HOTEL AND CATERING STUDIES

Ursula Jones

To Helen,

Wishing you every success in your future career.

Best Wishes.

Christine Bert Joy & Kaye.

OCTOBER 1993.

Hodder & Stoughton
LONDON SYDNEY AUCKLAND

ACKNOWLEDGMENTS

The author and publishers gratefully acknowledge the contribution of the following people and organisations: Boss Secretarial Services; Philip Burch, Head of Department, Tresham College; Franz Kreuzer, Walla Maschinen Gesellschaft MBH & Co, photos (b) and (c) p. 228; Roddy Paine for the cover photography; Bill Smith, Chrissons Journey Ltd.,(Authorised Corian Fabricator); Eric Tune, Director of Marketing, Diversey Ltd., Fig. 1.19; Barbara Warrior, Dietician, Kettering General Hospital.
The publishers would also like to thank the following organisations for permission to include copyright photographs and material: Anchor Industries, p. 133; Anglo Manufacturing., (b) p. 227: Bass Inns and Taverns Ltd for Fig. 4.7; Bartlett Harmony, p. 43 (right) and 44; BERG (Autek Control System Inc), Miami, 228 (a); Britannia Inter-continental Hotels for Fig. 4.5; China craft Hotelware, p. 189; Contico Manufacturing Ltd, p. 52, 54, 55 and 58; Crown Management Systems, p. 226; Edwardian Hotels, p. 3 (bottom); Fortronic Ltd, p. 224; Health Education Authority for Fig. 3.5; Gardner Merchant Ltd., p. 3 (top and middle); Hobart Manufacturing Co. Ltd., p. 35; Optech International, (a) p. 227; Primo Furniture Ltd., p. 189; Trusthouse Forte Hotels for Fig. 4.2; Keith Turnbull, p. 43 (left); Wedgewood Hotelware for Fig. 4.2;

Every effort has been made to trace copyright holders of material reproduced in this book. Any rights not acknowledged here will be acknowledged in subsequent printings if notice is given to the publisher.

British Library Cataloguing in Publication Data

Jones, Ursula
Hotel and catering studies.
I. Title
647.94068

ISBN 0-340-52749-8

First published 1992

Typeset by Taurus Graphics, Abingdon, Oxon.
Printed in Great Britain for the educational publishing division of Hodder & Stoughton Ltd, Mill Road, Dunton Green, Sevenoaks, Kent by St Edmundsbury Press Ltd.

CONTENTS

PART 1

Hotel and Catering Industry

This section will give you an understanding of the:

- diverse nature of the hotel and catering industry.
- paths and methods of communication used within the industry.
- legislation that is relevant to the catering industry and an understanding of its necessity.
- methods and systems employed to ensure that foods are stored, handled, prepared and served in a safe and hygienic manner.
- methods to ensure that a safe and clean working environment is maintained.

CHAPTER 1 Types of establishment

The hotel and catering industry provides for many types of consumer and it is necessary to appreciate their different needs and expectations if an efficient service is to be offered for sale. Whether it is a simple meal, a single drink or a complete accommodation package, the need to identify customers' requirements is the same. It is for this reason that the catering industry is so diverse in its composition: each individual establishment setting its own character and quality of service according to its size, the location, and capabilities of its key staff.

Accommodation and catering facilities are available at hotels, university and polytechnic hostels and halls of residence, boarding schools, hospitals, HM prisons, holiday camps, holiday chalets (within a leisure complex), time share apartments, and on cruise liners and ferries, to name but a few. From this list it is easy to see that there will be large differences between many of the establishments readily identified as being suitable for a particular type of guest or consumer. For example, halls of residence and hostels are designed to accommodate and provide self catering facilities or cafeterias for students; holiday camps, on the other hand, offer chalet accommodation with self-catering and have restaurants, bars and entertainment to suit many age groups and tastes.

THE PROVISION OF CATERING UNITS IN THE COMMUNITY

In any area, there will a variety of catering units selling or providing food and beverages to customers, guests or restaurants. For analytical purposes these catering units are described as either commercial units, which sell services for profit, or welfare units, which provide a service for residents or customers at a breakeven point or sometimes with a subsidy from a managing agent. Some examples are listed in Table 1.1.

A catering unit profile for a town can be created by drawing a diagram of the commercial and welfare units (see figure 1.1). The size of the circles indicates

Table 1.1 Examples of commercial and welfare units

Commercial units	**Trade groups**
Hotels	Residents or travellers requiring meals and beverages, served in a formal manner
Restaurants	Persons requiring a formal meal
Bistros	Persons requiring a light, informal meal
Wine bars	Persons requiring wine by the glass or bottle, and light snacks
Alcohol bars	Persons requiring alcohol and non-alcoholic beverages and light snacks
Fast food	Young persons requiring instantly prepared food
Airports	Travellers requiring light meals and beverages
Railways	Travellers requiring beverages and light meals while travelling
Roadside services	Travellers requiring beverages and snacks
Welfare units	**Target groups**
Schools	Pupils and staff, requiring light meals
Industrial canteens	Employees requiring beverages, light meals and snacks at work
Hospitals	Patients and staff requiring meals and beverages
Homes for the elderly	Residents and staff requiring meals and beverages
Prisons and remand homes	Inmates and staff requiring meals and beverages
Hostels	Residents and staff requiring meals and beverages
University halls of residence	Students requiring meals, beverages and snacks

Examples of commercial and welfare units. From top: a staff cafeteria, school cafeteria and a hotel restaurant.

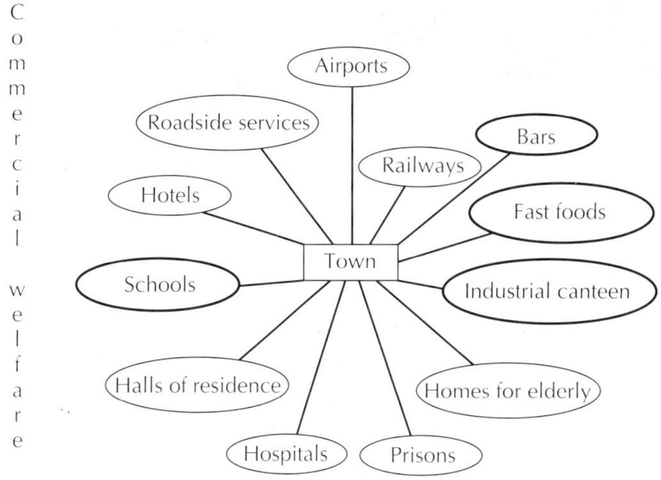

1.1 Catering unit profile. This size of the circles indicates the number of commercial or welfare units.

the number of catering units but not the number of meals that they serve. However, from the profile one can obtain information such as whether there are more commercial units than welfare units, more fast food units than any other type of unit, or an equal number of school and industrial units.

ACTIVITIES

Hotels,
public houses,
restaurants,
in-store restaurants,
coffee shops,
fish and chip shops (eat in and take away),
pizza houses (eat in and take away),
hamburger restaurants (eat in and take away).

All these businesses are different, yet they all have some things in common, such as the provision of food and drink and an area in which to eat the food.

- Examples of the types of hotel guests and their expectations are listed in Table 1.2. In groups, discuss the facilities that you would like a hotel to provide on a weekend break.
- Draw a catering unit profile for your town (see figure 1.1). Vary the size of the circles to indicate the number of each units in each. Consider the reasons why a town has many different types of catering units.
- Select an establishment of your choice, and draw a diagram to show the department and staff structure within your chosen catering unit.

Table 1.2 Types of hotel guests and their expectations

Customer	Type of establishment	Purpose of visit	Facilities/services expected by customer
Company	Large city hotels Exclusive country hotels Business parks University halls of residence	Meetings, exhibitions, conferences, sales, conventions, product launches	Efficient, unobtrusive service Conference rooms Quiet, comfortable surroundings Bedrooms with *en suite* facilities and room service Leisure facilities – sauna, swimming pool, jacuzzi, gym, etc. Access to main arterial routes Car park and garage facilities
Family	Hotels, chalets and holiday camps in historical towns, seaside resorts, etc	Holidays, leisure and recreation	Leisure facilities Trips to areas of interest Child facilities Entertainment Friendly atmosphere Provision for dogs
Students	Halls of residence or hostels attached to universities, polytechnics and colleges of higher education	Safe, economical accommodation while studying away from home	Single study bedrooms Cafeteria facilities Self-catering facilities Communal lounge and TV areas Leisure facilities
School pupils	Boarding houses attached to schools	Safe, controlled accommodation	Dormitories or study Common rooms TV rooms Sports facilities Communal dining facilities

CHAPTER 2 Staffing patterns

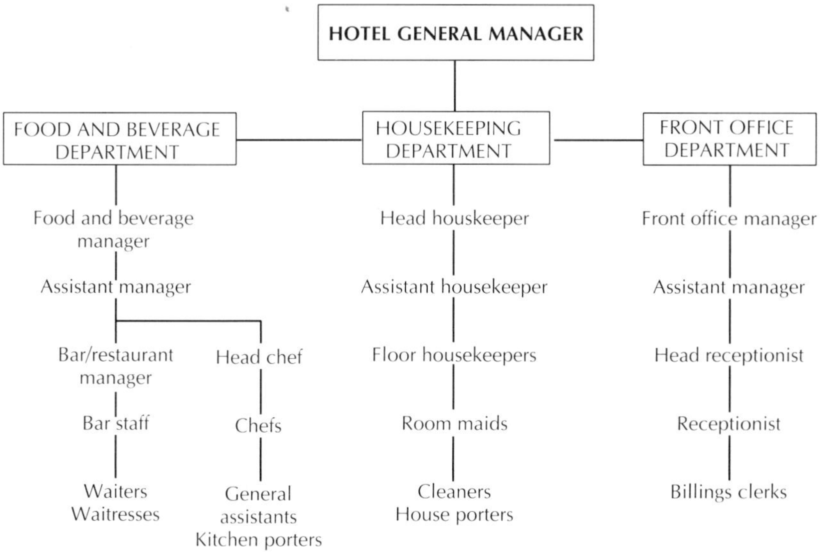

1.2 *Example of an hotel's departmental structure.*

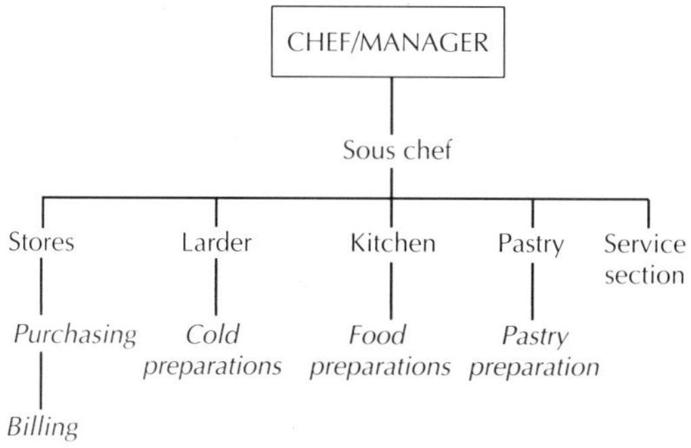

1.3 *Organization of production for a catering unit.*

The staffing and organization of a business will vary according to its type and size. An example of a departmental structure in a hotel is illustrated in figure 1.2. Each department works as a team within the hotel, all aiming for the same goals, that is, to create a comfortable, pleasant environment for the hotel's guests so that business will increase and thus the hotel will be more profitable. The organization of production in a catering unit is shown in figure 1.3.

INFLUENCING FACTORS

Factors that affect employment and the provision of facilities or services offered to the public in the hotel and catering industry may be summarized as follows:

1. General economic variables:
   - (a) Personal disposable income and higher real incomes give rise to the opportunity to spend more money on holidays and leisure activities.
   - (b) Business activity - this results in more travel and company business, including conferences, seminars, sales conventions and exhibitions.
   - (c) The exchange rate of sterling affects a person's decisions about the viability of a holiday abroad.
2. The relationship between employment and relative pricing structures in the industry.

Pricing decisions are usually taken, first, after due consideration has been given to the effect that an increase in price will have upon demand for that service or product and, second, the effect that this will have upon turnover and profits. Pricing structures are closely related to sales volumes and employment levels.

One factor that the hotel and catering industry has no control over is the excise duty charged on such items as alcohol and tobacco products. When these taxes increase, the price of the products has to rise accordingly. This may cause a drop in sales, and the resulting reduction in staff employment hours. Another factor is the seasonal nature of much of the UK's hotel and catering business. There are peaks of trade in some sections, for example, in December, for Christmas, and in the three summer sections (Easter break, spring bank holiday and school holidays in August), for holidays and the tourist trade.

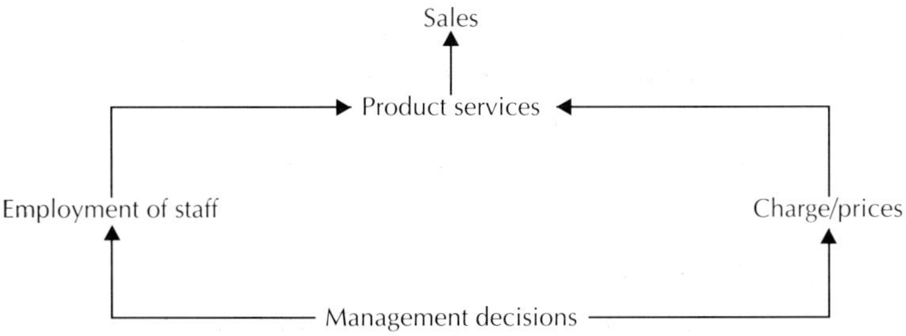

1.4 Sales flow diagram.

To cope with these changes in demand for services and fluctuations in sales, the work force in the hotel and catering industry comprises:

- permanent, salaried key personnel in each department/section,
- seasonal personnel who may be employed on fixed-term contracts, that is, for example, for twelve weeks, and paid on a weekly or monthly basis,
- part-time personnel who are contracted to work a set number of hours per week, and usually paid weekly at an hourly rate,
- casual workers are employed as and when required, to cope with a particular function or event, for example, receptions, conferences, banquets, conventions and seminars.

Another factor that affects the nature of employment within the industry is the necessity to operate 24 hours per day, 7 days per week, 52 weeks of the year. In other words, some establishments never close to the public.

The roles, skills and activities of staff are affected by ever changing developments in the industry - staff must be flexible and adaptable to cope with such developments.

It is possible to see from the few examples given in Table 1.3 that there have been many changes in the hotel and catering industry over the years. Everyone employed within the industry must be aware of these changes and be trained to become part of an adaptable, flexible team of operatives of all skill levels.

When decisions are being made about changes to the systems being employed, consideration must be given not only to the quality of the end result, product or service, but to the ability of the staff to work to the proposed changes and the suitability of the existing premises. It must be remembered that the service is only as good as the staff who provide that service.

STAFFING RATIOS

The number of staff employed in relation to the number of meals to be served is known as the staffing ratio.

The staffing ratio will be affected by the following factors:

- kitchen efficiency,
- overhead costs,
- complexity of the menu and type of establishment,
- staff capabilities,
- kitchen equipment and conditions,
- preparation and service time available,
- customer profile/expectancy.

Sometimes part-time staff are employed to increase the staffing ratio to cover peak time periods such as meal service times, special functions and maximum sales periods. Part-time or casual staff are employed to cover such eventualities as staff sickness, staff

Table 1.3 Examples of developments in the hotel and catering industry

Traditional activities	Intermediate developments	Fastfood activities
Washing up by hand	Washing up by machine	Disposable cartons
Silver service	Family service	Cafeteria service
Afternoon tea service	Cup of tea from cafeteria	Vending machine
Cooking with traditional equipment	Use of deep freeze and microwave equipment	Use of specially designed, semi-automatic cooking systems

holidays, and extra duties associated with renovation and redecoration schedules. There are some employees who only make themselves available for part-time or casual work, and many of these staff work through an employment agency. Two of the main reasons for their choice of this type of employment are:

Pay – it is not always necessary to earn a full wage if there is already a full wage earner in the family.

Time – family commitments mean that part-time employment can be fitted around other activities.

ATTRIBUTES OF EMPLOYEES

Employees need to be aware that the following attributes are essential if they are to be valued members of the team.

Punctuality

Punctuality is essential as all catering employees are part of a team of staff: if one member of the team is late or absent, all the other members will have to do a proportion of that member's work in order to complete the team's objectives. Thus, if a member of staff continually arrives at work late and without the correct uniform, there is every likelihood that he or she is unsuitable to work in the industry. For example, an early morning call ordered by a guest for 7.30 am means just that, and not earlier or later. A certain cause of guests becoming dissatisfied is if the staff call them at the wrong time.

Meals and services are offered for sale to the public between specified times – if staff are late it may mean that the meals and services are of a lower standard or are offered late (even a couple of minutes late is an indication of a slackness on the part of the establishment offering the service).

Self-presentation

How one values oneself is evident by the way one presents oneself to others – if an employee comes to work with low standards of personal cleanliness and in torn and unclean clothes, it will be obvious that he or she has a low opinion of him or herself. Even if an employee comes from a humble home, there is no reason why he or she should not arrive for work washed and wearing a clean, well-pressed set of clothes. In addition, the member of staff who arrives clean, bright and alert is more valuable than one who comes to work in an outfit that looks as if it has come straight from a fashion show.

Integrity

Integrity is an essential quality in the industry. If team work is to be successful, all employees should be able to trust one another at all times. Qualities that are closely allied to integrity are, honesty, trustworthiness, dependability, reliability, truthfulness, and loyalty. Loyalty to one's employer and colleagues is the secret of a good working relationship – arguments and mistrust do not enhance the working atmosphere and should be avoided.

Responsibility

The expression 'able to take responsibility' is frequently used in staff appraisals. It is the first stage of a member of staff being considered for a higher position within an organization and can be applied to every member of staff at every level of every job in the industry.

Each member of staff will agree upon a written job description (see figure 1.5) when they are appointed to a position but a caring attitude towards others also makes for a responsible employee.

JOB TITLE	Sous chef.
PLACE OF WORK	Regal Hotel Bristol. Food and beverage department.
PURPOSE	To assist in the preparation of food. To train junior staff in the department. To ensure that work is carried out in a safe and hygienic manner.
HOURS OF WORK	37 hours per week.
RESPONSIBLE TO	Head chef.
RESPONSIBLE FOR	Trainee junior staff; kitchen porters and cleaners.
DUTIES	The production of meals/dishes as specified by the head chef. Assist in the production of food consumption figures. The maintenance of a clean and safe working environment. Assist in the production of staff work schedules. The training of junior staff. Undertake any task delegated by the head chef.

1.5 An example of a job description.

Effort

It may be possible to carry out specified tasks or duties with the minimum amount of effort, in which case it is likely that only a minimum acceptable standard will have been reached and the minimum level of job satisfaction will result. On the other hand, if staff put effort into their work and really try to improve their standards, then not only will the tasks or duties be completed satisfactorily, but the level of job satisfaction will increase.

THE MANAGER

The manager is the leader of a team of staff and the decision maker for the business or section. The manager will take the responsibility for the success or otherwise, of the business or section. The aim of every management team is to create an environment which will enable others to perform their duties effectively.

To be a successful manager, it is necessary to possess the qualities to:

- interview, select and employ suitable staff,
- supervise and organize teams of staff,
- predict and interpret business trends,
- control financial expenditure,
- make valid business decisions,
- use discretion and tact in difficult circumstances,
- interpret sales analysis data,
- set and keep to business targets,
- delegate tasks to subordinates,
- comply with relevant legislation.

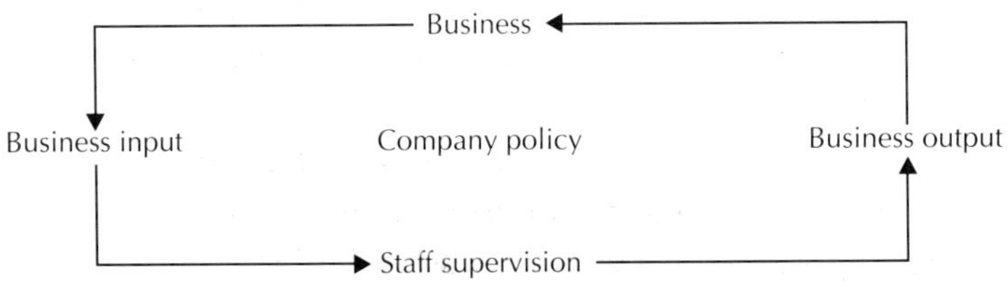

1.6 Company policy flow diagram.

From this list you will be able to understand how difficult and complex a manager's position really is. The successful manager should be able to:

- accept and shoulder responsibility,
- demonstrate a firm, positive attitude towards work,
- treat staff and problems in an unbiased manner,
- show a respectful regard towards others,
- work with and for other people of all types,
- demonstrate a caring attitude towards other staff,
- have a sense of humour in a crisis.

There are many different management positions in the hotel and catering industry. The hotel manager may be responsible for:

- control of the hotel and its staff,
- financial control of the entire hotel,
- promotion of sales and marketing,
- sales analysis and forecasting,
- creating a comfortable, inviting establishment.

The front office manager may be responsible for:

- promotion of room and function sales,
- control of all room and function reservations,
- creating successful customer relations,
- compiling and keeping customer accounts,
- keeping records of guests,
- creating a good business image,
- employment and training of staff.

The food and beverage manager may be responsible for:

- promotion of food and beverage sales,
- planning of functions and their execution,
- planning the provision and service of food and beverages,
- control of stock, food, alcohol and tobacco,
- purchasing of stock,
- budget forecasting and analysis,
- employment and training of staff.

The domestic services manager may be responsible for:

- control of cleanliness,
- liaison with the hospital infection control committee,
- maintenance of hygienic premises,
- disinfection and sterilization of equipment,
- employment, control and training of staff,
- budget forecasting and working within a budget,
- ordering and controlling equipment and stock.

STAFF MOTIVATION

Staff should be committed to work with the interests of their employer in their minds, in other words, they will do all that they can to promote and increase the standards and business wherever they are employed.

To ensure that members of staff are enthusiastic and committed people, they must be able to:

- feel that they belong to a team of staff in a secure job,
- realize how important and necessary their job is in relation to the rest of the organization,
- appreciate the consequences of their job and its timing in relation to the rest of the team and staff,
- use their knowledge and skill to their full potential,
- see a clear progression path for their career, and recognize that promotion can be earned,
- use the necessary equipment in a safe and efficient manner,
- complete the job in the required time and to the required standard,
- gain job satisfaction upon completion of the task and recognize a job well done.

TEAM APPROACH

A high proportion of work carried out in the hotel and catering industry is dependent upon a team or group of staff, not individual effort (see figure 1.7). The team of staff is a group of persons employed in a common area with a common aim between them. Within the team it may be found that certain people have definite responsibilities, but the common aim remains. The various team members will have a definite set of tasks to execute in order that the common aim may be achieved. For this reason, each team member is to some extent dependent upon all other members. Successful team work will depend upon such things as:

- a clear line of authority,
- respect for the team leader,
- a definite ladder of progression,
- knowledge and skill necessary to complete assigned tasks,
- sufficient time and stock to complete tasks,
- a caring attitude towards other team members,
- good team spirit, an essential part of the hotel and catering industry.

The following example illustrates the effect of one team member not performing his or her designated duties correctly. Team A in the kitchen is dependent upon team B in the restaurant. The restaurant customer is dependent upon both teams A and B. If a member of team A who is responsible for ordering the food stock omits an item, the entire team will be affected and, as a consequence, so will the customer. If a member B omits to present the menu to the restaurant customer, then food sales will be reduced.

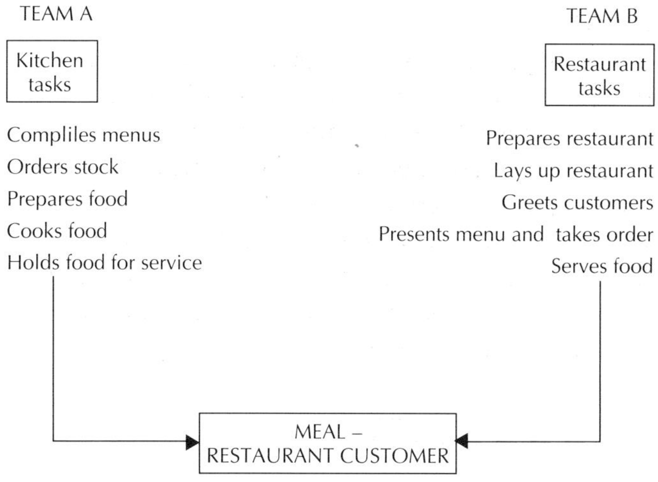

1.7 Tasks for restaurant and kitchen teams.

There are, however, many other factors that will affect the teams' performance. These include:

- absenteeism resulting in the other team members having to complete the absent members duties between them,
- laziness resulting in the other team members feeling resentful as there is an uneven work load in the team,
- antiteam spirit resulting in a poor team performance through members of the team working against one another.

ACTIVITIES

- In groups answer the following questions:
- Which types of hotel and catering establishments operate for 365 days a year, 24 hours per day?
- Describe the type of services that roadside service areas offer to the travelling public. Are some only offered at particular times? What is the reason for this?
- Look in trade papers and journals, and make a collection of those products that will cause changes in staff roles, skills and activities.
- Discuss the reasons for part-time work and complete a list of jobs in a hotel that could be covered adequately by part-time staff. Give reasons for your listed jobs.
- Discuss the importance of proper self-presentation in the hotel and catering industry and how one can ensure a smart personal appearance. Discuss the relationship between self-presentation and good performance levels at work.
- List the responsibilities that you have towards the college where you are studying. In groups, discuss

the responsibilities that you would expect the following staff in the industry to take:
1. room maid,
2. school caretaker,
3. counter assistant at MacDonalds,
4. table clearer in a canteen.

Identify any common responsibilities from the four staff listed above. Discuss the reasons why there are common responsibilities even when the type and level of jobs are very different.

- There are common tasks that all managers have to undertake. Can you compile a composite list of a manager's main duties from the summaries given in the chapter?
- Discuss staff motivation and then list the factors that enhance staff motivation.
- List the qualities that staff in the following departments should possess in order to work successfully in team situations:
  1. housekeeping department,
  2. food provisions department,
  3. food and beverage service department,
  4. front office department.

In groups, select a department in a hotel and list the possible causes of poor team work.

In groups, compile a chart to illustrate the staffing structure of the four departments listed above.

In groups, discuss the areas where team responsibility is dependent upon another team or teams within the same organization.

CHAPTER 3 Communications

Communications simply means the passing of information from one point to another. In order to communicate effectively it is necessary to analyse the information systems flow path (see figure 1.8).

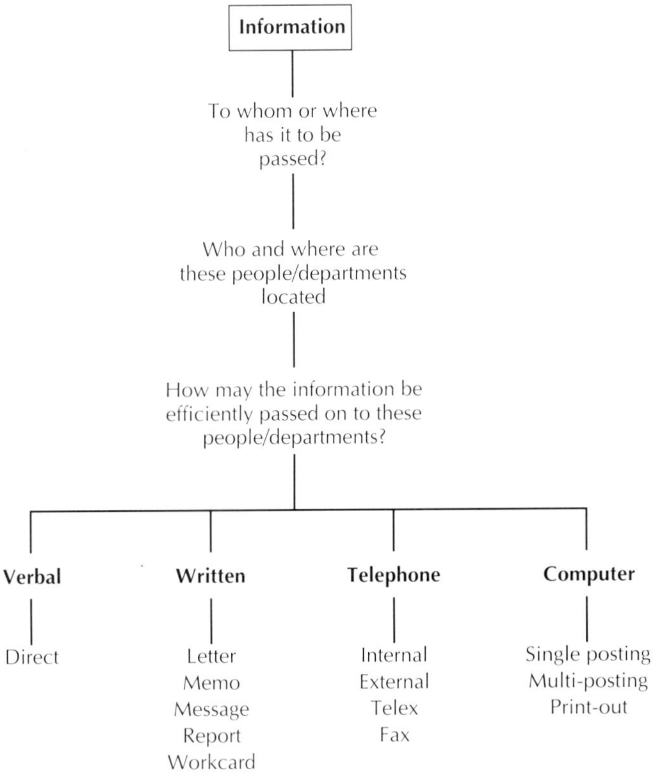

1.8 Information systems flow path.

In order to make the correct decision, the most efficient communication method should be chosen. It does not necessarily follow that the same method is always the most appropriate for similar information. For example, when confirming a hotel reservation for a booking three months ahead, a letter would be efficient, whereas if confirming a hotel reservation for the next day, either a telephone, telex or fax message would be more efficient unless direct verbal communication was possible.

DIRECT VERBAL COMMUNICATION

Direct verbal communication is the giving and receiving of information in person to and from customers, guests and staff. It is essential that staff are made aware of the impact of their responses to customers' queries as poor verbal communication can lead to a loss of business (see table 1.4). It is just as easy to give an adverse impression as it is to give the impression that one is interested and keen to please the customer.

Table 1.4 Verbal communication skills

Positive skills	Negative skills
Politeness	Impoliteness
Clarity of speech	Unclear, soft speech or mumbling
Informative statements	Giving extra, unnecessary information
Direct, correct answers	Not totally answering a question
Applying tact in awkward situations	Proving the customer is incorrect
Uncontroversial	Expressing one's own opinions
Listening attentively to the guest	Interrupting before the guest has finished
Looking at the guest	Gazing around or over the head of the guest
Answering so that the guest can understand	Answering using jargon
Conveying information impartially	Use of emotive language
Accurate pronunciation	Poorly pronounced words causing confusion
Using the correct form of address	Not using the formal manner of address

INDIRECT VERBAL COMMUNICATION

This is communication by telephone or intercom. Telephones are used to generate sales and to give information to potential guests and customers. It is therefore of paramount importance to ensure that the impression created is one of efficiency, politeness and a willingness to help and serve. More sales are lost through a poor response to an initial inquiry than by any other reasons. It is imperative that all queries are handled with care and that no caller is left at the end of a blank line - this is a sure way to lose business. Likewise, callers expect a quick response when they telephone and do not like to hear the telephone ring again and again. It gives the impression of an inefficient organization: 'If they cannot even answer the telephone they will not be able to do much else.'

The following points should be noted about answering the telephone:

- calls should be answered promptly giving the name of the establishment,
- calls should be handled promptly or quickly transferred,
- messages should be taken with care and read back to the caller to check their accuracy,
- callers should be thanked for their call and a polite 'good-bye' phrase should be used at the end of the conversation,
- clear, concise method of speech should be used,
- an even, slow tone of voice should be adopted,
- polite, conventional terms and expressions should be used.

Table 1.5 Expressions that create favourable and unfavourable impressions

Business-like impression	Unfavourable impressions
Good morning/afternoon	Hello or hi
Can I help you?	Yes or OK
Please wait a moment	Hang on or wait a mo
I will find out	I don't know
Thank you for calling	Cheerio or So long
I will forward the . . .	I'll send it
I will transfer you to . . .	. . . will deal with you
I look forward to meeting you	See you soon

WRITTEN COMMUNICATION

Wherever an employee works or whatever level of work or series of tasks are to be completed, communication in the written form will be used. Figures 1.9 to 1.12 are examples of the four most commonly used written communication methods: a letter, a memorandum, a message form, and a work card. Reports are also common and a description of these is included.

A Letter

Letters should be laid out formally and include the address, telephone and fax numbers of the sender; the date the letter was sent and a reference; and the name, position and company name and address of the addressee. All relevant information should be included in the letter.

A Memorandum

Memos are used to communicate information from one person to several other persons within the same organization. Many organizations have preprinted memo forms to enable information to be passed from one section or person to others, quickly and efficiently. The use of memos is frequently encouraged because it provides a written record of information, something that a telephone conversation does not do. The content of the memo should be concise as it is a short note, not a long explanation.

A Message Form

A message form is frequently used to record and pass on messages that are taken either by the telephonist, receptionist or duty member of staff. A printed form is usually used; this has the dual purpose of reminding the receiver to take down all the caller's details as well as their message, and to ensure that messages are passed on in a neat and legible manner for staff and guests.

A Work Card

A work card is designed to be a source of reference for the employee. It gives a short list of instructions and safety points for a specific task. Work cards are used in most sections of the catering industry and form an integral part of employee training and supervision. The work card may be either wall-mounted, slotted into a wall-mounted rack, or given to the employee so that it may be in his or her pocket for instant reference. Examples of tasks that are

Our ref: KJ/CSB

The White Hart Hotel
Cliff Top Road
Douglas
Isle of Man
2AI 1IM

Address telephone and fax number of sender

Tel: 0624 7732
Fax: 0624 8196

Name and position of addressee:
G.S. Freeman
Sales Manager

Company name and address:
ICP Food Products Ltd
23 Eastland Road
Pillrocky Trading Estate
Pillrocky
Kent

20.7.91 *Date*

Dear Mr. Freeman

I am pleased to be able to confirm your booking for a Sales Promotional Conference and exhibition in the Buchanan Suite on the 16th and 17th of January 1992 .I am also able to confirm the booking of 5 executive en-suite bedrooms for the nights of 15th to 18th January 1992 inclusive.

Please find enclosed details of our conference charges, sample menus and conditions of sale together with details of the hotel's facilities.

Should you require any further information do not hesitate to contact me.

Yours sincerely

C. S. Black

C. S. Black
General Manager

enc.

1.9 A business letter.

Name of Firm

Memorandum

From — To

Date — Ref

Subject

1.10 A memorandum.

Message

Date: — Time:

For:

From:

Address:

Tel No.: — Ext:

Message:

Taken by:

1.11 A message form.

Front

TASK:	Cleaning mixer
USING:	Stiff brush, detergent
SAFETY:	Switch off
	Unplug
	Remove bowl and paddle

Reverse

METHOD

1 Prepare detergent solution in bowl

2 Using cloth, clean around mixer including stand.

3 Using the brush, clean around the central spindle over the feet.

4 Wipe entire mixer dry. Replace bowl and paddle.

1.12 A work card.

frequently supported by the use of work cards include: the operation of a cash till; the use of electrical food preparation machines; the cleaning of specific pieces of equipment or complete areas; and the preparation of simple sweets, snacks or salads. The main advantage of using such cards is that a standard procedure is followed by all staff for a particular task, thus ensuring quality control.

A Report

A report is used when presenting information to one person or a group of people. It is a concise document listing the information in a manner that is easy to read. It will have a title or terms of reference, the procedures, the findings and, finally, the conclusions and recommendations. Before compiling a report, one should decide who will receive the report and how it will be used so that one can include only essential information.

The terms of reference includes the name of the person who requested the report, the precise area that the report is to cover, and what the intended outcome is to be. For example:

'This study, requested by the area manager, investigates the present marketing strategies for function trade and presents proposals for reorganization and improvement.'

The procedure states how the study was conducted, lists those consulted, and details any surveys completed or research carried out.

The findings are listed in an easy to read format. Tables and charts are sometimes included, either in the text or attached as appendices.

The conclusion and recommendations are numbered and the solutions are derived from the findings of the report. Naturally, they should address the points raised in the terms of reference.

INDIRECT MEANS OF COMMUNICATION

Telephone

Much communication is carried out by the use of telephones, either by the use of an internal system or an external system.

Internal Telephone System

This will link all areas within the organization thus enabling employees to hold conversations without having to leave their work areas.

External Telephone System

This will consist of one or more telephone numbers that a caller may use to contact the company, and will enable employees to contact any other person or company outside the organization.

Telex

Telex messages are transmitted from one telex user to another by typing in the message on the telex machine (which has a keyboard) and using an individual telex code number to identify the receiver. The message will be automatically printed on the telex receiver at its destination. Provided the telex receiver is switched on, messages may be received at any time, ready to be dealt with as soon as the office is opened. Telex messages are costly to send, and are therefore brief and concise (See figure 1.13).

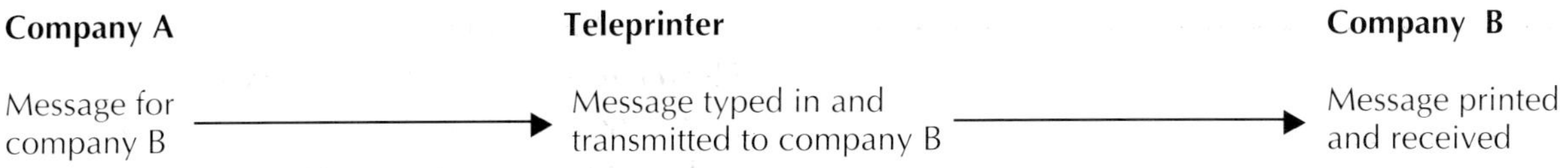

1.13 Sending a telex message.

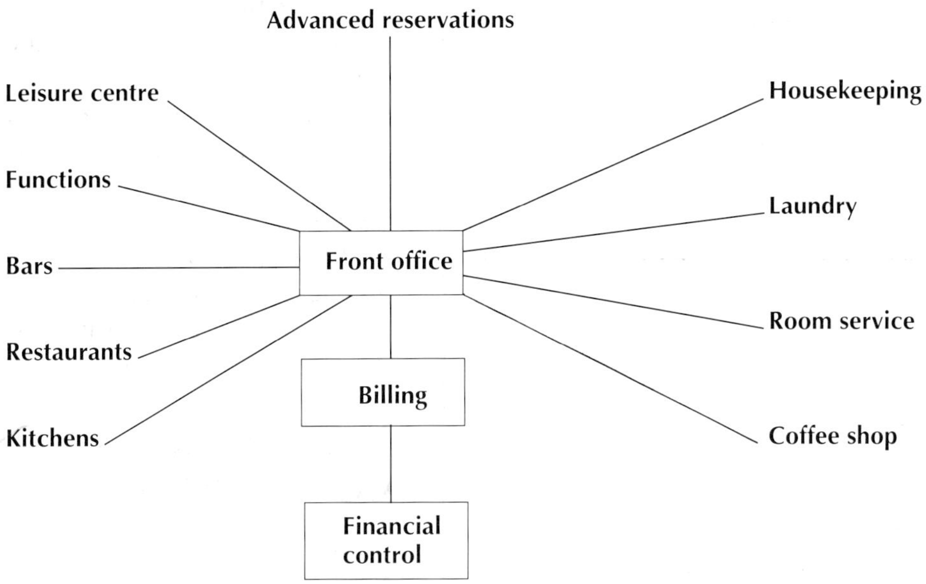

1.14 Links via computer between departments in a hotel.

Facsimile Machine

It is a copier which produces its copy at another place by using a telephone line. The most valuable use is for documents/papers that have to be seen by a distant person immediately, e.g. end of day trading figures from point of trading to head office. The sending party has proof that the document has been sent and received the other end. After a brief message has been sent the fax machine prints out a brief report confirming that it made contact with the recipient. Because the fax machine uses a telephone line the charge is exactly as it would be for a telephone call.

Fax machines can be programmed to make the calls during cheap rate hours. If a document needs to be transmitted to more than one destination the machine simply dials each number in turn and transmits to each one after the other. It is important to remember to leave the fax machine switched on over night so that it can receive messages.

Computerized Communications

A computerized communication system is frequently used in both small and large organizations. The computer will be used to link various customer departments of the organization, and to relay and store information. Figure 1.14 shows a possible link between departments in a hotel.

Computer terminals, comprising a keyboard and monitor, will be situated in each of the work areas. Guest bookings, charges and details, can all be transmitted via these terminals to the guests' accounts. This computerized system replaces the old paper and pen guest charge system. It is also possible to communicate between work areas. For example, the housekeeping and advanced reservations frequently need to communicate as do the Functions Suites, bars, kitchens and restaurants. The same computerized system is used to exercise financial control, stock control, equipment control, guest records, guest accounts and sales predictions.

INTERDEPARTMENTAL CO-OPERATION

Close co-operation between various departments and sections of the hotel is essential to ensure the smooth running of the business, its success and profitability.

Communications and Sales

Sales of any kind will not be made without some form of communication. Whether it is simply a sale of a meal or the sale of a complete reference package, communications will have played their part.
What is the meal being sold?
Of what does the meal consist?
What time is the meal being sold?
How much does the meal cost?

These four basic questions could be answered by the customer simply looking at a menu board outside a cafeteria, fast-food establishment or restaurant. Thus you can see that the menu is a prime form of communication that enhances sales as well as being a legal document. Alternatively, they could be answered by the customer reading an advertisement in a newspaper or journal, or by talking to a previous customer.

Hotel sales are achieved by many means, these include the use of advertisements, by offering special discounts and weekend prices, by mounting special events and activities, by encouraging company bookings and accounts, by bookings through travel agents and tour companies, and, more importantly, by returning guests. Large hotel groups employ sales and marketing specialists to ensure that these aspects of their business are efficient and effective.

Every employee is, in fact, a sales person, and can play his or her part in ensuring that guests and customers are served well so that they leave satisfied and will return again in the future.

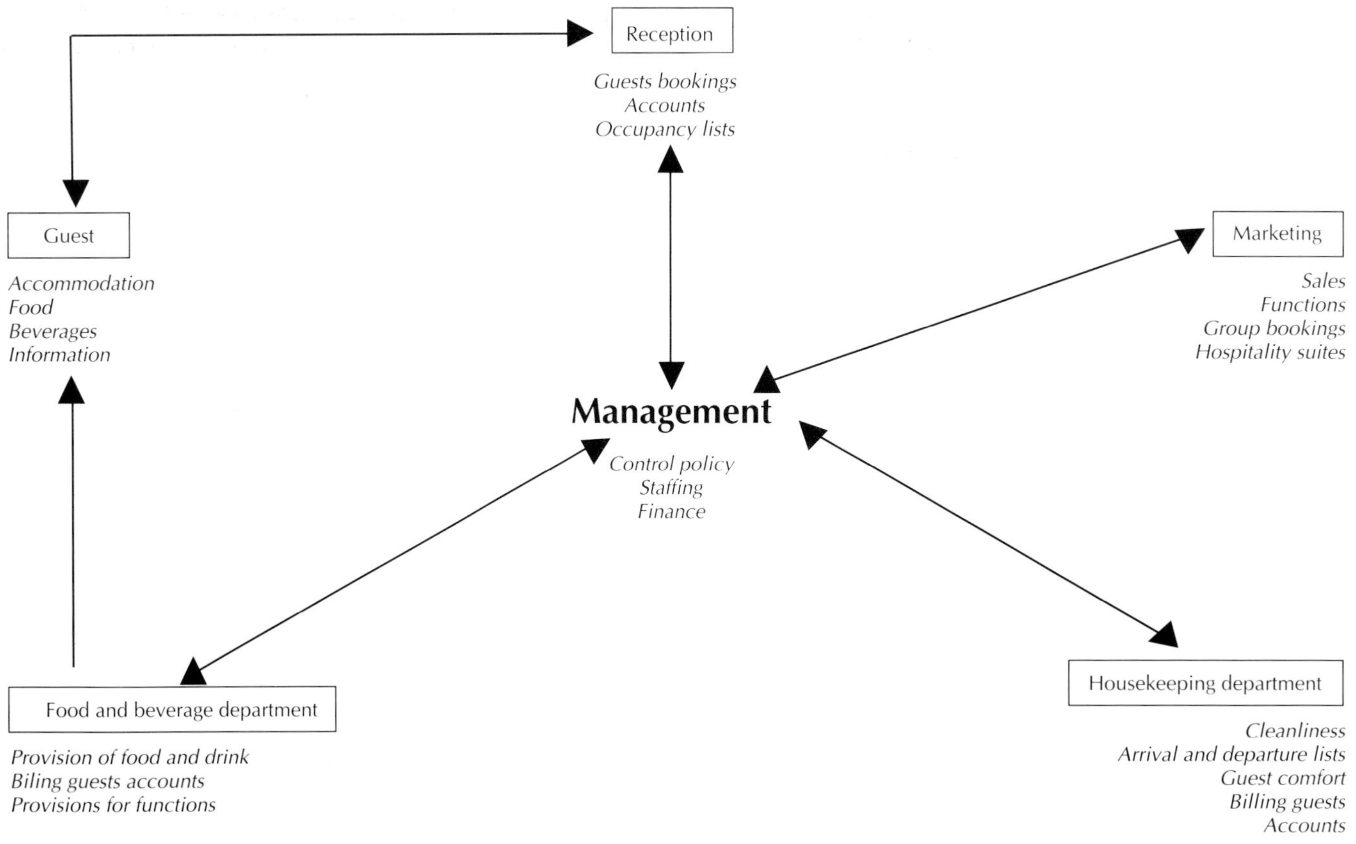

1.15 Pathways of communication in a hotel.

ACTIVITIES

- Wherever an employee works or whatever level of work or series of tasks are to be completed, communication with other persons or departments is essential. Below are some examples of written communications that may be used in the various areas of hotel and catering industry. Look at these examples and then add your own under each heading.

Kitchen
Orders of supplies
Restaurant orders
Function menus
Equipment operating cards

Restaurant
Menus
Guest reservations
Service instructions for functions
Restaurant bills

Housekeeping
Equipment operating instructions
Laundry lists
Occupancy sheets
Cleaning schedule

General
Safety notices
Fire drill procedures
Statutory legal notices
Pay/salary details

Front Office
Guest requests for accommodation
Diary bookings
Housekeepers reports
Guest accounts

Management
Memos
Reports
Contracts
Letters

All written information should be in a clear concise format giving all the details necessary to act upon the instructions or facts. It is always advisable to ascertain whether the written information can be understood - for example, foreign visitors.The blind or the elderly, may need assistance.

Take one item from each of the areas above and produce a clearly written example of each item chosen.

- Study the paths of communication illustrated in figure 1.15 (page 16), and list the ways that the information from each department may most efficiently be communicated. For example, occupancy lists may either be in the form of a printed sheet from the computer's monitor, information displayed on a typed list, or a hand-filled-in form.
- Look in trade and local papers and make a collection of promotional material and advertisements that are designed to increase sales in local hotels and restaurants.

  Design a quarter page advertisement to be placed in a newspaper to increase sales in the hotel restaurant.

CHAPTER 4 Legislation

THE HEALTH AND SAFETY AT WORK ACT 1974 (HASAWA)

This act of parliament was put on the statute book in 1974 to update many older pieces of safety legislation. The main updated aspect is that of employee responsibility.

The HASAWA is concerned with the health, safety and welfare of persons at work, by enforcing the protection of persons against harm caused by machinery or thoughtless actions of others.

Under this legislation, all employers must promote the health and safety of their employees by providing and maintaining:

- safe premises, equipment and work conditions,
- safe facilities for handling commodities,
- instruction, training and supervision of employees,
- safe means of entering and vacating premises: normally and in an emergency,
- comprehensive welfare arrangements for employees.

Furthermore, employers are required to compile a written 'safety policy' statement to substantiate their arrangements for the safety and welfare of their employees. It is also possible for a safety representative to be appointed from a recognized staff association or trade union, who will ensure that the safety policy is being maintained by the employer.

The HASAWA also contains a code of practice for the employee. This new aspect of legislation means that every employee is now responsible for his or her own actions, and must:

- take care of the health and safety of themselves and others,
- work in a safe manner and according to the employer's instructions,
- use safety guards and equipment,
- wear safety and protective clothing and footwear whenever appropriate,
- never use commodities or equipment in a manner likely to cause harm to others.

The HASAWA legislation covers the state and maintenance of working areas. It contains clauses relating to the:

- cleanliness of the work area, recommending a minimum of 40 sq ft floor space be available, and where the ceiling is under 10 ft, at least 400 cu ft per person,
- temperature of the premises which should be maintained at not less than 16°C where persons are employed in a still position for a considerable length of time,
- ventilation should be efficient and effective,
- lighting, which should be adequate in every part of the premises,
- sanitary conveniences,which should be supplied and maintained, the numbers being dependent upon the number of employees,
- washing facilities, which should be well kept and clean, and dependent upon the number of employees and their work areas,
- provision of drinking water, which should be adequate and wholesome,
- arrangements provided for employees' clothing that is not worn while working (they should not be kept in a food area),
- seating, which should be provided for workers whose work is of a sedentary nature,
- provision of adequate accommodation should meals be provided at work,
- floors and stairways, which must be maintained in a safe manner, and remain free from all obstructions,
- dangerous machines, which must have safety guards fitted and used at all times.

The HASAWA is enforced by the local authorities, and the Environmental Health Officer is usually the inspector who ensures that it is being adhered to.

The inspectors have various powers under the HASAWA:

- to enter premises to exercise their duties,
- to take a police constable with them if they think fit,

- to take any technical support personnel with them,
- to isolate any part of the premises for further examination,
- to photograph, measure or record evidence of malpractice,
- to test or dismantle any item or items that are deemed unsafe,
- to take possession of any item for further testing or analysis,
- to require any person to give evidence or relevant information,
- to inspect relevant books, records and documents,
- to require persons to assist with the investigation.

Should an inspector identify a contravention of the HASAWA, an improvement notice will be issued which will give details of the matters to be improved, altered or corrected. The improvement notice will allow a minimum of twenty-one days for an appeal before coming into effect.

If an inspector finds premises or activities that are likely to cause serious injury, a prohibition notice may be issued which will prohibit further activity or trading until matters have been rectified. A prohibition notice may have immediate effect.

Penalties are in the form of a monetary fine - maximum £2,000 plus a fine of £100 per day for a continuing offence, such as failure to act upon an improvement or prohibition order.

HEALTH AND SAFETY FIRST-AID REGULATIONS 1981

These regulations make it obligatory for the employer to provide equipment and facilities for first-aid treatment for employees while at work. There should also be an adequate number of persons who are trained to give first-aid treatment to employees at work. It is common practice to provide first-aid boxes at suitable points around the workplace. In a hotel these could be kept in the chef's office; the front office; the housekeeper's office; the linen room; and the food and beverage manager's office. All boxes should be easily recognizable - green with a central white cross. The number of first-aid boxes in the workplace should be adequate to ensure that all sections or teams of staff have ready access in cases of emergency; they should not be locked away in offices or cupboards.

These are minimum contents guidelines for first-aid boxes; it should be realized that extra items may be required (see Table 1.6).

Table 1.6 Minimum contents for first-aid boxes

Contents	No. of employees	
	6–50	51–100
Guidance card	1	1
Industrial wrap sterile adhesive dressings	20	40
Sterile eye pads with attachments	2	6
Triangular bandages	2	6
Sterile coverings for wounds	2	6
Safety pins	6	12
Medium sterile dressings	6	10
Large sterile dressings	2	6
Extra large sterile dressings	2	6
Sterile water if water is unavailable	1	6

A record should be kept of all injuries to employees and the treatment given. This record should include the following facts:

- the name of the injured person,
- the nature of the injury and where it took place,
- the time it occurred and when it was treated,
- an indication of the cause of injury,
- the treatment given to the injured party,
- any action taken after the initial first aid was given.

THE PRESCRIBED DANGEROUS MACHINES ORDER 1964

No person under the age of 18 years may clean, lubricate or adjust any machine if it exposes them to risk of injury from any moving part or adjacent machinery.

No-one may work at any machine specified in the order unless they have been fully trained to use a particular piece of machinery, and are to work under adequate supervision by a person who has a thorough knowledge of the machinery to be used.

The following machines that are used in catering units are covered by the order:

- Power-driven machines - catering: worm-type mincing machines,

rotary knife bowl chopping machines,
dough brakes,
dough mixers,
food mixing machines when used for cutting, mixing and slicing,
pie and tart making machines,
wrapping and packing machines,
machines of any type equipped with a circular saw blade

- Power-driven machines - laundry:
  hydro extractors,
  calenders,
  washing machines,
  garment presses.
- Power-driven or manual machines:
  circular knife slicing machines used for cutting bacon and other foods,
  potato chipping machines,
  platen printing machines and cutting and creasing machines,
  guillotine machines.

FOOD HYGIENE (GENERAL) REGULATIONS 1970

The regulations apply to any premises in or from which a food business is carried on. A food business must not be carried out in insanitary premises, or those infested with vermin or pests. The premises should be in good repair, clean, tidy, well ventilated and adequately lit. The food should never be exposed to any risk of contamination. Equipment should be well maintained and scrupulously clean at all times.

Food handlers should ensure that they maintain high standards of personal hygiene and at all times are as clean as is reasonably practicable. Adequate supplies of protective clothing must be available and worn by all food handlers. Materials for wrapping or containing open food must be clean. Personal clothing and possessions should not be kept in the food preparation area.

Any person engaged in the food business that is suffering from any of the following diseases must notify their local Medical Officer of health: typhoid, paratyphoid, and other salmonella infections; amoebic dysentery; bacillary dysentery; any condition likely to cause food poisoning, for example, septic cuts, boils, spots, burns, throat or nasal infections.

Conveniences should be designed with an outer lobby for hand washing, and should never lead directly into a food preparation area. They should be clean, well ventilated and have 'now wash your hands' notices prominently displayed. Hand basins in kitchens should have an adequate supply of drying facilities. These hand basins should never be used for any purpose other than the washing of hands.

The food rooms should never lead directly to a sleeping area. All food rooms, whether for storage or preparation purposes, should be in a good state of repair, and built of suitable materials to facilitate good standards of hygiene. Refuse should never be allowed to accumulate in the food areas, or even outside, as this will attract vermin and pests.

The Environmental Health Officer (EHO) has power to enforce the regulations. He or she has the power to inspect food premises at any reasonable time. If the officer is refused entry, or feels that there is reasonable need to enter, he or she may apply to a justice of the peace for a warrant to enter and, if needed, take a police officer with him or her.

THE FOOD ACT 1984

The act states that:

- No person shall sell, or have in possession for sale, any food for human consumption which has been rendered injurious to health.
- A person shall be guilty of an offence if he or she sells to the prejudice of the purchaser any food which is not of the nature, or not of the substance, or not of the quality demanded by the purchaser. For example, a person would be guilty if the following were sold:
  smoked cod instead of smoked haddock,
  a nail in a jar of pickle
  frozen chicken instead of fresh chicken

THE FOOD SAFETY ACT 1990

This act strengthens and updates the law on food safety and consumer protection. The aims of the act are to:

- continue to ensure that all food produced for sale is safe to eat and not misleadingly presented,
- strengthen existing legal powers and penalties,
- enable this country to fulfil its role in the European Community,
- keep pace with technological change.

The act came into force on 1 January 1991. From April 1992, Crown premises will be covered by the Food Act.

The main offences under the act are:

- selling or possessing for sale, food which does not comply with food safety requirements,
- rendering food injurious to health,
- selling, to the purchaser's prejudice, food which is not of the nature, substance or quality demanded,
- falsely or misleadingly describing or presenting food.

The food safety requirements are that food throughout the food chain must not:

- have been rendered injurious to health,
- be unfit, or so contaminated that it would be unreasonable to expect it to be eaten.

The food law provides that:

- officers (EHO) can enter the food premises to investigate possible offences,
- officers can inspect food to see if it is safe,
- officers can detain suspect food or seize it and ask a Justice of the Peace to condemn it.

Enforcement officers must be given reasonable information and assistance, and they have the power to enter any business premises in the country. Officers may inspect records and take samples of food for later analysis. They may also take their own visual records, such as still photographs and videos. Officers must carry evidence of their identity, and may only enter premises at reasonable times.

The powers of the enforcement officer

If an officer believes that a food business does not comply with the hygiene regulations or food processing regulations, he or she may issue an improvement notice which requires the proprietor to put matters right.

If a business is convicted of breach of hygiene or processing regulations and the court feels that public health is at risk, it will impose a prohibition order. This prohibits the business from further trading until the faults have been eliminated and the enforcement authority lifts the prohibition order.

If an officer believes that a business gives rise to imminent health risks he or she may close it by issuing an emergency prohibition order through the courts.

Legal proceedings

The principal defence under this act is the defence of due diligence. Food traders will be able to plead a modified form of 'due diligence' where defendants can be deemed to have exercised 'due diligence'. These deemed defences are not available to manufactures and importers, nor do they apply to the offence of rendering food injurious to health.

The Food Act has increased the maximum penalties available to the courts. For most offences, Crown Courts will be able to send offenders to prison for up to two years and/or impose a fine of unlimited size. Magistrates courts will be able to impose a fine of up to £2,000 per offence and a prison sentence of up to six months. For the first three of the main offences, a magistrates court may set a maximum fine of £20,000: there are additional penalties for obstructing an enforcement officer.

Regulations

From April 1991, new regulations require that premises, including vehicles used for food businesses, be registered with their Local Authority so that the authority can identify the number and type of food premises in its area and be able to target enforcement action more effectively. Registration of short-term events, where catering is provided over several days, will be necessary, as will permanent premises used by charities. Food handlers will all be required to have appropriate hygiene and food handling training.

Regulations will cover the sale and importation of 'novel foods', that is, foods which are new or have only rarely been eaten in this country. Premises where food is irradiated will be subject to licensing. Inspectors appointed by central government will check food irradiation plants before approval is given and regularly thereafter.

Many of the key provisions in the Food Safety Act are contained in The Food Labelling Regulations 1984, The Food Hygiene Regulations (General) 1970, and The Imported Food Regulations 1984. The new act will strengthen and aid enforcement of these important regulations, while legalizing the registration of food premises and the training of all food handlers.

ADVERTISING STANDARDS AUTHORITY (ASA)

The ASA lays down a set of rules to which advertisers must conform. All advertisements must be legal, direct, honest, and not be offensive.

FOOD LABELLING REGULATIONS 1984

According to the these regulations, the labels on food products must give:

- the name of the food,
- a list of ingredients with the largest component first,
- an indication of the minimum durability (life of the food).

It is an offence to give a label to any food which falsely describes the food or is calculated to mislead as to its nature or quality. However, it is not an offence if it can be proved that the person who did so, did not, or could not, know, or with any reasonable diligence have found out, that the label was false or misleading.

ACTIVITIES

- Design an accident report form using the list of facts to be recorded listed on page 19.
- Using the following headings, complete a list of work conditions and facilities that employees should find at their place of employment:
  machinery,
  first aid,
  staff facilities,
  premises,
  training.
- Within the legal framework provided by the Health and Safety at Work Act 1974, both the employer and the employee has obligations towards each other and the general public. Discuss the obligations in relation to a busy kitchen and restaurant. Give specific examples of how these aspects of health and safety can be incorporated in the work place.
- Employers will create a reliable, willing and able work force if they can appreciate how their employees feel about their jobs. Figure 1.16 indicates the aspects of the job that concern the employee. Discuss the figure and list the points that you think the employer should be concerned about. Do you think that there are other aspects of the job other than those given that are of concern to the employee?

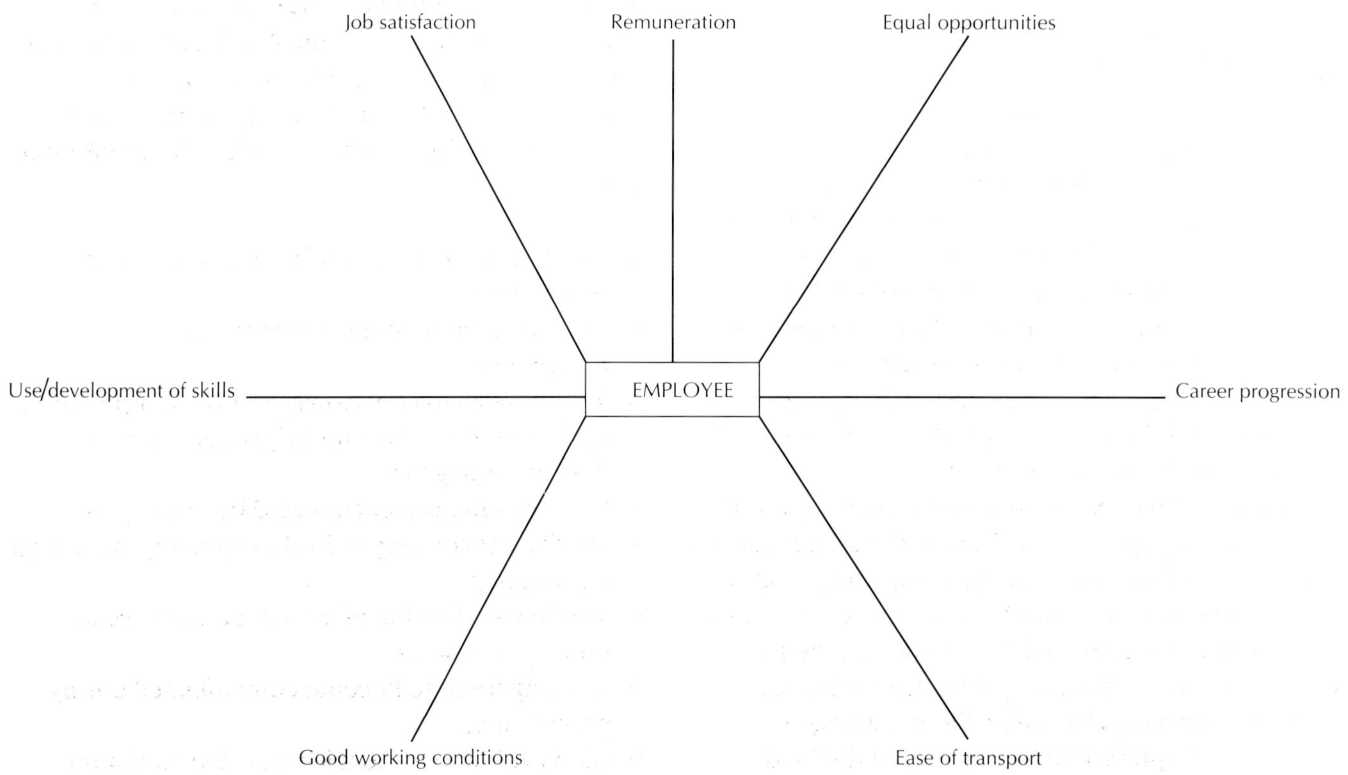

1.16 Employee's concerns.

CHAPTER 5 Food hygiene and safety

A kitchen and the staff should be clean, and the possibility of food poisoning should be minimized through proper preparation and storage of food. It should never be forgotten that food poisoning can kill.

Under the Food Safety Bill, basic food hygiene training is now a statutory obligation for all persons who are engaged in the storage, preparation and service of food. It does not apply to those establishments that only handle wrapped, canned or bottled food, packaged confectionery, fresh fruit and wet fish.

It is essential that all food handlers understand the ways that bacteria multiply, the effects of food poisoning, food contamination, personal health and hygiene, cleanliness of premises and equipment, and their responsibilities under the law.

MICROBIOLOGY

Microbes are organisms that are microscopic in size. Bacteria are a group of microbes (micro-organisms) that are mostly unicellular, and may be spherical, rodlike, spiral, comma-shaped or corkscrew-shaped. They reproduce by dividing in half once every 20 minutes. Therefore in 6 hours 1 microbe will have multiplied to over 130,000 microbes. In order to live and grow, many microbes need to have access to time, food, oxygen, water and warmth together with a degree of acidity pH 4.5 to pH 8 (vinegar has a pH of 4 and caustic soda a pH of 10).

Pathogens are microbes that will produce poisons (toxins) as they grow in the food, and these toxins will cause human beings to become ill when the food is eaten. An example of these is *Staphylococcus aureus*. Some pathogens grow in the intestine after being eaten in the food where they produce toxins. An example of this is *Salmonella*. Other pathogens produce heat-spores that lie dormant in the food. Once eaten, the spores become active bacteria that produce toxins. *Clostridium perfringens* is a spore-producing bacterium that is able to grow in the centre of foods, such as minced meat or vacuum packed foods, and without needing oxygen (an anaerobe).

Pathogens have to be present in large numbers in order to cause illness – it is impossible to prevent small numbers being present in food. Growth of these pathogens may be prevented by keeping food at a low temperature (below 5°C), or at a high temperature (above 64°C). The food itself can prevent their growth if it is too dry or too acid, or if the organism is prevented from absorbing water in the food because of salt or sugar.

Food spoilage is caused by the growth of moulds on the food or by foods being bruised or otherwise damaged.

POISONOUS FOODS

The common symptoms of food poisoning are diarrhoea, vomiting, abdominal pain and sometimes, headaches and fever. Symptoms may become evident after six hours and up to three days after eating the infected food. The common causes of food poisoning are:

- food being prepared too far in advance of the service time,
- food stored at room temperature, not in a refrigerator,
- food allowed to cool slowly at room temperature,
- food not reheated adequately/sufficiently to destroy pathogens,
- the use of foods contaminated by pathogens,
- insufficient cooking of foods, especially meat, eggs and poultry,
- insufficient thawing of frozen foods, especially poultry and meats,
- allowing foods to become contaminated during preparation,
- allowing hot foods to fall below the minimum temperature of 63°C,
- allowing infected food handling staff to work in food areas.

Food poisoning cases are being reported in larger numbers due to increased public awareness, changes in eating habits, increased numbers of outside and function caterers, the popularity of retail chilled foods, and the move towards foods that contain no preservatives. *Salmonella* organisms are responsible for 70 per cent of reported cases, *Clostridium perfringens* for 20 per cent of cases and *Staphylococcus aureus* for the remaining 10 per cent. *Salmonella* may be found in many foods and, in particular, poultry, eggs, milk and beansprouts. Great care should be taken to ensure that sufficient heat is applied in order that *Salmonella* organisms are destroyed (see Table 1.7).

There are some foods that cause severe allergic reactions and violent diarrhoea. These foods include poisonous mushrooms, toadstools and raw, red kidney beans; some fish from the Caribbean, including tuna, mackerel, sardines and pilchards at certain times of the year, or in certain weather conditions. Scombroid toxin poisoning occurs if the fish is kept at a temperature that permits the growth of bacteria which break down the fish flesh and produce a poison. To prevent this action the fish must be kept at a temperature of below 4°C. Shellfish and mussels can become infected by algae. Spring and early summer are the most likely times and fish are regularly inspected by the local authorities for this type of contamination. If eaten when infected, they can be the cause of paralytic shellfish poisoning.

THE LARDER

Since much of the food used in a kitchen is prepared in the larder, special care must be taken in this area to avoid contaminating the food. Equipment must be kept clean and used only for its specific purpose; food must be stored appropriately at the correct temperature and away from likely contaminants.

It is essential that all members of staff are trained in all aspects of hygiene so that they know how to avoid the risk of food poisoning, particularly through cross-contamination. The likely areas in the larder where contamination may occur are given in Table 1.8, together with the means of preventing such an occurrence (see also Table 1.9).

FOOD HANDLER'S RESPONSIBILITY

1. If an employee develops an illness it is his or her duty to notify the employer. This includes skin rashes, boils, septic spots, nose and throat infections, bowel trouble, nausea or vomiting. Should a member of his or her immediate family become affected by any of the above symptoms, the employer should also be notified.
2. An employee should notify his or her own doctor that he or she is a food handler if medical advice is sought. An employee should also seek medical advice if he or she has fallen ill while on holiday at home or abroad.
3. Employees should wash their hands after using the toilet, upon entering a food handling or storage area, and after handling any food.
4. Employees should wear clean protective clothing including protective shoes and a head covering when preparing and cooking food.
5. Employees should refrain from licking their fingers or utensils when tasting foods. A special tasting spoon should be used once and then washed.
6. Employees should take every precaution to minimize cross-contamination of foods by proper storage and handling of all foods and equipment.

Table 1.7 The sources and prevention of food poisoning

Contaminant	Source	Cause of poisoning	Prevention
Salmonella species	Raw foods of animal origin, especially poultry	Foods inadequately cooked	Cook foods properly
	Human intestines	Hands contaminated	Hands washed after using toilet
	Insects, vermin, animals	Bacteria carried on their bodies	Clean premises, keep out animals
Staphylococcus aureus	Unclean hands	Bacteria passed onto foods during preparation	Wash hands after each task

Table continued

Table 1.7 continued

Contaminant	Source	Cause of poisoning	Prevention
	Septic cuts, spots, boils	Bacteria passed onto food during preparation	Clean skin, waterproof plaster, good health
	Hand touch face, nose, ears, neck	Bacteria passed onto foods during preparation	Clean skin, good health, adequate uniform and hats
Clostridium perfringens	Soil on vegetables	Foods inadequately prepared, bacteria not destroyed	Separate vegetable preparation area, wash vegetables carefully
	Raw meat	Food inadequately prepared, bacteria not destroyed	Raw meat prepared in separate area and cooked adequately
	Human intestines	Hands contaminated	Hands washed after using toilet and between tasks
Chemicals	Chipped enamelware, worn copperware	Foods contaminated	Use stainless steel and plastic/ polythene ware
	Lead pipes	Water contaminated	Know water supply source
	Insecticides on fruit and vegetables	Fruit and vegetables contaminated	Wash carefully before use
	Rodenticides (rat poison)	Foods contaminated	No rodents on premises
	Chemicals in water	Fish and watercress contaminated	Know source of water supply
Campylobacter species and *Listeria monoytogenes*	Unpasteurized milk, gut of cow or pig, poultry and other animals, dairy products	Food contaminated	Foods prepared properly Foods stored at low temperatures
Bacillus cereus	Rice and cereals	Foods contaminated	Purchase from reliable supplier
Clostridium botulinum	Fish, canned foods	Foods contaminated, blown cans	Purchase from reliable supplier

Table 1.8 Contamination in the larder – danger areas and means of prevention

Contamination danger area	Means of prevention
Refrigerators	Separate refrigerator for fish, meat and cold items Never allow different items to touch each other
Cutting surfaces	Butcher's blocks for carcass preparation Coloured boards for cutting: blue: fish red: meat yellow: cooked meats green: vegetables white: dairy produce
Washing areas	Separate sinks for fish, meat and vegetables
Preparation equipment	Complete stripping down and cleaning between each task: mincers, bowl choppers, slicers, mixers, tenderizers

Table continued

Table 1.8 continued

Contamination danger area	Means of prevention
Waste	Waste from raw and cooked items should be disposed of via a waste disposal unit Bones, skins and other items that are difficult to dispose of should be bagged and placed in outside lidded bins Collection of bins should be frequent and regular
Utensils	Knives, choppers and all other utensils should be washed thoroughly after each task
Cloths	All cloths and muslins should either be boiled after use or be of a disposable nature (bacteria can be harboured in damp cloths)

Table 1.9 Contamination of foods

Food	Cause and prevention
Fish	Has a short shelf life and distinctive odour that will contaminate other foods
Meat	Bacteria present in raw meats are *Salmonella* species and *Clostridium perfringens* Cooked and raw meats must never be allowed to come into contact with each other Cooked meats should be kept very cold and covered until ready to be eaten
Pork	*Trichinia* (a parasitic worm) may be present in the flesh, so very thorough cooking is essential
Poultry	Liable to contain *Salmonella* bacteria, and so should be kept separate from all other foods and as cool as possible. Poultry should be thoroughly cleaned on the inside and only cooked when completely thawed out
Game, feathered	Should be hung for 2–3 days to tenderize and improve the flavour. When drawn, game should be thoroughly washed inside to remove all traces of bacterial contamination
Offal	Susceptible to contamination and has a very short shelf life – for this reason it is usually purchased frozen. It must never be refrozen after defrosting
Meat products	Avoid excessive handling and chopping as this generates heat, thus increasing the growth of bacteria present in the meat
Eggs	*Salmonella* bacteria are present in some eggs, but it is impossible to determine in which eggs they are present. Special care should be taken when using raw eggs in preparations that will not subsequently be cooked, especially mayonnaise and Hollandaise sauce
Sauces	Sauces must be kept either very hot or very cold because they provide an ideal place for bacteria to multiply
Stocks and glazes	Should be maintained at boiling point or strained, cooled and refrigerated for 24 hours maximum. Never keep warm stock in the larder as it is an ideal place for bacteria to multiply
Vegetables	Soil contains *Clostridum perfringens* so vegetables must be washed thoroughly and prepared in a separate area away from all other foods
Cream, fresh and artificial	If allowed to, cream will quickly reach room temperature and then provide an ideal area for bacteria to multiply. It should be kept refrigerated until it is to be eaten

FOOD SAFETY

Food safety is of paramount importance to the caterer. Figure 1.17 illustrates the main types of danger likely to affect the safety of food being prepared and served for human consumption.

One of the most important areas of food safety is the area in a large, traditional kitchen where food is prepared - the larder section. In a smaller unit, it may simply be an area of the main kitchen where cross-contamination can easily occur unless all food handlers are consistently careful as they work.

Food storage is another area where care must be exercised at all times. Table 1.10 indicates the main types of foods, their storage method and quality points. Food is of no value to the caterer if it is of poor quality when it is delivered. Efficient catering establishments will ensure that all foods are checked for quality as they are delivered to their unit.

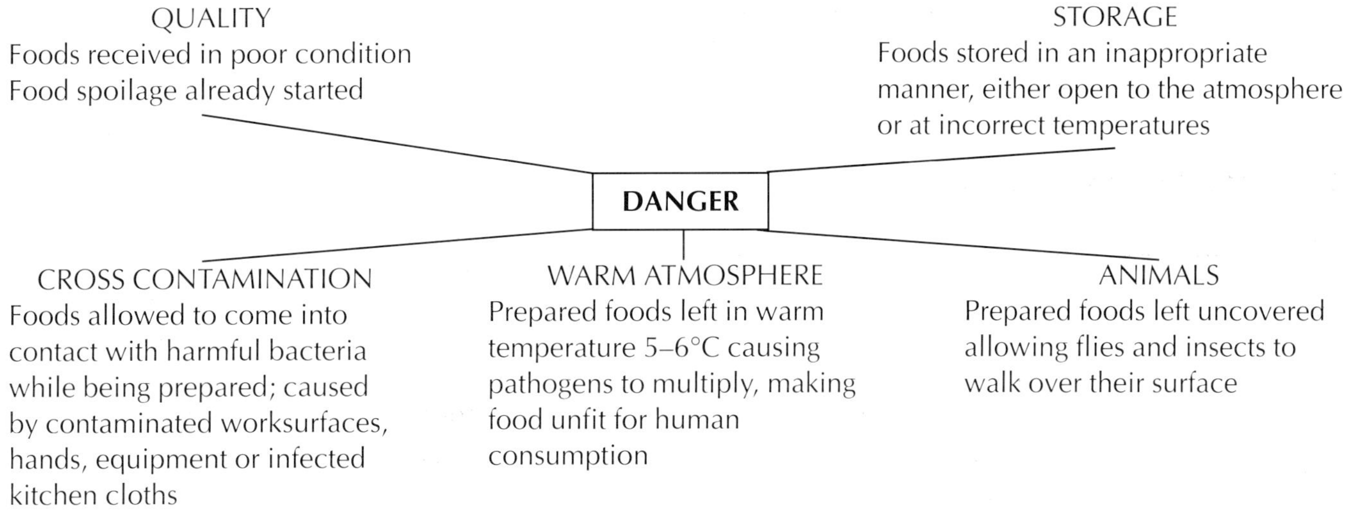

1.17 Areas affecting the safety of food.

Table 1.10 Food storage and quality

Category	Food	Storage method	Quality points	Safety points
Fresh	Meat	Hung in Cold Room 1–4C or placed on clean trays in refrigerator	Firm, moist fat and flesh No dried areas or discolouration Pleasant smell	Keep separate from other foods Do not allow blood to drip onto other areas
	Bacon	Hung in cold room 1–4C Wrapped in muslin or when sliced placed on tray and covered lightly	Thin rind Fresh smell Moist but not slimy	Can easily dry out Will deteriorate quickly
	Vegetables	In bins or on racks in cool, well-ventilated area	No insect infestation Natural colour Crisp	Use within 24 hours to ensure crisp and fresh
	Fruit	In trays or punnets Cover with muslin Keep cool	Natural colour No sign of mould or insect infestation	Will bruise or squash easily Handle carefully
Dairy	Milk and cream	In refrigerator	Fresh smell Sealed containers	Use within 24 hours Excessive heat and sunlight will cause deterioration

Table continued

Table 1.10 continued

Category	Food	Storage method	Quality points	Safety points
	Cheese	Wrap in muslin Place on shelf in cool area	Natural colour No mould growth No fat globules or cracks on surface	Keep strong cheeses away from more delicately flavoured varieties
	Eggs	On their trays in a cold room 1–4C, pointed end downwards	Smooth shell No cracks in shell Check use-by date on crate	Use in rotation Eggs deteriorate easily if allowed to become stale
	Fats	On trays in refrigerator	Check use-by date No broken wrappers Fresh smell Natural colour	Will deteriorate if stale Use in rotation Will go rancid if past sell-by date
Dry goods	Cereals and pulses	In sealed bins or containers in cool, dry area	No sign of insect infestation Natural colour and smell Dry	Oats will go rancid if stale Keep dry Use in rotation
	Grains and pasta	In sealed containers in dry area	Natural colour Dry No sign of damage	Pasta is brittle and will break if carelessly handled
	Flour	In lidded bins in cool, dry areas	Natural colour and smell No sign of insect infestation	Use in rotation Wholemeal flour in particular will go rancid
	Sugar	In lidded bins Icing sugar in sealed containers	Should not be lumpy Natural colour and smell	Will go lumpy if allowed to get moist
	Fruits	In sealed containers in cold, dry area	No sign of stalks or mould Should be moist inside	Can dry out too much if stale Use in rotation
	Tea	In sealed containers	No sign of stalks Natural smell	Use in rotation Will lose flavour if stale
	Ground coffee	In sealed containers or vacuum sealed in easy to use containers	Dry and even Fresh smell	Use in rotation As soon as ground, the flavour will begin to deteriorate

HIGH RISK FOODS

Some foods are more susceptible to pathogen growth and/or contamination than others. This is because they provide the ideal growing conditions for the pathogens or because they have to be handled frequently during their preparation. The following foods come into this high-risk category. All these should be treated with even more care than usual. Attention should be paid to minimal handling or to the use of disposable gloves and the fast cooling and refrigeration of foods:

Stocks, soups, sauces, gravies, savoury dips, mayonnaise, Hollandaise sauce, cream, custard, synthetic creams and cold desserts. Trifle, cream cakes, meat rissoles, fish cakes, Scotch eggs, fish pie, shepherd's pie, burgers, meat pies, pasties, sausage rolls, cold water crust pies, pâté, and cold, sliced cooked meats. Cooked meats in sauce or gravy. Chicken portions, peas, pancakes, and cooked, cold poultry.

The cooling of foods that have been cooked and that need to be served at a later time or served cold is a high-risk process. There are two safe methods of cooling foods. These are either immersing the food in cold, running water, or by placing on shallow trays, covering and keeping in a cool place until cold enough to refrigerate.

Those foods that have a liquid element, such as soup or braised steak, should be cooled by transferring to shallow trays. This has the effect of increasing the surface area that is exposed, thus enabling the heat from the food to disperse quickly. The trays of food should then be covered and placed in a cool atmosphere to speed up the cooling process. Once cool, the food may be poured into a deeper, clean container, covered and refrigerated until it is to be used.

The 'dry' types of foods, such as rissoles, Scotch eggs and pastry products, should be cooled rapidly by placing on cooling racks, covering with muslin and then placed in a cold atmosphere until they are cool enough to refrigerate.

REFRIGERATORS AND DEEP FREEZERS

The refrigerator should be set at a temperature of below 5°C, and the deep freeze at - 18°C- -22°C so that all the foods are maintained at a safe storage temperature. Refrigerators and deep freezers should be defrosted and cleaned regularly. While this cleaning is in progress, any foods that were in the cabinets should be stored in other refrigerators or freezers. Foods should be covered before placing in a refrigerator and always in clean, suitably-sized containers. Foods should never be permitted to touch other commodities or surfaces, and should be labelled and dated so that they are never allowed to be stored for longer than is essential. In a deep freeze, all foods should be firmly sealed in suitably-sized bags and labelled so that they may be clearly and easily identified. Once thawed, raw frozen foods must be cooked (see Table 1.11).

It is very dangerous to refreeze foods once they have been thawed as pathogens grow readily while food is thawing out and freezing does not destroy them. Pathogens must be destroyed by heat before the food is safe to eat (Above 63°C).

Care should be taken when thawing freezer foods. They should be placed in clean, deep trays, covered and placed in a refrigerator to allow them to thaw safely at a controlled, low temperature.

Table 1.11 Examples of minimum thawing time

Fruit	6 hours	Joints of meat 8 hours per 2kg weight
Fish	6–hours	Poultry 24 hours per 5 kg weight
Chops, sausages	6–8 hours	Cooked casseroles 10–15 hours

KITCHEN HYGIENE

The food hygiene regulations contain basic recommendations about the premises used for storing, cooking and preparing foods. The site should show no obvious sources of contamination. The size and layout of the premises should be planned to facilitate the maintenance of hygienic conditions, incorporating a logical work flow pattern with the minimum of back tracking in the process. There should be adequate means of ventilation with a minimum of fifteen changes of air per hour. All extraction systems should be regularly cleaned and maintained, and windows should be screened to reduce the risk of flies entering. Surfaces should be impervious to water and designed to eliminate dirt traps, and floors should have a non-slip surface.

Doors should be self-closing, made of metal, rubber or laminated. External doors should be close fitting to prevent the entry of vermin. Windowsills should be slightly sloping to prevent them being used as additional shelves. Lighting should be adequate and designed to eliminate dark corners and shadows. There should be an adequate, continuous supply of hot and cold water. The nondrinking water taps should be labelled. Hand basins should be supplied in all food preparation, storage and service areas; these should have hot and cold running water, soap, a nail brush and hand drying facilities, and should not be used for any other purpose. Floors should be well drained, gullies should be covered with suitable grids and well maintained. All refuse should be dispersed of either via a waste disposal unit, or in proper lidded refuse containers, which should be cleaned regularly and emptied daily. Food stores must be well maintained, adequately ventilated and organized so that food is issued in rotation. Cooked and raw foods should never be allowed to come into contact with each other. Kitchen equipment should be kept clean and well maintained, laminates should not be used for chopping purposes. There should be separate boards for vegetable preparation to prevent cross-contamination.

PESTS ON FOOD PREMISES

Pests are a cause of food spoilage and contamination. Brown rats will enter premises via sewers, drains, lift shafts and open doorways. The house mouse will enter through cracks in walls and mortar, and ill-fitting doors and windows. The signs of rat and mouse infestation are easily recognized – look for droppings, gnaw-marks on food stock packages, damaged stock and grease marks from their bodies along skirting boards and walls. Cockroaches emit a foul smell as they run around the premises. They are nocturnal and so are not often seen during working hours. They breed in warm areas where there is a good supply of food, which they nibble and contaminate. Cracks in plasterwork and poor quality wood skirting boards create the ideal home for cockroaches. Flies, wasps and birds also contaminate foods. Screens should be fitted over all openings so that doors and windows may be opened for ventilation without the risk of flies and wasps entering the premises. Wall-mounted, electrically operated fly traps may also be installed. If flies land on food, they contaminate it by either passing on bacteria from their bodies, or by regurgitating their previous meal onto the food. Blow flies are attracted to meat. They lay their eggs in clusters, which hatch out to form maggots. Birds will enter food premises more by accident than desire, they will spread *Salmonella* bacteria and will foul food with their faeces.

In severe cases of infestation the Environmental Health Officer should be contacted. However, there are many steps that the caterer can take to prevent infestation by pests. These are as follows:

- keep premises in a good state of repair and check external mortar,
- seal up any holes around pipework and drains,
- clean the lift shaft regularly and keep it in good repair,
- fit removable screens over windows against flies, birds and wasps,
- check for and repair any cracks in skirting boards and walls,
- install self-closing doors,
- store food in well-organized storerooms, above floor level (never leave open or uncovered food out),
- clean the entire premises regularly and thoroughly,
- use a waste disposal unit for food waste and keep refuse in clean, lidded bins outside the premises and empty them regularly,
- train employees to be vigilant and aware of the dangers of pests on premises.

THE IMPACT OF A CLEAN, HYGIENIC, SAFE ENVIRONMENT

Dust consists of organic and inorganic substances. It is generally termed 'dirt' when it becomes trapped in moisture or grease. Inorganic substances acting as dust include grit, soot, dry earth, sand and chalk; organic substances are such items as leaves, pollen and grass, tiny animals, hair, fur, skin, droppings, bacteria, yeasts and moulds.

When an area said to be 'clean', it means that as far as is possible these inorganic and organic substances have been removed by cleaning procedures. It should be realized that, in addition to these substances, much debris is created by the people using the premises.

Cleaning is carried out to create a safe, clean environment and to maintain the condition of furniture and fittings. Bacteria can multiply in dirt, and dirty premises attract vermin.

It is essential to provide clean and attractive surroundings for the guests and for the staff. The cleaner and more attractive environments will tend to encourage staff to maintain high working standards, thereby increasing their job satisfaction. Poorly maintained establishments, on the other hand, will only become worse as staff have no incentive to keep them clean. Needless to say, guests will only frequent the clean and attractive establishments.

FOOD PREPARATION AREAS

In areas such as restaurants, bars, kitchens, cafeterias and store rooms, it is less easy to divide the cleaning into daily and weekly tasks. The prime consideration is the necessity to maintain hygienic conditions for the preparation and service of foods and beverages. It is therefore necessary to employ a cleaning rota that will ensure that every area is maintained in a hygienic condition.

Where a kitchen, restaurant or bar is used for twelve or more hours a day, it will be necessary to thoroughly clean the area at the end of each working shift, so that it is ready for the incoming team of staff. This means that these areas have a daily clean or a special clean (when all areas are stripped and cleaned). This special clean often has to be completed overnight so as not to interrupt the daily business.

HYGIENE SCHEDULE

WEEK ENDING ________

	ITEM	PRODUCT CODE	FREQUENCY	M	T	W	T	F	S	S	C.T NO
	BAIN MARIE & HOT CUPBOARD	4A	DAILY								16
	BOILER & STEAMER	4A	DAILY								15
	CAN OPENER	4A	AFTER USE								24
	CANOPIES	5A	WEEKLY								30
	COLD ROOM	3K/4A	WEEKLY								38/58
	DISHWASHER	4A	DAILY								13/29
	DISPLAY CASE	3K/4A E	DAILY								18/53
	FAT FRYER	4B	WEEKLY								27
	FLOORS	4B	DAILY								11
	FOOD PREP SURFACES	4A	AFTER USE								10
	FOOD STORE	3K/4A	WEEKLY								1/50
	FRIDGES/FREEZERS	4A	WEEKLY								3
	MIXERS & SCALES	3K/3A	AFTER USE								25/56
	OVENS & GRILLS	4C/4N	WEEKLY								26/59
	SHELVES	3K/4A	WEEKLY								32/57
	SINKS & BASINS	4A	DAILY								9
	SLICERS & MINCERS	4A	AFTER USE								23
	TABLES & CHAIRS	5A	DAILY								20
	VENDING MACHINES	4A	DAILY								54
	WALLS & DOORS	5A	DAILY								12
	WASTE BINS	4A	WHEN EMPTY								4
	WASTE DISPOSER	4A	TWICE DAILY								7

1.18 Hygiene schedule.

To ensure that cleaning is completed to the required standard, work cards are used. These will specify exactly how each item of equipment or area is to be cleaned. Teamwork is an essential component of keeping the work areas clean and the reasons for the thorough cleaning of all food handling areas should be familiar to all staff:

- to eliminate the risk of cross-contamination,
- food debris provides a growth area for pathogens,
- food scraps provide food for pests,
- stale food crumbs can accidentally be incorporated into fresh foods,
- cleanliness encourages cleanliness among employees,
- cleanliness is a requirement under food hygiene regulations,
- cleanliness will encourage trade.

CLEANING OF FOOD HANDLING AREAS

Water and detergent are used to clean effectively. First, a suitable detergent must be selected. Detergents are made from varying combinations of soap, carbonate, silicates and phosphate. Some will have a low pH value and therefore have acid properties, for example, toilet cleaners have a pH of 2 or 3. Some will have a high pH value and therefore have alkali properties, for example, bleach has a pH of 10, caustic soda a pH of 12, and oven degreaser a pH of 13. General cleansing detergents, such as washing-up liquid, will have a neutral or very slightly alkali property and will, therefore, have a pH value of 7 or 8. Some detergents have disinfectant properties - these are known as sanitizers - and they have the ability to clean away food debris and grease as well as destroying micro-organisms. To be effective, a detergent should be able to:

- reduce the surface tension of the water (wetting properties),
- emulsify fats, thus enabling them to be rubbed away freely,
- soften the water thus reducing the formation of scum,
- be used safely, that is, be harmless, nontoxic, odourless and tasteless,
- be rinsed away freely,
- be used easily and economically.

Within a food handling area it will be necessary to use more than one type of detergent: a good purpose sanitizer, an acid toilet cleanser, an alkali degreasing agent, an acid descaling agent, an abrasive cream or powder, a dishwashing cleaning agent and rinse aid, and a detergent suitable for floor cleaning.

The term 'disinfect' means to rid something of micro-organisms. Water at 80°C is an effective disinfectant and is used to disinfect dishes after the washing and rinsing process.

The cleansing programme for all food handling areas should be compiled after the following points have been considered:

- standard of cleanliness required, for example, a detergent will be used for visible cleanliness whereas a sanitizer of disinfectant will be used for microbiological cleanliness,
- identify the areas and equipment to be cleaned and decide on the frequency of cleaning necessary to maintain the standard previously identified,
- specify cleaning agents and equipment suitable to obtain the standards required,
- identify any safety factors connected with either the equipment to be cleaned or the cleaning process itself,
- decide upon a cleaning rota that will specify when

each task has to be carried out, and identify responsibility for its standard and completion,

- train and supervise employees so that they are not only competent to carry out the cleaning to the required standard, but appreciate the reasons why cleanliness is of paramount importance.

Regular monitoring is essential to ensure that standards of cleanliness are maintained. Work cards for staff to use are a valuable training aid, and also serve as a constant reminder, both of the standard to be achieved, the methods and agents to be used, and the actual surface areas and items of equipment that are to be cleaned.

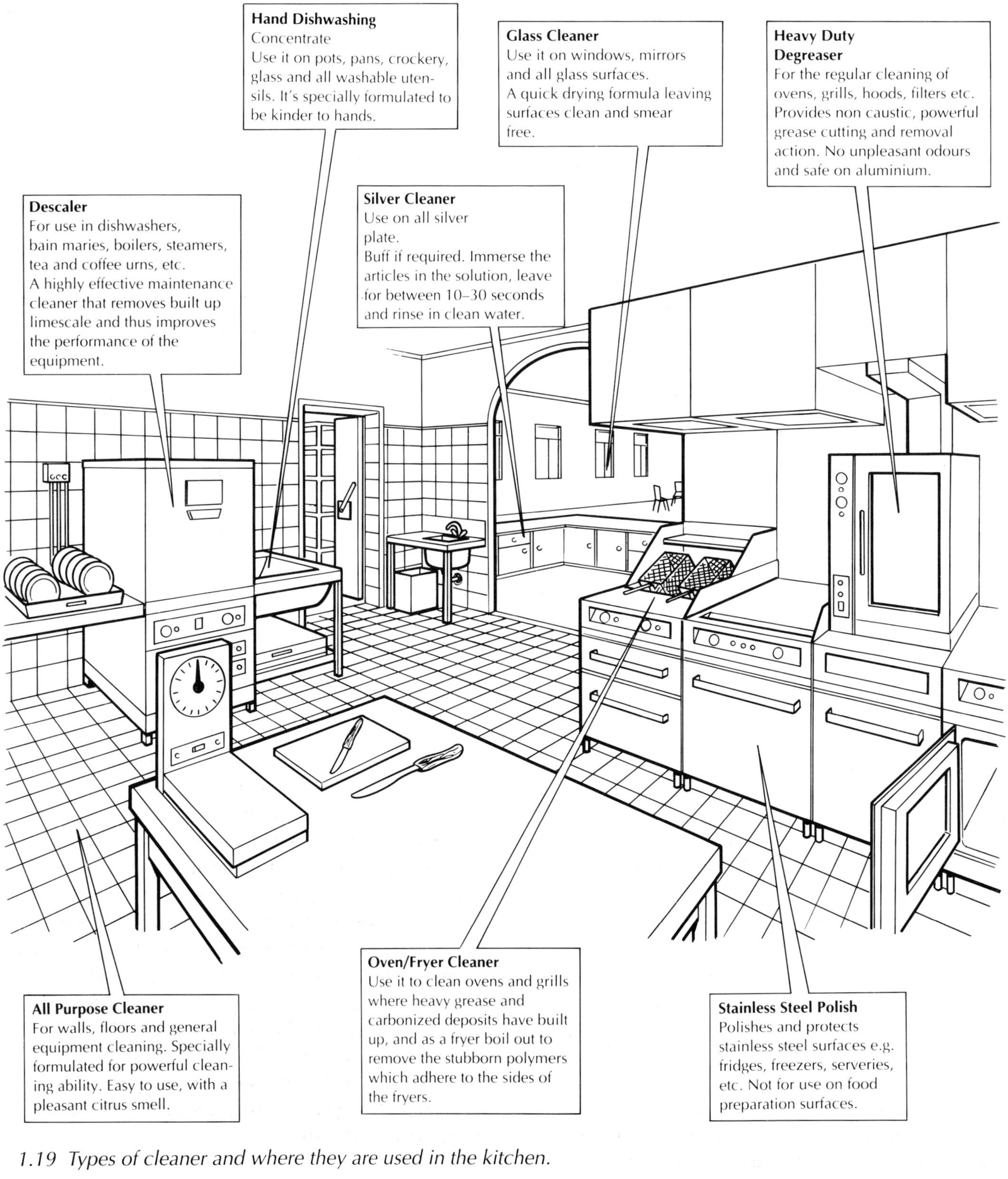

1.19 Types of cleaner and where they are used in the kitchen.

PRODUCT	EQUIPMENT	CLEANING METHOD	STANDARD
DIVERSOL BX 4A Pink Powder 	Bucket Sponge cloth 	After defrosting 1. Make up a solution of DIVERSOL BX crystals. 1/2 cup per bucket of water. Concentration 1-2oz/gallon, 6-10g/litre. 2. Rinse sponge cloth in the solution and wash. 3. Rinse with clean water. 4. Allow to air dry. Take care that door seals and tray channels are clean and dry. Keep clear of motor housing.	No food crumbs No smells Door seals clean

FRIDGES AND FREEZERS

PRODUCT	EQUIPMENT	CLEANING METHOD	STANDARD
R4 FURNITURE POLISH 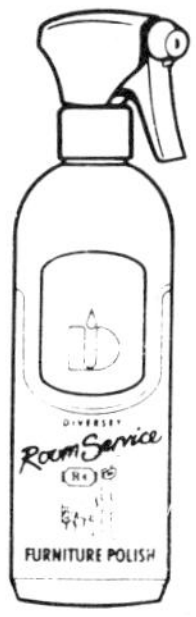	'D' cloth 	1. Shake the bottle well. 2. Spray sparingly onto a clean dry 'D' cloth. 3. Apply to the surface and polish to a shine. 	Dust free Lustrous sheen Clean

1.20 Work cards for cleaning fridges and freezers and tables and chairs.

METHODS OF WASTE DISPOSAL

There are many methods commonly used to make the disposal of waste products safer, more hygienic and easier. Within the catering industry there is often a combination of more than one of the following methods used for waste disposal.

Waste bins may be wheeled or static with a hinged or free lid and are often used with strong polythene liner bags. These bins are emptied by the Local Authority.

Waste disposal units are installed with a cold water supply and outflow to the waste water drainage system. Normally the grinding action cannot commence until the safety plate is in position to safeguard the operator's fingers. These units are useful for food scraps, vegetable peelings, stalks and outer leaves. They will not grind up solid items such as bones, cans, skin and polythene.

Compactors come in various capacities. There are small units that may be fitted neatly into swill and preparation areas. An electrical supply is required to develop the compaction force necessary to reduce the volume of waste by 75 per cent of its volume. The compactor will crush bottles, flatten cans and cardboard and crush food and vegetable matter. The use of a compactor will eliminate the problems caused by numerous overflowing waste bins. Once compacted, the refuse is removed in a plastic bag, sealed and placed in a waste bin ready for collection by the Local Authority.

Shredders are used to cut up such items as paper, card, metal and other waste in order to reduce large volumes of waste to a more manageable size. However, unless there is likely to be a large quantity of waste to be shredded, such a method may not be economical to employ.

Paper balers cope with large volumes of paper and cardboard, compressing and baling up these products into neat 30–50 kg bales. Unless there is a large volume of paper and card, it is not an economical method to use.

WASHING UP METHODS

Washing up is an essential part of any catering process. There are three areas where the washing process is used to break the chain of infection by the destruction of harmful micro-organisms and by the removal of waste particles of food and, in some cases, human saliva: the busy pot wash area in the kitchen itself, the crockery and cutlery wash area adjacent to the restaurant and, finally, the glass wash area in the bar. Each area has to be systematically worked and logically planned so that no cross-contamination can occur between the soiled items and the cleaned items. The following points should be considered:

- washing up should be completed away from food preparation and storage areas,
- the washing up sinks and facilities should never be used for any other purpose because of the risk of cross-contamination,
- greasy kitchen pots, pans and utensils should always be washed in a separate designated area,
- a sanitizer should be used to ensure that grease is removed and micro-organisms are destroyed,
- to ensure safe results, the recommended washing and rinsing temperatures must be adhered to – wash 60°C, sterilize 82°C,
- after sterilizing, air drying will ensure minimum recontamination.

A machine is normally used to wash crockery and cutlery. Machines are of many different capacities and designs, small, manually-operated models right through to vast, fully-automatic versions. They all work on the same principles (see Table 1.12).

Cutlery needs special attention, between the fork prongs and the spoons that could have traces of lipstick on them. These are checked at stage 1 so that no problems occur later on.

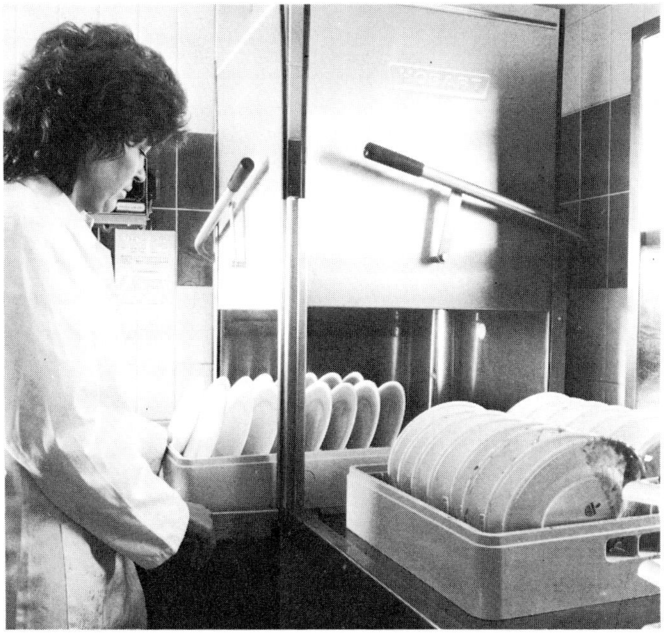

An automatic, door-type dishwasher.

Table 1.12 Principles of dishwashing machines

Process	Action	Result
1. *Preparation*	Scraping, soaking, prewash Items placed into racks	Food debris removed
2. *Washing*	Water at 60°C, with sanitizer Agitation with bristles or jets of water	Complete removal of all items of food particles
3. *Rinsing*	Water at 60°C, with rinse aid Water sprays onto racks	Complete removal of greasy deposits
4. *Sterilizing*	Water at 82°C sprayed onto racks	Complete destruction of micro-organisms
5. *Drying*	Racks removed and left to allow air drying	The rinse-aid at stage 3 and the heat at stage 4 ensure quick air drying

A washer for pots and pans.

STORAGE OF CLEANSED EQUIPMENT

It is of little value to cleanse items of equipment, crockery and glassware only to handle and store them in such a manner that they are recontaminated almost immediately. The transportation and storage of all cleansed items is therefore a major consideration.

In the kitchen, equipment should be allowed to air dry and then be stacked on racks in an inverted position or, where appropriate, with lids on so that dust will not settle on the surfaces that will be in contact with food. All racks should be constructed of impervious material, preferably stainless steel, so that they can be cleaned effectively and will create an unattractive surface for micro-organisms. Cooking utensils should be stored in plastic or stainless steel trays. The practice of keeping dishes, trays and utensils on racks above the cooking ranges and hot plates is not encouraged, as they will soon be covered with a thin film of grease from the cooking atmosphere – just right for dust to stick to.

Air dried items in the crockery and cutlery area should be stacked in such a manner that minimum further handling is necessary. For example, cups may be stacked directly onto their trays and carried back to the still room or beverage service area ready for further sales. This will ensure that they are as clean as humanly possible as they will only have been picked up once (by their handle) before they are reused. This principle of minimum handling and suitable stacking between uses is one that should be applied to all types of crockery and cutlery.

In the glass wash area, glasses are washed by machine in a similar manner to the crockery. The main problems here are the traces of lipstick that need to be removed at this stage, and the water marks that form on glasses with an uneven, exterior finish. Some glasses have a dip under the stem base which will hold water when the glass is inverted. Care should be taken to avoid such 'pools' of water.

The selection of the style of glassware should be made only after consideration to its care, cleanliness and storage. Glasses are washed and dried in an inverted positon in racks. They should be stored on impervious shelves or trays still in the inverted position as the rule of minimum handling applies as much to glassware as to other items. The rims should never be handled because of the risk of contamination. If the design of the glassware makes hand drying essential, then clean linen cloths should be used. Care must be taken so that the interior and rims are not touched by the hand. These linen glass cloths will leave the glass sparkling and dry, but will absorb moisure themselves; therefore a plentiful supply fo clean, dry cloths is essential. They will be of no use once they are themselves wet.

A compact glasswasher.

ACTIVITIES

- Study figure 1.21 and then compile a hygiene code of practice for food storage and handling in the larder.
- Describe what is meant by cross-contamination in the kitchen.

List precautions to be taken in the following areas: Bar areas
Restaurants
Salad bars
Sandwich preparation areas

- Discuss why a logical work flow pattern should be enforced in catering premises.

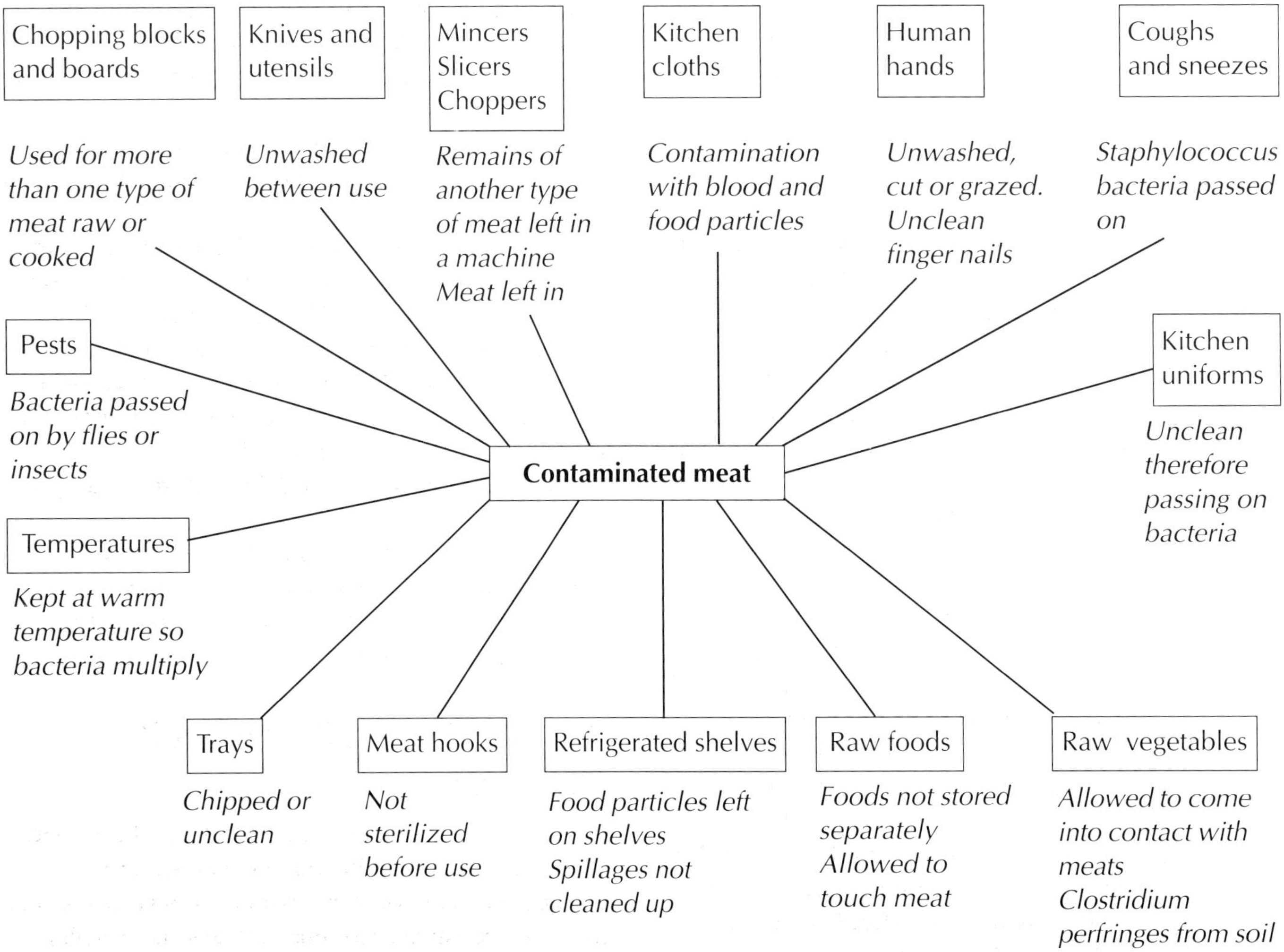

1.21 Cross-contamination causes.

CHAPTER 6 Catering premises

KITCHEN PLANNING

Kitchens should be planned so that the best use is made of the space available and so that each of the different areas is situated in the most convenient position. Thought should also be given to the type of food being prepared, cooked and served, the type of equipment that will be required, and the materials that will be used. However, questions of hygiene and safety will play a large part in the planning process.

Best use must be made of the area. It must be large enough to enable staff to carry out their work without accidents, or crossed work-flow lines.

- Floors must be non-slip, impervious, easily cleaned, resistant to heat damage and stains from foods and cleaning agents.
- Walls and ceiling must be light coloured, washable and nonabsorbent.
- Natural light should be used wherever possible. Flourescent lighting may be used to create as natural a light as possible, without causing any shadows or false colours on the food. There should be no dark, unlit corners as these attract rodents.
- Natural ventilation should be used as long as it will not create draughts. Extractor fans and hoods, fitted above the main cooking areas, will remove excess steam and cooking odours. There must be adequate and quiet ventilation in all service areas.
- The supply of drinking water should be checked. Determine the quantity of hot water that will be needed and then select the most suitable heating method.
- Decide whether to have one type of fuel for the whole unit, or two types in case of overloading or emergencies. The running costs, storage and delivery arrangements of the types available should be compared.
- Measure the areas to be heated, and check the availability of fuels and systems. Compare the costs and suitability of the systems.
- The number of staff who will be in the unit at any one time will need to be estimated, also the total number of staff that will be employed. It will be necessary to plan the type and size of toilet and cloakroom facilities that will be built for the staff. There will need to be a locker for each member of staff, and a rest room area with first-aid facilities. Hand washing facilities will be required in all parts of the unit.
- Kitchens should be planned so that work is completed in a logical sequence from food delivery, through preparation and cooking to service (see figure 1.22).

KITCHEN MATERIALS AND SURFACES

The choice of materials and surfaces should be made with safety in mind. The following aspects to consider are: ease of maintenance; imperviousness to moisture; resistance to damage (cracking, warping etc.) and to growth of pathogenic organisms.

Goods inwards →	Checked and stored →	Issued and prepared →	Served and sold
Easy access Unloading parking areas	Secure area Trained staff to check goods Adequate store area	Storekeeper to issue and record stores Kitchen staff to prepare foods to their best advantage	Foods garnished packaged, sold in an appetizing manner Clean, appealing surroundings

1.22 By examining the flow of work, the importance of a well-organized kitchen becomes clear.

Table 1.13 Material uses, cleaning care and cleansing equipment

Material	Uses	Cleansing care	Cleansing equipment
Wood	chopping blocks, rolling pins, and spoons or spatulas	carefully remove food debris; wash with warm detergent water; dry away from direct heat,	wire brush, detergent with a sterilizer, and a cloth.
Stainless steel gloss matt	sinks, splash guards, tables, shelves, trays, bowls, hotplates, ranges, ovens, grills and water boilers	brush away or soak off any food debris taking care not to scratch the surfaces; wash with detergent solution, or cream cleanser; rinse and dry,	soft brush, plastic scraper, cloth, dry cloth, and detergent with a sterilizer.
Enamelled steel	refrigerator cabinets, ranges and ovens,	wash with hot detergent water; rinse and dry; do not scratch surfaces,	cloth and a detergent with a sterilizer.
Steel	baking trays or sheets, oven shelves, ranges and hotplates,	wipe clean while still hot and oil sparingly to avoid rusting,	damp cloth, kitchen paper roll and edible oil.
Cast iron	frying pans, griddle plates and pans,	wash with hot detergent water; rinse and dry; do not scratch or scrape surfaces,	nylon brush, dry and wet cloths and a detergent with a sterilizer.
Chrome	handles on equipment, oven runged shelves and refrigerator shelves,	wash with hot detergent water; soak to remove food stains,	neutral detergent, soft cloth (wet and dry).
Aluminium	pans, trays and cooking equipment,	wash in hot water with scouring powder and wire wool; soak stains; rinse well,	scouring powder or pad, wire wool, still brush and a detergent with a sterilizer.
Plastics	handles on equipment, light equipment, bowls, chopping boards, shelves and work surfaces,	soak off any debris of food; wash in hot water with a detergent; rinse well and dry,	neutral detergent, soft brush and a wet and dry cloth.

SAFETY IN THE KITCHEN

A kitchen is a very hazardous place. Not only is much of the equipment dangerous, but many people are often working in a busy environment, all concentrating on their own tasks. Accidents do happen, but many can be avoided.

It is everyone's responsibility to ensure that they abide by the simple safety rules that are laid down in the kitchen. Each person is responsible for his or her own and everyone else's safety.

Table 1.14 summarizes some major causes of accidents and methods of prevention.

Any accidents to staff or guests should be reported to the management immediately. A clear statement of events leading up to the accident must be written in the accident book, which should be kept in the manager's office. This report is essential even if the person does not appear to be badly injured. An insurance claim may be filed at a later date and a detailed record will be required.

FIRE

There are strict regulations concerning the provision of escape routes, emergency exits, fire doors, hoses, sprinklers and alarms in the event of fire, as well as customer numbers in relation to the size of the premises. The fire brigade are obliged to regulate and inspect premises under the Fire Precautions Act 1971. Under these regulations, a fire certificate must be granted by the fire authority to all hotels sleeping more than six persons, and all catering establishments. The certificate ensures that the

Table 1.14 Summary of some major causes of accidents and methods of prevention

Areas	Causes	Prevention
Floor	Wet or damp floor Food debris left on floor Brooms and mops lying in corner Boxes or cartons left by refuse sack	Clean, tidy kitchen Cleaning rota Staff training Staff training
Light equipment	Knives and mandolins incorrectly used or handled, or broken equipment Equipment not stacked on shelves properly	Proper stacking of cleaned equipment Staff trained to use sharp implements
Heavy equipment	Oven doors left open or items placed on the edge of ranges Incorrect lifting procedure	Proper, and safe use of heavy equipment Staff training
Electrical equipment	Misuse of equipment Incorrect attachments used Not disconnected from power before being cleaned or damaged plugs or flexes	Staff training Attachments coded for each machine Warning notices
Heat/dry	Solid top ranges left on heat Thermostats not working on hot plates or handles too hot	Equipment maintenance
Heat/wet	Incorrect use of steamers, boiling pans, water boilers and coffee machines	Staff training
Clothing	Cuffs, buttons, apron strings caught in equipment Unsafe footwear	Adequate, correctly-sized clothing Safe, nonslip shoes
Foods	Contaminated foods or foods incorrectly stored Foods and meals incorrectly carried	Staff awareness of food poisoning causes Use of lifts/trolleys
Personnel	Staff running about Staff not concentrating upon the task to be completed or staff unaware of others	Staff training Staff supervision and motivation

premises are safe, have sufficient fire fighting appliances and that staff training is a regular feature.

Possible causes of fire

The two main causes are accidents and carelessness. Accidents that could result in a fire include the incorrect use of equipment, particularly when frying in the kitchen, and the poor maintenance of electrical equipment. Fires can also be caused by the careless extinguishing of cigarettes and emptying of ash trays, not guarding open fires in a grate or in the kitchen, by allowing cloths to come into contact with direct heat causing them to ignite. A further cause may occur outside the buildings, by allowing combustible refuse to accumulate, by leaving outbuildings unlocked thereby attracting vagrants and vandals, and by the inability to control lit bonfires and incinerators.

Employee's role in fire prevention

All employees should be trained continually to be aware of the dangers of fire, and work in a manner that will not be the cause of a fire. They should always be aware of the dangers of faulty electrical appliances and equipment, and of the dangers of overloading electrical sockets by using multi-plugs and extension leads. Damaged flexes and plugs should always be reported immediately. When cleaning ashtrays, an empty metal container should be used, ash and stubs should never be emptied into waste paper bins. Ideally, ash trays should be emptied down a waste disposal unit, and at the end of the service time and never left to smoulder over night. Employees should also be alert to potential dangers, and trained to report any suspicious circumstances immediately.

In the event of a fire

There are two points to be remembered when a fire alarm is rung. One is that the safety of human life is more important than the buildings, therefore the immediate evacuation of both employees and customers is essential. The second point is that the fire brigade are specialists at fire fighting, and so unless the fire is really very small, all fire fighting should be left to the professionals.

If a very small fire should occur, then the extinguishing may be attempted by a unit or team supervisor, and all instructions given by this employee should be followed by the remainder of the team. For example, should a kitchen cloth catch fire on the side of a solid top kitchen range, it would not at this stage be necessary to raise the alarm and cause the entire building to be evacuated. It can be extinguished quickly and easily by smothering the cloth with a fire blanket, or by dousing it with water.

It should always be remembered that fire spreads quickly, and so speedy actions are essential. Any fire, however small, must be reported in the accident book; it may mean that further training is required, or that maintenance checks have not been thorough enough.

Fire fighting should never be attempted by an untrained or junior employee, and never be continued if the fire is continuing to spread in spite of efforts to extinguish it, or if the escape route is threatened.

Should it be necessary to evacuate the premises, doors and windows should be closed where possible, as draughts will spread the flames further afield. All gas and electricity mains switches should be turned off - in kitchens, these are normally near the main exits for safety.

Fire fighting equipment

Fires will only burn if they have a supply of oxygen, and so all extinguishers are designed to eliminate oxygen from the area of the fire. Table 1.15 is a summary of the types of equipment and their purpose.

Evacuation of the premises

Once the fire alarm has been activated, the building should be evacuated. The alarm will alert the fire, police and ambulance services. Once evacuated, it will be necessary to take a role call of all persons that were in the building, to ensure that no-one is missing, presumed trapped. For this reason, the guest list, hotel register and list of staff on duty should be taken out of the building, this will normally be the responsibility of the front office manager in a hotel or the manager in a restaurant or other catering outlet. In large, residential premises, there will be predetermined staff with duties to aid and escort the guests from the premises, and to keep them together and look after them once outside. No member of staff or public should re-enter the premises until the 'all clear' is given by the police or fire brigade.

BOMB SCARE OR ALARMS

In these cases all persons should be evacuated from the building and accounted for as for a fire alarm. For bomb alarms there is normally a different sounding alarm signal, and in this instant the army bomb disposal unit will take control of the situation with the police and fire brigade assisting. To minimize the blast damage, it is advised to leave all doors and windows open, and guests may pick up their light, hand-held luggage if this is possible, to eliminate too many small bags and brief cases having to be searched before the premises are declared safe.

No person should re-enter the building until the premises are declared safe by the army bomb disposal personnel. It is the staff's responsibility to look after the guest in as comfortable a manner as is possible under the circumstances.

Table 1.15 Types of fire fighting equipment and their purpose

Type	Colour	Type of fire
Water	red	wood, paper, fabrics
Powder	blue	all risks, flammable liquids and gases
Foam	cream	flammable liquids, oils, fats, spirits
BCF	green	electrical and flammable liquid fires
CO_2	black	electrical and flammable liquid fires
Fire blanket	red	use to smother a small fire
Sprinklers	—	automatically start when the fire alarm is raised (positioned in ceilings)
Smoke detectors	—	should the smoke level rise, they will sound the alarm

CHAPTER 7 Catering equipment

CATERING EQUIPMENT

There are many aspects to be considered when choosing catering equipment for the kitchen. Below is a summary of the main points:

- The choice of fuel to be used is governed by the *type of fuel supply* available and the *estimated fuel consumption*.
- The choice of equipment must be suitable for the proposed menu. The number of meals to be produced will also dictate the type of equipment to be purchased.
- The space available in relation to the size and height of the equipment is important.
- Will the total cost of the equipment be worth the use to which it will be put?
- The equipment may be too heavy for the structure of the premises; a special floor may be needed to help spread the weight evenly.
- Is there an adequate supply of water and drainage facilities for such equipment as steamers, boilers, and combination ovens?
- It is necessary to estimate the capacity of the equipment in relation to the menus being prepared, and to compare this with the time it will take to prepare and cook the food.
- The heat up time and ease of heat control should be considered in relation to the preparation time and staff capabilities.
- The necessity to use an air extractor system with the equipment should be considered. This may dictate where the equipment can be positioned within the kitchen.
- The equipment should be well designed so that cleaning is a simple operation. All handles and knobs should be robust and heat resistant.
- There should be adequate arrangements for maintenance and a spare part replacement service provided by the manufacturer or local dealer.

EQUIPMENT SAFETY

The following points need to be considered:

- adequate staff training is necessary before the equipment is used,
- staff should be tall enough to use equipment safely - grills are often too high,
- waste food particles will burn easily and may ignite if they are not removed,
- splashes of grease and fat may ignite if they are not cleaned away,
- drip trays and grill troughs should be kept clean to prevent fire,
- large trays will be heavy and hot and so care is needed when removing them,
- the tops of grills, salamanders, ovens and steamers have air vents, and these must be kept clear and clean,
- the tops of equipment are often hot and so nothing must be placed on these surfaces,
- solid top ranges may be switched off but the heat will be retained for a considerable length of time,
- gas safety taps must always be switched off when equipment is not in use,
- drainage water taps may be hot and will splash when opened,
- electricity switches should be turned off when equipment is not in use,
- the door seals on ovens and, particularly, microwave ovens, should be regularly checked.

REFRIGERATORS

A refrigerator is a cabinet with well-lagged walls and a door, which closes tightly to insulate the inside. Inside, an evaporator surrounds a smaller compartment used for ice making. The evaporator absorbs heat from the foods and the whole compartment. In doing so, it sets up convection currents in the air enclosed in the cabinet and also draws moisture out of the foods. This moisture condenses and forms 'frost' on the outside of the evaporator.

Using a refrigerator

- open the door as seldom as possible,
- arrange foods so that the air can circulate, enabling convection currents to be set up to draw the heat from the food and pass it up to the evaporator,
- keep all foods covered to prevent them from becoming dried as the heat is extracted, and to prevent smells and odours from passing from one type of food to another (if food is left uncovered, it produces extra moisture and then frost, which reduces the refrigerator's efficiency),
- always keep raw and cooked meat and meat products and poultry completely separate from each other to prevent cross-contamination,
- all foods placed in the refrigerator should be in clean containers and be cold,
- one person should have the responsibility of checking the contents of the refrigerator on a daily basis (it must never be used to keep foods that are stale and unwanted).

HEAT IN COOKING

Radiation

With radiation, heat passes from its source in direct rays until it falls on an object in its path, for example, when grilling food.

Conduction

This occurs when heat is transferred through a solid object by contact with that object. Conduction is the principle involved in the use of a solid electric plate or gas hot plate.

Convection

This occurs through the movement of heated particles of gases and liquids. On heating, the particles expand, become less dense and rise. The colder particles sink to take their place, thus causing convection currents which distribute heat. Convection is the principle involved in the heating of an oven and in the heating of liquids.

Microwaves

Microwaves are high-frequency, electromagnetic waves, similar to radio waves. When the food is placed in a microwave oven and the power is switched on, energy waves cause friction in the molecules of food, producing heat. The cooking time is greatly reduced, and the shrinkage and moisture loss of the food is minimal.

Microwaves are generated by a vacuum tube called a 'magnetron', which converts electrical energy to microwave energy and transfers it to the oven where it cooks the food. Microwaves are reflected by metal, and so it is not possible to cook food in metal containers. However, microwaves pass through containers that are made of paper, high density plastic, china and glass. As heat is produced within the food itself, only the food gets hot - the container and the air inside the oven remains cold.

Microwaves penetrate food surfaces to a depth of several centimetres from all angles and start the rapid heating activity of the water, fat and sugar molecules present in the food. Because of the heat is conducted below the surface, the food remains moist, and does not become dry and crisp.

Modern catering equipment is designed to save fuel and examples of these are induction hobs, combination ovens and pressure cookers.

Induction cooking

Induction cooking hobs are vitroceramic solid plates that are not heated in themselves. A two-way magnetic field exists at the hob level. When a ferrous metal based cooking pan is placed on the hob, the magnetic energy is immediately transmitted through the pan and into its contents. Unlike convention cooking hobs, there is no need to wait for the hob to heat up. Only the area covered by the cooking pan will be generating heat, thus the surrounding hob area should remain cold. This has the advantage of reducing the heat in the atmosphere to an acceptable level.

Independent tests have shown that cooking by induction uses, on average, 49 per cent *less* energy than a traditional electric hob, and for certain methods of à la carte style cooking, the savings can be as great as 64 per cent. The induction hob produces these savings because it only consumes energy when there is a pan on the cooking plate. The empty hob can be switched on all day without any heat being generated. In addition, cooking times can be as much as 15-20 per cent faster than traditional hobs, as it is possible to control the intensity of the heat just as swiftly and accurately as with gas.

The cool-running induction hob means that there should be no more kitchen fires from accidentally touching the hot hob surface with a dry cloth.

Induction cooking hob on a mobile trolley for maximum versatility.

A combination oven capable of cooking various foods simultaneously without cross-tainting flavours.

The induction hob is easily cleaned by turning off the electricity and then wiping the hob with a damp cloth and neutral detergent; the food spills will not cook as there will be no heat.

Induction hobs are available in a variety of styles to suit different operations - they can be mounted on a heavy duty oven, fitted into a work bench area, stand mounted with storage space below, or mounted on a mobile trolley.

COMBINATION OVENS

Combination ovens are designed to be used either as a convection oven or as a steamer and, for this reason, they need a cold water supply. If these two methods are combined in one cooking mode, hence the name combination oven, then the food may be cooked by convected hot air and steam being circulated evenly throughout the oven. With this combination, foods may be successfully roasted, baked, poached, braised or steamed. There will be no danger of cross-tainting of foods, and the mode of cooking can be adjusted during the process. For example, a joint of meat may be placed in the oven set at convection mode to seal the juices into the joint, then, after 10 minutes, the mode may be changed to combination mode - the steam will prevent the food from drying out and shrinking while the convection heat continues to cook the joint. Combination ovens are ideal for the cooking of fresh and frozen vegetables, which may be placed on gastronorm trays and cooked using the steam mode, virtually as they are required, as they will only take a couple of minutes to cook to al denté texture. In addition, the fast cooking time will ensure that the vegetables will retain their nutrients and colour.

Combination ovens come in a variety of sizes and because of the speed of cooking, are becoming increasingly popular in kitchens of all types. They mean an end to the traditional batch cooking process for vegetables and small baked or poached items, which may now be cooked to order.

PRESSURE COOKER

This item of equipment is far quicker, more efficient and easier to clean than a vessel used to cook food by steam (atmospheric steamer). It will require a cold water supply and may be either heated by gas or electricity. The pressure cooker will take three, full-size gastronorm trays at one time, and no cross-tainting of foods will occur. It has a simple control system and is able to run on high pressure or, for more delicate foods, on a reduced low pressure. There is an automatic timer with a time elapse indicator. Like atmospheric steamers, pressure steamers help foods to retain their natural juices, colour and flavour.

Foods cook quickly in pressure cookers. For example, 54 individual steamed puddings will take 35 minutes to cook; 40 portions of chicken will take 30 minutes to cook; and 56 portions of frozen peas will be cooked in 11 minutes. Pressure cookers are useful in catering establishments where large quantities of food are required at the same time; also they will be found in use in welfare catering units.

Table 1.16 is a summary of modern catering equipment, showing the method of heat transfer and the methods of cooking that are used.

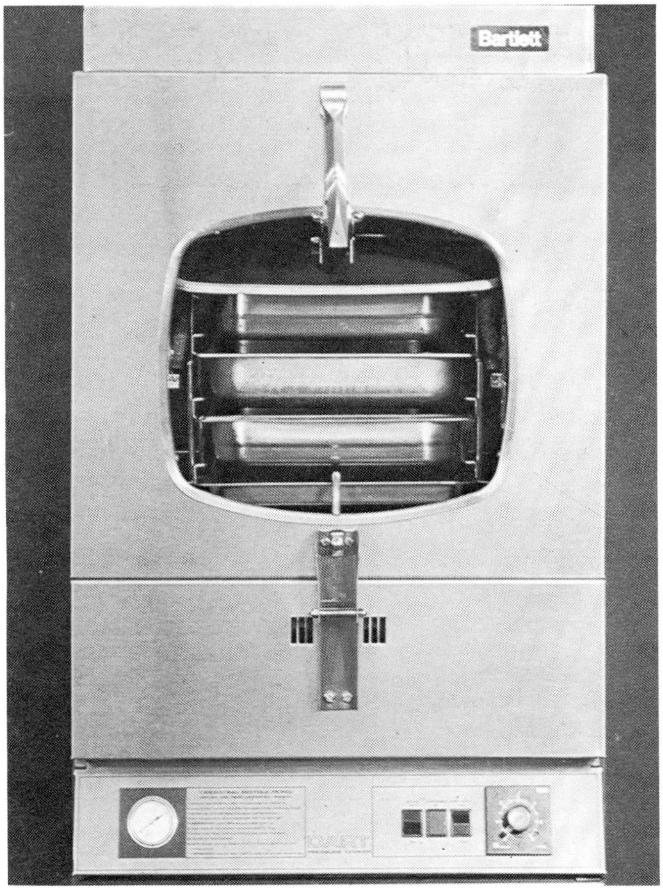

Right: a pressure cooker.

Table 1.16 Summary of modern catering equipment

Equipment	Method of heat transfer	Examples of methods of cookery
Boiling pan	Conduction	Boiling, direction or in steam jacket
Boiling top	Conduction	Boiling, simmering, shallow frying
Bratt pan	Conduction/convection	Stewing, shallow frying, braising
Induction hob	Induction	Boiling, simmering, shallow frying
Atmospheric steamer	Convection	Steaming
Pressure cooker	Convection	Steaming, stewing, boiling
Deep fat fryer	Conduction/convection	Deep frying
Griddle	Conduction	Griddle or shallow fry
Char grill	Conduction/convection	Grill
Salamander	Radiation	Grill
Oven	Convection	Roasting, baking, braising, oven poaching
Cook and hold oven	Convection	Roasting, baking, oven poaching
Convection oven	Convection	Baking, roasting, braising, oven poaching
Combination oven	Conduction	Baking, roasting, steaming, poaching, braising
Microwave oven	Friction	Microwave cookery, defrosting, reheating
Microwave oven	Friction/convection	Fast baking, oven poaching, of small quantities only, reheating

CHAPTER 8 Cooking methods

METHODS OF COOKERY

There are eleven different methods of cooking foods (see Table 1.19 pp 46–48). Each method has its own characteristics and advantages and is particularly suited to specific foods (others, such as pot roasting and en papillot (cooking food in oiled greaseproof paper or foil) are sometimes added to the list).

APPROPRIATE METHODS OF COOKING DIFFERENT FOODS

Foods can be prepared using moist or dry methods of cookery. Tables 1.17 and 1.18 divide the main methods into these two categories and give examples that illustrate which are appropriate for various foods.

Table 1.17 Examples of moist methods of cooking foods

Food	**Boiling**	**Poaching**	**Steaming**	**Stewing**	**Braising**
Potatoes	Pommes persillées		Pommes vapeurs		Pommes boulangère
Cabbage	Boiled cabbage		Choux vapeurs		Choux braises
Topside beef	Boiled beef			Ragout de boeuf	Paupiettes de boeuf
Gammon joints	Boiled gammon		Steamed gammon		Braised gammon
Trout		Poached trout	Steamed trout		Braised stuffed trout
Eggs	Oeuf mollet	Oeuf poché	Oeuf en cocotte		

Table 1.18 Examples of dry methods of cooking foods

Food	**Roasting**	**Baking**	**Shallow fry**	**Deep fry**	**Grilling**
Sirloin beef	Roast beef	Boeuf en croute	Entrecôte steak		Minute steak
Potatoes	Roast potatoes	Jacket potatoes	Sauté potatoes	Chips	Gratinated
Bread		Bread rolls	Croutons	Fried bread	Toast
Puff pastry		Vol au vent			
Batter		Yorkshire pudding		Fritters	
Plaice fillet			Plaice meuniere	Plaice in batter	Grilled fillet

Table 1.19 Eleven methods of cooking foods

	Definition	*Method*	*Temperature*	*Liquid*	*Equipment*	*Advantages*	*Safety rules*
Boiling	Food is placed into liquid, brought to and maintained at boiling point; simmered with gentle agitation.	Food is placed into cold liquid and then brought to the boil and cooked through. Or, food placed into boiling liquid and maintained at boiling point and cooked.	100°C	Water, stock or milk.	Stockpots, saucepans, or boiling pans.	Simple method, quick, foods cook evenly, and tenderizes fibrous foods.	Pans are heavy and hot, the pan should be a suitable size for volume of food, and careful immersion/removal of food from cooking pans.
Poaching	Food is carefully placed into liquid that is just below boiling point.	*Deep* Food is placed into deep liquor, sometimes on a rack, e.g. fish. *Shallow* Food is lowered into liquor and removed individually. *Oven* Food is placed in a shallow dish, with minimum liquid; covered and cooked in the oven.	80°C	Water, stock or milk.	Bratt pan, fish kettle, shallow pans, or oven-proof dishes.	Allows fragile food to be cooked gently, and nutrients in oven poaching may be kept in liquor and used in a sauce.	Foods must be carefully immersed into the cooking liquor, foods should be removed from the liquor as soon as they are cooked, and precise timing is essential.
Steaming	Food is cooked by the action of steam (water vapour).	*Atmospheric* Low pressure food is placed in lidded containers and cooked on shelves in atmospheric steamer *High pressure* Small quantities of food are cooked under high pressure.	Steam	Steam	Atmospheric steamer, high pressure steamer, covered basin, or sleeves and dishes.	Foods and easily digested nutrients are retained in food, less likely to over cook, bulk capacity – low pressure, and quick cook/reheat in high pressure steamer	Check water level and temperature before use, use door as a shield from the steam and use oven cloths, cover food to prevent water penetration, and time foods carefully when using a high pressure steamer.

	Definition	*Method*	*Temperature*	*Liquid*	*Equipment*	*Advantages*	*Safety rules*
Stewing	Food is cut into even sized pieces, simmered in minimum liquid or sauce which forms part of the stew.	Food is cooked in a liquid which is thickened by its own unpassed ingredients. *Or* food is cooked in liquid from which a sauce is made. *Or* food is cooked in a prepared sauce.	Liquid, 180°C.	Well-flavoured stock or sauce.	Heavy based, lidded pans, lidded casseroles, bratt pan.	Nutrients retained, less expensive, and cuts of meat may be used.	Container of a suitable size for the volume of food; careful handling of the container in and out of oven. Oven cloths to be used. Avoid over cooking as this causes food to break up and discolour.
Braising	Food is placed on a mirepoix in a lidded container with just enough sauce/stock to cover the mirepoix.	Method preceded by:either quick fry sealing or blanching. *Brown braising:* the meat is sealed, and cooked on a mirepoix with appropriate sauce. *White braising:* the food is cooked on a mirepoix with white stock.	Oven 200°C.	Well-flavoured stock or appropriate sauce.	Lidded braising pans. Lidded casseroles, Bratt pans.	Nutrients and the flavour are retained in the sauce.	Suitable sized container for food. Careful handling of food in and out of oven. Use dry cloth. Avoid over cooking and adjust temperature while cooking as necessary.
Roasting	Food is cooked with oil or fat, either in the oven or on a spit over a heat source.	Joint is placed on a trivet of bones in a roasting tin and placed in a hot oven to seal the joint. The temperature is reduced to cook the joint through and it is basted frequently, placed on a rotating spit over a heat source and basted regularly.	Oven 230°C.	No liquid, fat or oil.	Oven: general-purpose or convection. Spit, roasting tin and trivet.	Excess fat is lost, natural juices are used in the gravy and enhances natural flavours.	Roasting tray should be deep enough to allow for safe use. Adjust the temperature while cooking as necessary. Care is required when basting. Use dry oven cloths. Leave the joint to 'rest' before carving. Drain off excess before making the gravy.
Grilling	Cooking small quantities of food by radiant heat.	Food is placed on a greased bar over a heat source. *Or* food placed on a tray or rack under a heat source. *Or* food placed between heated bars or plates.	Variable according to food.	Clarified fat, baste.	Salamander top heat. Under fire grill. Infra-red grill. Or toaster.	Quick method for prime quality foods. Product is visible while cooking. Excess fat is lost. Quick adjustment of heat while cooking. Colour/Crisp finish may be created.	Careful preparation of grill bars or trays. Never place items on the top surface of a heated salamander. Clean regularly and remove excess fat before it ignites.

	Definition	*Method*	*Temperature*	*Liquid*	*Equipment*	*Advantages*	*Safety rule*
Deep frying	Food immersed in preheated fat or oil.	*Partial:* immerse in hot oil or fat until softened, not coloured. *Complete:* immerse in hot fat or oil until cooked through and coloured and crisp on the outside.	Partial 160–175 °C Complete 170–190°C	Oil or fat.	Friture, basket/ spider. Deep fry unit: manual or automatic. Pressure fryer.	*Partial* frying enables items to be finished to order. *Complete* frying of foods are quickly sealed in hot fat or oil. Fast cooking, easy handling and serving. Pressure frying for large volumes.	Correct volume of fat or oil to food. Food should be free from excess moisture. Careful loading of fryer. Operatives should be trained in use of the equipment and fire drill procedures.
Shallow frying	Food cooked in a small quantity of fat or oil.	*Shallow:* fried both sides in clarified fat. *Sauté:* tossed in hot fat to cook quickly. *Griddle:* cooked quickly on a lightly oiled hot plate. *Stir-fry:* tossed in hot butter or oil over a fierce heat.	Variable according to the food to be cooked.	Oil or fat.	Shallow frypan, round or oval, sauté pan, bratt pan, griddle, omelette pan, pancake pan, wok.	Quick method for prime quality foods cooked to order. Colour and crisp coating may be created. Flavour of frying fat or oil may enhance the foods.	Pan should be of a correct capacity to allow for safe practice. Careful movement of pans over the heat. Never leave a pan unattended over the heat. Fat or oil will ignite when too hot.
Baking	Cooking of prepared foods by convected dry heat in an oven using natural moisture.	The cooking of foods in a heated oven usually without any significant amount of fat or liquid. *Or* the placing of foods into a heated oven with humidity introduced (steam) or within a water bath (bain marie).	Variable according to the requirement of the food.	None.	Specialist tins and trays for specific baked goods. General purpose ovens, pastry ovens, pizza ovens, forced air convection ovens.	Products cooked easily in specially designed ovens. Effective control of the temperature. Specially designed trays and tins are available. Acceptable presentation or finish is easily achieved.	Oven should be loaded within its capacity. Foods should not be placed directly over heat source. Base plate in the oven should be cleaned regularly, spilt food particles will ignite.
Microwave	Cooking with energy created by the emission of microwaves at high frequency. Waves activate water molecules in the food and the agitation created produces heat which cooks or reheats the food.	*Prime cooking*: total preparation from raw food. *Partial cooking*: foods prepared in conjunction with the microwave and conventional equipment. *Defrosting*: raw or cooked foods are safely defrosted. *Reheating*: cooked or prepared foods are safely reheated to order.	Variable according to the require–ment of the food.	This will vary according to the food.	Microwave oven, specialist containers made from earthenware, glassware, plastic, cardboard, and cling film and roasting bags.	Can save up to 75% of the time taken by conventional cooking methods. Fast method of prime cooling or reheating. Safe, rapid method of defrosting. Cooking odours kept to a minimum. Minimizes shrinkage of meats or fish. Useful for small quantities of food.	Never operate the oven when it is empty. Limited capacity for cooking. Some foods will not achieve an acceptable finish. Keep the interior spott-lessly clean. Never use any metal containers. Check the door seal regularly – if damaged, do not use the oven and call a service engineer.

THE EFFECT OF HEAT UPON NUTRIENTS

Once heat is applied, the nutrients in the food are affected by the method of cooking as well as by the actual heat. Heat generally improves eating quality and can reduce the effect that harmful bacteria can have upon food. Some nutrients (vitamins B and C and minerals) are destroyed by the application of heat or dissolve into the liquids. Heat improves the digestibility of food, but too fierce a heat or a prolonged heat caused the food to burn or dehydrate so it becomes indigestible.

It should be remembered that there is a difference between applying heat to cause physical changes to take place in the food, and just warming food after the actual changes have taken place.

PRE-COOKED CHILLED FOODS

Cook-chill is the term used to describe food that is prepared in the conventional manner, cooked, and then quickly chilled, followed by controlled low temperature storage (0–3°C). The food is reheated immediately before consumption.

The cook-chill kitchen may be an existing kitchen which has been modified and chilling, packaging and chill storage facilities added. Alternatively, a purpose-built unit may be constructed which will be used for the preparation, cooking, portioning, packaging, blast chilling and chill storage of foods. The production unit acts as the supply centre for smaller satellite units where the chilled foods are reheated immediately prior to service, which eliminates the necessity to keep foods hot for any length of time.

Foods that are prepared for the cook-chill process are packaged, chilled and served from suitable containers. Disposable containers are used for vending machines and other less formal outlets, but ceramics, stainless steel and aluminium ware may also be used. The final choice will depend upon the standard and type of service that is to be provided. Labels for each dish should give precise details of contents, date and reheating instructions.

Guidelines for precooked foods:

- Pathogenic organisms that can cause harm to consumers, will grow and multiply freely at temperatures above 10°C. This temperature is the danger temperature and should be remembered.
- The food should be of sound quality and fresh. It should be prepared following the requirements of the Food Hygiene (General) Regulations 1970.
- The cooking process should ensure the destruction of any pathogenic micro-organisms present in the food.
- The chilling process should immediately follow the completion of the cooling and portioning process, within a maximum time of *30 minutes.*
- The food should be chilled to 3°C within a maximum of 1.5 hours. This low temperature will reduce the growth of organisms that, if permitted, will spoil the food.
- A maximum of *five days* chilled storage is permitted. Any longer time will result in the foods being damaged by food spoilage organisms.
- The chilled foods should be distributed to the reheating area in refrigerated containers to maintain the chilled temperature - maximum 3°C.
- Reheating should take place as soon as the food is removed from the chilled storage area, and the method used should permit the chilled foods to rise in temperature from 3°C to 70°C as rapidly as possible. Remember that the centre point of the food must reach 70°C to ensure that the food is safe to eat.
- The reheated foods should be consumed as soon as possible after reheating. They should *never* be kept hot for long periods of time, and any foods not immediately eaten should be disposed of in a safe manner.
- A critical point in the preparation of chilled foods is the period immediately following the cooking process, when the foods are portioned and placed in their appropriate containers - strict hygiene standards must be practised to reduce the risks of cross-contamination.
- Cook-chill foods are stored in specially designed chill cabinets, many on trolleys for ease of transport. They should be labelled adequately and used in strict rotation so that no foods are stored for longer the five days.

The Department of Health has issued guidelines on cook-chill and cook-freeze catering systems which state that control checks on the following items should be carried out for every batch processed:

- the quality of raw materials to be used in the cooking process,
- the temperature at which the perishable raw materials are stored prior to preparation for cooking,

- the centre temperatures of meat and poultry during cooking,
- the period which elapses during the portioning process,
- the period which elapses during the cooking process of large pieces of meat and poultry,
- the time of the chilling process for portioned foods and the temperature of the food at the completion of the chilling process,
- the temperature of the food during storage and the air temperature of the chill store as recorded on the automatic recording instrument,
- the rotation of stocks within the store,
- the temperature of the food at the completion of its distribution from the refrigerated store,
- the centre temperatures achieved during the reheating process for each type of food which is subjected to a treatment prior to consumption.

All records should be kept for a minimum of three months.

Caterers may wish to keep a sample of each item of food for a further two days after the batch has been eaten. This sample will be used to help with any investigation in the event of suspected food poisoning.

PRE-COOKED FROZEN FOODS

There has been a significant increase in the use of precooked frozen foods (cook-freeze) in all sectors of the catering industry. Cook-freeze foods are foods that have been prepared and cooked, and then frozen rapidly by a blast freezer and stored at 18°C or below until ready to be reheated and served. Great care should be exercised when using this system because of the possibility of contamination – the two critical stages in the process are the preparation and cooking stage and the storage, reheating and serving stage.

Guidelines for the preparation and cooking stage:

- the quality of the raw foods should be high,
- all foods should be stored correctly and for minimum periods of time,
- there should be minimum handling of foods to reduce risks of contamination,
- employees who are unhealthy and likely to cause food contamination should not handle the food
- all equipment should be clean and in good repair,
- raw foods should be kept apart from prepared foods or cooked foods,
- when being cooked, the foods should be submitted to sufficient heat to render them safe for human consumption,
- a high standard of both premises and personal hygiene should be maintained,
- foods should be cooled *immediately* after cooking,
- as soon as foods are cooked, they should be blast-frozen within *15 minutes* of being cooked,
- foods should be correctly weighed and arranged on trays to facilitate even and thorough freezing and, later, reheating,
- great care should be taken with cooked meats to be sliced as they may easily become contaminated while the joints are cooling. Joints should be cooled to *below 30°C* in *two hours* then sliced and frozen, with a further *30 minutes*. the longer meat is kept at a temperature of between 30°C and 50°C, the greater risk of bacterial multiplication.

Guidelines for the storage, reheating and service stage:

- the pre-cooked frozen food should be stored at -18°C or below,
- a shelf-life of *8 weeks* is recommended,
- a clear system of stock control and marking of products should be used,
- stocks should be used on a rotation basis: first in, first out,
- some foods, such as cold desserts, need to be thawed before service and should be placed in a refrigerator and allowed to thaw safely and slowly,
- foods that are to be reheated by microwave oven may need top be prethawed,
- most foods should be reheated direct from their frozen state, thus eliminating risks of bacterial multiplication,
- foods should be clearly labelled to indicate their optimum cooking and temperature times,
- final heating should take place as near to the service area as possible,
- the heating process should be controlled so that the food is not spoilt by under- or overcooking,
- food that has been reheated must *never* be allowed to cool and then be served to a customer or reheated again as bacterial multiplication will have taken place,
- if frozen food has to be transported to its reheating and serving point, insulated containers should be used to maintain the food at -18°C,

- food handling personnel should be trained in the skills required to operate the reheating system and should understand the reasons for care when operating this process.

BLAST-FREEZE PROCESSING

In this process a machine blasts air at high velocity (6 m per second) over refrigeration coils and then over the foods to be frozen. The freezing process is completed quickly so that the foods are kept in as good a condition as possible. However, the freezing of food does *not* destroy harmful bacteria, it only halts their multiplication.

SOUS-VIDE

This is a method of cooking food in a vacuum. Its main application is in restaurants with a wide à la carte menu; in slack trading periods, foods can be prepared and packed.

Good hygiene is imperative since two types of bacteria are particularly successful in surviving this process, *Listeria monocytogenes* and *Yersinia enterocolitica*, both of which can cause food poisoning.

When preparing foods, sulphurized paper is used to help prevent bacteria from collecting on the food's surface. This paper is disposed of after each use to prevent contamination. The prepared food is then packaged in cryovac bags with heat resistant and antitoxin layers. Care must be taken when placing the food into the bag, since if the neck of the bag is not clean, then it will not seal properly.

Once the bag is filled and sealed, the air is removed slowly and the food labelled with the date and cooking temperatures. Sous-vide foods must be cooked in steam convection ovens at temperatures ranging from 70–100°C. Because of the low temperature, the actual cooking time is two to two and a half times longer than with conventionally prepared foods. After being cooked, the sealed bags must be chilled down to a temperature of 0–3°C within *90 minutes*, and then kept in a storage refrigerator at the same temperature.

Foods prepared this way must be carefully controlled and used in rotation so that they are only stored for a maximum of *five days*. they can be regenerated for use in a steam convection oven at a temperature above 65°C.

WORK ORGANISATION

If customers are to leave a catering establishment satisfied, then not only must the food have been to their liking, but they should have received an appropriate level of service. It is imperative that the correct order be delivered to a table and that the food arrives on time. Although the waiter may be held to blame by the customer, if these things do not occur, such failures usually reflect poor work organization behind the scenes.

KITCHEN WORK PLANS/ SCHEDULES

Whenever food is being prepared, all members of the kitchen staff should work to a previously agreed schedule. It is essential to decide the following when compiling a work schedule:

- What work has to be done?
- How much time is there?
- Is there any natural sequence to be followed?
- Is there a definite service/completion time?
- Are the foods to be hot or cold?

GENERAL PRINCIPLES

The following stages can be applied to the preparation of most foods:

- *Stage 1*
  Prepare foods such as peeling vegetables,
  start sauces that need time to simmer,
  make pastes that need time to rest before being cooked,
  start items that need to be cold or set for service.
- *Stage 2*
  Work through the main part of the food preparation, cooking dishes at appropriate times.
- *Stage 3*
  Complete dishes and items,
  garnish appropriately ready for service.

ACTIVITIES

- Look at the guidelines on precooked foods and discuss their relevance to the storage of foods in an industrial kitchen. State how these guidelines could apply to general food preparation.

CHAPTER 9 Keeping premises clean

The accommodation industry has to practice and be aware of the necessity of maintaining a clean, hygienic and safe environment for its customers and staff.

A dust control mat in the foyer.

DUST CONTROL

Dirt of many types may be prevented from going beyond the entrance to a building by the use of barrier, dust-control carpeting. This is used because it has antistatic properties that attract the dirt from a person's feet. It is strong and easily cleaned and can be fitted at the entrance to the floor area in place of an ordinary carpet. When laid it is difficult to distinguish the difference between the barrier carpet and an ordinary carpet. Dust control mats may also be used. These should be large enough to take a minimum of two complete strides so that the soles of persons feet are cleaned as they walk on the mat.

In addition, many establishments use doormats and scrapers in porch areas to remove the heavy dirt from shoes before people enter the premises. Circular doors may also be fitted to prevent draughts, thus preventing air-borne dirt from entering the premises.

HYGIENE

All of the many different areas within a catering establishment have to be cleaned on a regular basis. A summary of these areas and the likely waste or refuse to be found in each are given in Tables 1.20 and 1.21.

To ensure that all of these areas are maintained in a clean and safe manner, it is necessary to be able to identify the types of surfaces and materials to be cleaned and, therefore, to know which method of cleaning is the most appropriate to use. However, every area presents its problems to the cleaning personnel. Careful planning is therefore important to ensure that each area is appropriately cleaned.

The general principles of cleaning are:

- to remove loose dust and dirt,
- to use the simplest method first, using the least harmful cleaning agent first,
- to restore surfaces to their original state at the end of the process,
- to remove and collect dust, dirt and debris in such a way that it cannot cause harm to any additional surface,

Table 1.20 Specific areas and cleaning problems

Area	Features	Likely problems
Halls and foyers	Dust control matting/carpet; heavy-duty floor covering; welcoming atmosphere.	People constantly walking through with luggage; leads directly to the outside causing draughts and litter.
Stairs and corridors	Heavy-duty floor covering; quiet and nonslip; stain resistant; curved corners.	Constant numbers of people walking about; crevices on stair rises and sides, will harbour dust.
Passenger lifts	Dust control floor covering; kick guards and handrails/bars around the walls.	In constant use; a small, confined space to clean; needs to be aired regularly; damp clean walls daily.
Offices and reception areas	Desks, chairs and shelves; hard-wearing floor coverings; well-lit and ventilated; venetian blinds or curtains; plants.	Files and documents must not be moved; telephones need to be cleaned and disinfected; keyboards and VDU need antistatic dusting; waste or shredded paper.
Computer rooms	Atmosphere and temperature is constantly controlled; dust-free fittings; floor is sealed.	Double filter specialist suction cleaner, special training and supervision of staff; high security area.
Lounges and common rooms	Comfortable, extensive upholstered seating; television; coffee tables; periodicals; plants; ash trays.	Needs to be well–ventilated when being cleaned; ash trays; newspapers litter; seating will need regular suction cleaning and damp dusting.
Kitchens and cellars	Impervious surfaces for walls, ceilings, floors and work surfaces; equipment permanently fixed in position; standards set by the food hygiene regulations.	Restricted floor areas, under and around equipment; heavy soiling and grease to be cleaned away; floor gullies and drains; special staff training and supervision; security areas.
Bars and restaurants	Impervious surfaces for floors walls, and shelves; equipment fixed and some small in size; carpets and seating in public areas; standards set by the food hygiene regulations.	Food particles and beverage stains; litter, bottle tops and empty crate bottles; beer pipes and pumps, cooling equipment; will need to be well–ventilated; should look clean and inviting; no odours of agents or food.

Table 1.21

Amenity	Area	Waste
Accommodation	Bedrooms	Fluff, tissues
	Bathrooms and toilets	Hair, skin, hairspray, toothpaste
	Lounges	Newspapers, dead flowers
Preparation	Linen room	Fluff
	Kitchen	Food, polythene, cans
	Store rooms	Crates, boxes
	Cellars	Crates, cartons
Service	Restaurants	Bottles, paper, corks
	Cafeterias	Polythene, cling film
	Bars	Bottles, bottle tops, cigarettes
	Coffee shops	Paper napkins, polythene
Staff	Rest Room	Cigarettes, cans, snack wrappings
	Cloakrooms	Hair, tissues
	Residential area	Fluff, paper

- to use the minimum possible equipment, cleaning agents and time,
- to create an airy, light atmosphere that will help to discourage the growth of bacteria or the habitation of vermin and pests.

Cleaning is carried out in order (1) to maintain a safe and clean environment, (2) to preserve furniture and fittings in good condition, and (3) to maintain an attractive environment in which to live and work.

The methods used to maintain clean premises will depend upon (a) the type of soiling to be removed, (b) the material or surface to be cleaned, and (c) the standard of cleanliness required.

Wherever cleaning is required, the aim should be to create an acceptable level of cleanliness and leave a pleasant, tidy, fresh room at the end of the cleaning process.

METHODS OF CLEANING

Dry dusting

A method of cleaning used to remove dust from dry, smooth surfaces; it is normally completed daily. It is very often the preliminary stage of the cleaning process.

Equipment

- Dry, soft dusters for horizontal dusting.
- Dry, small, soft mop or brush for vertical dusting.

Method

1. Remove dust from the highest areas first, gradually working down to the lower levels.
2. Work in a clockwise direction around the room.
3. Work in a draught free atmosphere so that the dust does not blow around.
4. Keep the duster/mop or brush in contact with the surface to eliminate dust being scattered around.

Safety

1. Bacteria live in dust and for this reason it is essential to wash equipment thoroughly after use.
2. Methods and equipment must not be permitted to cause cross-contamination.

Dust control mopping and sweeping

This is another dry method of cleaning, used on smooth, hard surface floors to gather up and remove surface dust and small items of dry litter and paper. It is used in many areas as it is a quick and efficient way to remove surface dust.

Equipment

- A long-handled sweeping mop, either straight headed or with a V-sweeper head with a scissor-like action. These sweeping mops are available in sizes that range from 40 cm length heads, suitable for confined spaces and staircases, to 80 cm length heads that are used in large, open areas. The V-sweeper heads are 100 cm in length and can be used in any unconfined areas such as gymnasiums, halls, shopping arcades, airport waiting areas, etc. The heads of these sweeping mops are static and are made from a synthetic fibre that resembles cotton. They attract and hold the dust particles and can be easily cleaned by using a suction cleaner or by washing. A disposable mop head should be used in areas where cross- contamination should be avoided at all costs.

Method

1. Keep the mop head in contact with the floor surface all the time, to allow the mop head to collect the dust as it is worked across the floor.
2. Each stroke should overlap the previous one so that the entire area is cleaned.
3. The floor area should be cleaned in a methodical manner, finishing the process at the doorway.
4. Excess dust and litter should be collected with a dustpan and brush.

A dust control mop designed to provide fast, efficient sweeping and dusting.

Safety
1. Always stand the mop upright when not in use to prevent accidents.
2. Mop heads should be removed and suction cleaned after use.
3. When soiled, the mop heads should be washed with a neutral detergent, rinsed well and then dried before further use.

Damp dusting

This method is used to remove dust from nonporous furniture fittings and shelves that can be cleaned by a moist method.

Equipment
- Soft, checked, cotton cloths.
- Plastic trigger spray bottles filled with a sanitizing solution.

Method
1. A sanitizing solution is sprayed onto a cotton, checked cloth. The moist cloth is used to remove the dust from the surface, leaving it with a reduced bacterial count.
2. Alternatively, a detergent sanitizing powder can be sprinkled onto the surface and then wiped with a damp cloth. This will leave the surface clean, with a reduced bacterial count.
3. A third method is to use a hard surface cleaner that is sprayed onto the surface and then wiped with a damp cloth. This will leave the surface clean and free from grease.

Safety
1. Always use the recommended quantity of detergent.
2. Rinse the damp cloth frequently to eliminate cross-contamination and to ensure a smear-free finish to the surface. The cloth should be washed and dried after use.
3. Remember that, in high infection risk areas, a disposable cloth may be used.

Damp dusting using a sanitizer in a trigger bottle and a soft cloth.

Floor mopping – damp methods

This method is used on soiled floor areas that are nonporous. For example, floors in kitchens, cafeterias, retail stores and schools are frequently cleaned by damp mopping. It is usually a daily task, with the exception of kitchens, where these floors are normally cleaned at the end of each working shift. The floor should be swept before the damp mopping is started.

Equipment
- Soft-headed brooms and dustpans for debris and litter removal.
- Floor cleaning warning notices or cones.
- Floor mops - either short tailed or long tailed.

Wet floor mopping using a wheeled wringer bucket and Kentucky mop.

- Mop buckets with wringers appropriate to the mop heads. A twin bucket is often used to allow for the rinsing process after the initial washing.
- Neutral detergent or sanitizer.

Method

1. Clear away all possible items from the area to be cleaned.
2. Sweep the floor if the level of soiling is excessive.
3. Using the hot detergent solution, mop the first section of the floor. Then using the clean rinsing water, rinse this section. Use the wringer each time to ensure that the minimum quantity of water is put on the floor.
4. Continue this sectional cleaning process until the entire floor has been cleaned.
5. Always work in a methodical manner so that the floor nearest the doorway is the last section to be cleaned.

Safety

1. Change the water regularly as the level of soiling dictates. This will help to eliminate cross-contamination.
2. Use the correct concentration of detergent.
3. Wash the equipment thoroughly after use, and hang the mop up to dry.
4. Remember that dirty, damp mop heads provide an ideal breeding ground for harmful bacteria.

Suction cleaning

A method of dust removal is used on carpets, upholstery, mattresses and any hard surface that requires maximum dust removal.

Equipment

- Floor cleaning warning notices.
- A suction cleaner with attachments as required.
- A small, stiff hand brush for edges of carpeted areas.

Method

1. Remove sharp, sticky items from the floor surface.
2. Remove small items of furniture.
3. Brush the corners and edges of the carpet with stiff brush if necessary.
4. Plug in the cleaner and work across the floor surface in a methodical manner keeping the flex behind the machine. Finish cleaning by the doorway.
5. Use the small brush attachment to suction clean any upholstery and curtained area.

Safety

1. Ensure that the machine is switched off before plugging it into the wall socket.
2. Ensure that the flex does not trail over a large floor area, as this can cause accidents.
3. Always empty the dust container after use and store the cleaner away correctly.
4. Remember to check the plug and cable regularly for signs of damage.

Central, built-in vacuum system

This system can be installed into any premises. It involves the installation of plastic ducting from a power unit and dust collection receptacle. The ducting is normally laid under the floor boards with inlet valves installed at strategic points so that the entire premises can be suction cleaned.

To operate, a lightweight hose is connected to one of the inlet valves; this automatically switches on the motor and the suction power. The motor is normally sited away from the main residential area, so the noise level is far lower than the traditional suction cleaner. Since it is fixed, the motor is more powerful than a normal suction cleaner, and therefore its suction power is greater.

The system is considered to be safer than traditional suction cleaners because there is no live flex trailing around, no fumes and dust are emitted back into the atmosphere, and the dust, dirt and bacteria are all collected in one central receptacle, which is emptied about once a month.

Staircase cleaning

Owing to the safety implications of staircase cleaning, it is essential to use only *dry methods* of cleaning during the day time. Should a wet method be required, this *must* be completed while the building or area is not in use, and adequate time allowed for the surface to dry.

Equipment

- Dust control mop for noncarpeted surfaces.
- Duster or small dust control mop for banisters and handrail.
- Suction cleaner with brush attachments for carpet areas.
- Safety warning notices.

Method

1. Place warning notices at the top and bottom of the staircase.
2. Start at the top of the flight of stairs, work across

the top tread and then the rise. Work across the second tread and then the second rise. Continue in this manner until the flight had been cleaned.
3. Starting at the top of the flight again, dust the banisters, handrails and skirting boards.

Safety
1. Place the warning signs where they may easily be seen.
2. Clear or replace equipment after use; empty the dust container.
3. Remember, never leave any equipment on the staircase.

Spray buffing or buffing

Buffing a hard floor surface will remove dull scuff marks and keep the floor looking glossy and clean between major stripping and repolishing. It is possible to use a floor cleaning machine and a buffing pad to achieve a good result. However, should it be necessary to increase the gloss, spray buffing may be used.

Equipment
- Polishing machine and buffing pad.
- Fine nozzle spray bottle with spray buff solution.
- Dust control mop.
- Floor cleaning notices.

Method
1. Assemble the machine.
2. Clear the floor area to be spray buffed.
3. Spray one section of the floor and then buff immediately.
4. Work in a methodical manner finishing at the doorway.
5. Dust control mop the floor to remove surface dust after the buffing process.

Safety
1. Wash the pads after use - it may be necessary to use several pads to complete the floor area satisfactorily.
2. Replace and store the floor cleaning machine.
3. Use floor cleaning notices.
4. Check the flex and plug of the machine.

Floor polishing

Before a floor can be polished it is necessary to strip off all the old polish. This process can be completed by machine scrubbing with an appropriate solution. The floor surface then needs to be rinsed and dried.

Points to note
1. The polish should be applied using a mop or applicator. One or more coats are normally applied evenly, at right angles to each other.
2. Care should be taken to ensure that the coats are applied evenly.
3. Each coat should be completely dry before the next one is applied.
4. Good ventilation will speed up the drying process (approximately 20 minutes).
5. Some polish manufacturers advocate buffing the surface between applications.

Furniture polishing

Points to note
1. A soft cloth should be used.
2. The polish should be applied to the cloth, which should then be rubbed gently, but firmly, over the surface. Buffing brings up a high gloss finish.
3. The polish should be used sparingly as too much will leave a sticky finish.
4. Wood that has not been sealed will take a wax-based polish and produce a gloss finish after the wax has built up after several applications.
5. Nonporous finishes such as melamine and varnished wood do not need polish as they have a natural gloss finish.

Wall cleaning

It is necessary to wash nonporous walls in food preparation and service areas, as well as in some other areas such as bathrooms, cloakrooms and toilets. Nonporous wall surfaces may be finished with smooth paint, or ceramic tiles. These surfaces may be damp mopped or wet washed to remove all traces of dust and food spillages.

Equipment
- Dry dust control or suction cleaner with a soft brush attachment.
- Wet wall mop or sponge.
- Bucket with bactericidal detergent.

Method
1. Use the dry mop or suction cleaner to remove dust and cobwebs.
2. Begin to wash the wall surface, starting at the skirting board and working upwards, cleaning a section at a time. Remember, the water running over a clean surface will not leave any marks.
3. Finally, rinse the wall surface, starting at the top and working downwards in sections as before.

Safety

1. Clear the area of equipment and food.
2. If wet washing, turn off the mains electricity before starting. Mask over any electrical sockets with nonporous tape.
3. Place warning notices near the base of any step ladder being used.
4. Change washing and rinsing water frequently so as to avoid cross-contamination.
5. Wash and put away all equipment after use.
6. Make sure that any water spilt on the floor is mopped up.

Interior window cleaning

Dirty windows will easily spoil the appearance of a room, and so it is necessary to ensure that they are maintained in as clean a condition as possible.

Most catering establishments use a window cleaning contractor to ensure that window exteriors are kept clean. This leaves the insides to the housekeeping section to clean on a regular basis. Interior glass panels and windows also need to be cleaned regularly, some, in fact, more than once a day, in order to keep them sparkling and clear free.

Equipment

- A soft cloth and glass cleaning fluid with ammonia, or vinegar added, or
- A neutral detergent solution, a bucket of hot water, a soft cloth and a chamois leather or synthetic leather cloth. Squeegees made with a rubber blade may be used for large windows in place of the chamois leather.

Method

1. Spray on the fluid and buff up with soft cloth till smear free, or
2. Wash the glass then rinse and dry it using either the chamois leather or the squeegee.

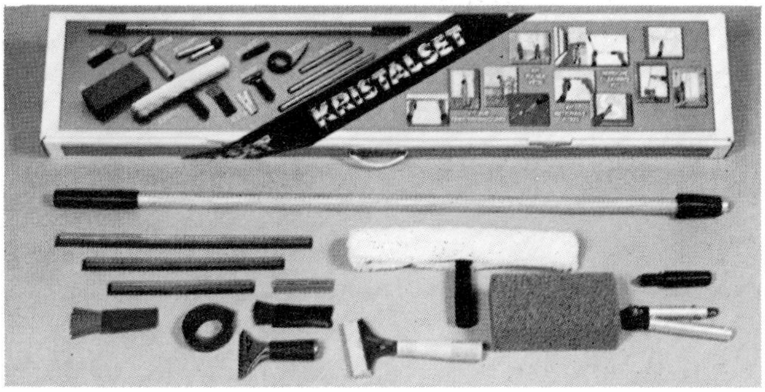

A window cleaning kit containing all the tools needed to wash, clean and squeegee any size of window.

Safety

1. The window sill should be dried completely after the window has been cleaned.
2. The equipment should be cleaned and stored correctly.
3. The chamois leather should be rinsed thoroughly and then stretched back into shape and hung up to dry.
4. Should it be necessary to use a step ladder, care should be taken to use it safely and a warning notice should be placed nearby.

THE CLEANING EQUIPMENT

Choosing equipment

Before choosing cleaning equipment the following points should be considered:

Task

- The size and type of area to be cleaned.
- The standard of finish required.

Economy

- The cost of the equipment, including cleaning agents.
- The amount of human energy required to use the equipment.

Maintenance

- The servicing arrangements and daily care necessary for the equipment.
- The storage facilities available or required for the equipment.

Training

- The training that will be required before staff can use the equipment.
- The suitability of the equipment for the existing staff to use efficiently.

Safety

- The safety of the equipment especially electrical items.
- Is there a BEAB mark of safety?

Cleaning equipment storeroom

The cleaning equipment storeroom should be situated near the work area to minimize loss or damage to equipment and to encourage staff to use the correct item of equipment for each task. All the equipment should be clearly coded to allow tools to be used in the correct manner and to ensure efficient stock control.

Storerooms should have adequate shelving and hooks provided for the equipment to encourage orderly storage of items. They should be well lit and ventilated, and should have areas suitable for cleaning, washing and drying equipment. Storerooms should be the definite responsibility of a particular member of staff and should be kept clean, tidy and well organized at all times. They should be kept locked when not in use to ensure efficient control of stores and equipment.

CLEANING AGENTS

In order to clean surfaces effectively, it is necessary to understand the basic differences between the various types of cleaning agents and to be able to choose a suitable cleaning agent for a given surface. Obviously, it is possible simply to damp wipe over dirty surfaces using just water but, if there is any trace of grease or other deposit present, then water alone will not be an efficient cleanser.

It is also necessary to know whether a cleaning agent is an acid or alkali so that surfaces are not damaged or incompatible agents mixed by mistake, causing noxious fumes to be emitted.

The pH scale and cleaning agents

Some cleaning agents are acid, some are neutral and some alkali (see figure 1.23). They are measured on a numerical scale known as the pH scale. On this scale, cleaning agents between 1 and 6 are acid, those at 7 are neutral, and those over 7 are alkali. The nearer an agent is to 7, the weaker its acidity or alkalinity will be.

	1	
	2	toilet cleanser
ACID	3	acid agents
	4	
	5	distilled vinegar
	6	
NEUTRAL	7	soap
	8	liquid detergent
	9	mild alkali detergent
	10	bleach
ALKALI	11	ammonia
	12	caustic soda
	13	strong alkali detergent
	14	

1.23 The pH scale and some cleaning agents.

There are many reputable cleaning agent manufacturers who will offer a complete range of cleaning agents, but it is very often more efficient and economical to use one reliable manufacturer than try and save a few pounds by purchasing different agents from various sources.

Below some examples of the types of cleaning agents commonly used in hotels and catering establishments.

Soaps

These are manufactured from a combination of animal fats and caustic soda. Soap does not reduce the surface tension of water or grease, and has poor wetting properties. It is quickly used up and forms a scum, thus recontaminating the surfaces.

- Toilet soap is used for personal hygiene, and is provided by hotels in small tablet form for the guests use, often with the hotel or group name imprinted on the surface.
- Liquid soap is used in liquid dispensers in kitchens, cloakrooms and toilet areas. The nozzles should be cleaned regularly to prevent a build up of soap, which provides a sticky area for bacteria to breed.

Detergents

These are made from various chemicals blended together to create a balanced mixture which, when used with hot water, will have the ability to:

- wet soiled surfaces,
- suspend the soil throughout the solution,
- free rinse away the soil from the surface being cleaned.

Liquid detergent with a sterilizer

This is used in a washing up machine. The detergent removes the grease and food debris and the sterilizer destroys the harmful bacteria.

Mild alkali detergent

This is used in a floor cleaning machine to remove water-based polish from thermoplastic tiled floors.

Strong alkali detergent

This is used carefully with protective gloves to degrease kitchen equipment and fittings. It should be rinsed away thoroughly.

Neutral detergent

A mild, multi-purpose detergent that is used for general cleansing.

Sanitizer and detergent
A detergent containing a bactericide that will effectively destroy harmful bacteria. This is used for cleansing in high-risk areas such as hospitals, schools, old people's homes, and many kitchen and food service areas.

Abrasives
Abrasives are made from pumice, detergent and, sometimes, bleach. The scratching properties of the powder help to remove soil deposits, but will damage most surfaces.

Scouring powder
This is used sparingly to remove stubborn food deposits from aluminium pans and trays.

Scouring pads
These are made from cubes of steel wool impregnated with scouring powder.

Metal cleansers
These are used on unsealed, decorative, hard metals such as brass and copper. The cleanser is rubbed onto the metal with a cloth or soft brush, and then buffed off with a clean, soft cloth. The result is a shiny gloss finish to the metal.

Polishes
Polishes are created to produce a shine by providing a smooth finish which will reflect light and help to protect the surface from general wear and tear. Furniture polishes and floor polishes are applied differently and are composed of different proportions of wax, silicone and other materials. It is therefore necessary to be able to choose the most suitable polish for each item of furniture and floor surface.

- A water-based metallized floor finish may be a paste or liquid and is used to finish floors of wood or cork. It helps to protect the surface from water and provide a gloss, antislip finish.
- A paste furniture polish contains a high proportion of wax and silicones and is applied sparingly with a soft cloth and buffed until a gloss is achieved. It is suitable for wooden surfaces that have no other dressing or seal.
- A spray-on furniture polish or cleanser often comes in a trigger-action spray bottle; a soft cloth is used to produce a shine. They may be applied to various hard surfaces such as melamine, ceramic and glass. Different types are available - some contain wax while others are wax free. They are effective but expensive.

Bleach
Sodium hypochloride is a poisonous, clear liquid that should never be used in a food preparation or service area. It destroys harmful bacteria and is used to clean drains and gullies. It can also remove stains from washable fabrics when diluted (one part bleach, 40 parts water). Bleach may be purchased in liquid or tablet form.

Caustic soda
When dissolved in very hot water sodium carbonate has the ability to dissolve grease and is used to clean drains and floor gullies. It should never be allowed to come into contact with aluminium kitchenware as it will quickly pit the surface. Caustic soda is used with perforated aluminium plate and hot water to clean tarnished silver; this is called the polivit method of silver cleaning and is often employed where large quantities of cutlery are in daily use.

Distilled vinegar
Use diluted in water (1:20). A soft cloth is placed into the solution, wrung out well, and then used on wooden furniture as an effective way of removing grease and dirt prior to applying a wax-based polish. Undiluted, it can be used to remove mild water stains from baths, shower trays and hand basins. It is also used in some window cleaning agents.

The storage of cleaning agents

In most establishments, there should be a central bulk store for all cleaning agents, which are issued to individual work areas as required. For example, a room maid should have a working stock for daily use, which should be kept with the cleaning equipment in a small cupboard adjacent to the work area.

Cleaning agents that are used for cleaning within a food preparation or service area should be stored in a cupboard well away from the food storage area or any food preparation equipment. All containers should be clearly labelled so that no mistakes can possibly be made when identifying or using the cleaning agents. Remember that food can easily become contaminated if it comes into contact with cleaning agents or equipment.

The wet mop and the dish cloth or sponge are

items of cleaning equipment that are likely to be full of harmful bacteria. Every time these are used, bacteria will be spread around unless a bactericide detergent or sanitizer is used, and unless they are regularly and thoroughly cleaned.

THE CLEANING PROGRAMME

Frequency of cleaning

Some cleaning has to be completed daily, some weekly and some will be required at less frequent intervals; this may be monthly, quarterly or even yearly, depending on the standards to be achieved and the degree of soiling. Whatever the schedule, it is essential to remove stains and marks as soon as they occur, so that they do not become ingrained and permanent.

Residential areas

In residential areas in establishments such as hotels, hospitals, schools and halls of residence, it is necessary to clean daily to remove dust. The schedule should include such tasks as dry dusting, damp dusting, suction cleaning, dust control sweeping and mopping, and the cleansing of bathrooms and toilets. Obviously, some tasks will require less frequent (weekly) attention; these include interior window cleaning, high dusting, suction cleaning of upholstery and beds, extra attention to floors, and extra attention to all bathrooms, toilets and cloakrooms.

It is common practice to incorporate one of these weekly tasks into the daily routine so that they will be all completed once a week. For example, a daily routine might involve a daily clean plus the appropriate day's allotted weekly cleaning task:

Monday – extra attention to all floors
Tuesday – high level dusting and walls if necessary
Wednesday – extra attention to all beds and upholstery
Thursday – extra cleaning of all sanitary areas and fitments
Friday – cleaning of windows and paintwork if necessary

Work schedules

Work schedules should give the cleaning personnel the precise information necessary to clean a given area, including (1) the length of time that the cleaning should take, (2) the cleaning equipment and agents to be used, and (3) the correct sequence to be followed when carrying out the cleaning.

Before deciding on a work schedule the following need to be done:

- *Decide* which surfaces and items need to be cleaned,
- *Identify* the materials that the items are made of,
- *Select* the equipment and cleaning agents suitable for each task,
- *Decide* how frequently each cleaning task should be completed.

An example of a daily work schedule for a hotel bedroom with an *en suite* bathroom can be seen in figure 1.24.

FREQUENCY: Daily		**LOCATION: Bedroom and bathroom**	
Task	*Time*	*Equipment/agents*	*Methods*
1. Ventilate	$\frac{1}{2}$ min.		Adjust curtain - open window.
2. Remove litter	1 min.	Waste sack	Remove litter, used crockery. Empty ash-trays.
3. Strip bed	2 mins.	Soiled linen sack	Remove used linen.
4. Clean bathroom	3 mins.	Bactericide detergent Toilet cleanser Cloths	Flush and clean WC. Clean bath, shower and basin. Clean floor if necessary.
5. Adjust ventilation	$\frac{1}{2}$ min.		Close window.
6. Remake bed	3 min.	Clean linen	Remake bed. Place towels in bathroom.
7. Dust	1 min.	Soft duster	Work clockwise round room.
8. Replenish	1 min.	Guest beverage	Refill tea/coffee making tray.
9. Suction clean	2 min.	Suction cleaner	Remove dust from floors.
10. Check	1 min.		Check around room. Close and lock door.

1.24 Work schedule for a hotel bedroom with an en suite bathroom.

Using work schedules

When using work schedules to plan the tasks that should be completed by staff it is important to:

- predict the time which will be taken to complete each task,
- plan the sequence of tasks to be completed,
- itemize the equipment necessary to complete the tasks,
- list the cleaning agents to be used,
- estimate the cleaning costs for a given period - a week, a month, etc.
- estimate the total cleaning time, giving the number of hours needed to clean the entire premises, on a daily basis,
- work out staffing levels and staff rotas,
- purchase the required cleaning materials in bulk.

Work schedules may be compiled for every area within the premises, and the sequence of tasks will vary according to the type of area and the type of cleaning to be completed - daily, weekly or periodically. Work schedules are useful in the following situations:

- when staff are not managing to complete work in the time allowed,
- when there is excessive use of cleaning agents,
- when cleaning is not being completed to the required standard,
- when there are staff complaints concerning insufficient cleaning equipment,
- when cleaning equipment is not replaced in the correct store area,
- whenever staff become irritable and unhappy when they have finished their shift,
- when staff think they are expected to complete more work than others employed to do the same tasks.

Task/job identification

In the cleaning of a hotel bedroom (see figure 1.24), it is relatively easy to produce a short schedule. This schedule is made up from a series of individual cleaning tasks, and in this example there are four main task:

1. stripping and remaking the bed,
2. cleaning the bathroom,
3. dry dusting, and
4. suction cleaning the floor.

However, in kitchen cleaning there are many more individual tasks to be completed, and more than one member of staff is needed to complete them. This is where a cleaning rota or schedule should be used, together with work cards.

PREPARING ROOMS

The aim in preparing rooms is to create an acceptable level of cleanliness and a pleasant, tidy room. The following list will ensure that this will be achieved:

1. Visual
   (a) check the level of dust,
   (b) identify any stains or damage,
   (c) check heat and humidity level.
2. Remove
   (a) litter, old newspapers and dead flowers,
   (b) used crockery, glassware, trays and ashtrays,
   (c) additional furniture that is no longer required,
   (d) used linen which should be sent for laundering.
3. Prepare
   (a) equipment necessary to clean the room,
   (b) measure or prepare cleaning agents required to clean the room,
   (c) collect any clean linen required for the room.
4. Collect
   (a) clean ashtrays,
   (b) fresh flowers or plants,
   (c) giveaway products, notepaper, menu cards and newspapers.
5. Action
   (a) dry or damp dust or polish furniture, and check handles, knobs and castors,
   (c) clear small objects from the floor and suction clean or dry dust. Wet clean is appropriate.
6. Check
   (a) the position of furniture,
   (b) the positon of flowers, ashtrays, giveaway products, newspapers and linen,
   (c) the ventilation and heating balance,
   (d) the room is ready for use.

General points;

- leave the room in a condition in which you would like to find it,
- always check that all cleaning agents and equipment are removed,
- ensure that the room is locked if necessary after cleaning,
- check the lights and other electrical items and leave ready for use,
- adjust the curtains and blinds according to the time of day,
- report to the supervisor any damage or breakages that are found while cleaning the room.

STANDARDS OF CLEANLINESS

Standards of cleanliness must be clearly defined, and, naturally, different types of establishment will have different standards of cleanliness. The terms used to describe the standards of cleanliness are: high risk, prestige and normal.

'High risk' is used in areas where infection control is of prime importance, for example, hospitals, old people's homes, nursing homes, maternity homes, cloakrooms, toilets and all food preparation, storage and service areas.

'Prestige' is used at the luxury end of the market in high-priced hotels, restaurants and cruise liners. These businesses pay close attention to comfort as well as cleanliness.

'Normal' refers to the maintenance of an acceptable standard of cleanliness.

INSPECTION PROCEDURES

When a room is ready for occupation, it should be checked by the assistant housekeeper or supervisor:

- the room should be entered and the door closed,
- the first impression that the room creates is the most important,
- proceed slowly round the room in a clockwise direction and note any fault (see figure 1.25),
- look high up, then at eye level, then at table top level, and finally at low level,
- all faults or soiled areas should be noted on a form (see figure 1.26).

Regular failures in cleaning the same area need to be investigated, since it must be caused by either lack of staff training or lack of time, or lack of equipment or agents for that task. If room maids and cleaning staff are made aware of the room inspection procedures, they will be more likely to clean the rooms to an acceptable standard.

A room inspection form for a hospital patient's room may include quite a different list of points to be checked, because here emphasis on [I]standards [R]is the control of inspection. It is a high-risk area. A typical form migh include the following:

- floor free from dust and litter,
- walls clean and smear free,
- ledges free from dust and litter,
- furniture in place and free from dust,
- cubicle curtains and rails clean,
- hand basin clean and free from scum lines,
- mirrors, windows and glass panels smear free,
- rubbish bins cleaned and relined.

1.25 Room inspection procedures

Room no.: .... Date: .... Time: ....

Staff name: .... Supervisor: ....

Item area		Mark	Action to be taken
Door			
Light fittings			
Window-glass			
Window-frame			
Ceiling			
Walls			
Upholstery:	chairs		
	headboard		
Beds	mattress		
	made up ready		
Flat surfaces:	table		
	shelves		
	mirror		
Floor:	carpet		
	surrounds		
Cupboards:	chest of drawers		
	wardrobe		

Marks: 5 = excellent, 4 = good, 3 = acceptable, 2 = some soiling, 1 = unclean

1.26 Room inspection form.

MAINTAINING STANDARDS

The quality of the work carried out will relate directly to the cleaning standards achieved. There are many aspects of quality control that will all directly or indirectly contribute to the achievement of cleaning standards (see figure 1.27).

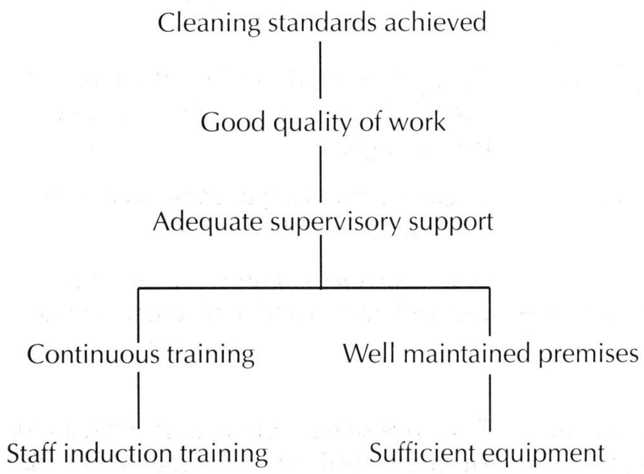

1.27 Factors that affect cleaning standards

HOSPITAL CLEANLINESS

Effective organization is an essential element of hospital cleanliness. An *infection control committee* works with the domestic services staff to give advice on infection control. Successful control depends on:

- interdepartmental co-operation,
- adequate training procedures,
- efficient supervisory staff,
- adequate selection and care of equipment,
- correct selection and use of cleaning agents.

Staff

Hospitals are high risk areas so it is important that the highest calibre of cleaning staff are employed. The following points are particularly relevant:

- staff should receive training in and be aware of the dangers of contamination and cross-infection,
- induction training and retraining schemes should be provided,
- members of staff should have high standards of personal hygiene,
- adequate protective clothing should be worn,
- there should be adequate procedures for the laundering and replacement of protective clothing,

- adequate daily staff supervision should be provided,
- members of staff should be made to feel proud of their role within the workforce and recognize its importance,
- members of staff should work in a consistent, methodical manner.

Cleaning agents and equipment

Care should be taken in hospitals that the correct cleaning agents and equipment are used and that they are stored in a safe and hygienic place. In particular, there should be:

- sachets of liquid agents to ensure correct dilution,
- buckets with capacity markings to ensure correct dilution,
- sufficient supplies of agents and equipment to ensure that an acceptable standard of work is maintained,
- adequate maintenance of equipment to ensure its efficient operation,
- adequate autoclave and pasteurizing facilities for equipment,
- sufficient storage areas for sterilized equipment,
- adequate areas for the washing, emptying and daily maintenance of cleaning equipment.

WASHROOMS AND BATHROOMS

Wherever or whatever the type, style or size of the washroom, the basic principles relating to the cleanliness of the fitments and fittings are the same. It should be remembered that:

- washrooms are used by members of the public, workforce and guests, and they present a potential health hazard,
- cross-infection risks are high as there are many shared surfaces in any bathroom or washroom,
- dirty or soiled rooms are unpleasant to use and will have the effect of reducing standards of cleanliness elsewhere in the premises,
- soiled, dirty or uncared for washrooms and bathrooms reflect a poor, unpleasant image to the general public.

Toilets should be supplied with:

- hot and cold water for hand washing,
- nail brushes to encourage clean hands and nails,
- soap to ensure bacteria are destroyed when hands are washed,
- hand drying facilities that do not harbour bacteria,
- toilet paper in a germ free container.

Furthermore fitments should include waste bins, sanitary bins, female toilets and 'now wash your hands' notices.

Table 1.22 The procedure for cleaning of toilets, washrooms and bathrooms.

Stage	Action	Reason
1. *Ventilate*	Open all windows, or activate extractor fans.	To eliminate unpleasant odours, and to reduce the number of airborne harmful bacteria.
2. *Litter*	Prop open doors. Put a 'cleaning in progress' sign on outer door. Sweep floor and remove refuse and litter.	Litter, refuse and dust will harbour bacteria and insects and cause unpleasant odours and infection.
3. *Water closet*	Flush, then brush inside vigorously. Apply an acid toilet cleanser.	To remove traces of paper, organic stains and lime scale stains.
Urinals	Flush, then stop the automatic flush mechanism while the walls and base trough are scrubbed thoroughly. Apply an acid cleanser.	To remove traces of urine and dust that harbour bacteria and unpleasant odours.
External surfaces	Wash all exterior surfaces of WC urinals and basins using hot water and a germicidal detergent.	To remove smears, urine, dust and harmful bacteria. Infection may be caused from the seat, handles and taps.

table continued overleaf

Table 1.22 continued.

Stage	Action	Reason
4. *Hand basins and drinking fountains*	Wash thoroughly using nonscratch cleaners. Pay attention to taps, plugs, chains, splash backs, soap dispensers, and the water outlet on drinking fountains.	Can cause cross-infection as hands are in direct contact with the basin. Remove harmful bacteria debris, scum marks and soap.
5. *Sundries*	Replenish soap dispensers, towels, toilet paper, contents of slot machines and litter containers. Adjust ventilation and doors.	To ensure an adequate supply for the staff, guests and general public to use.
6. *Surfaces*	Damp dust window ledges, doors and other surfaces. Use a germicidal detergent solution in the spray.	To remove dust, finger marks and harmful bacteria.
7. *WC and urinal*	Turn on the flushing mechanism for urinals and flush the WC. Polish up all seats, lids, cisterns and handles.	To spread acid cleanser evenly around the surfaces.
8. *Floor*	Clean the floor with appropriate notice.	To remove dust and bacteria.
9. *Final check*	Check and then remove cleaning notice.	To ensure the room is ready for use.

ACTIVITIES

- Find out which types of dust control procedures and fitments are used in your place of training or work and then discuss the following questions:
  How effective do you consider the dust control procedures and fitments to be?
  Are there any other areas where these preventative measures should be taken?
- Discuss the use of work schedules and consider the following questions:
  Why are work schedules useful in these circumstances?
  Are there any more instances where they could be helpful?
- Discuss the different priorities of cleanliness standards between a *public bar* in a public house and a *ward* in a general hospital. See if you can identify priorities in the following areas:
  infection control - specific priority areas;
  level of cleanliness - frequency and depth of cleaning;
  customer expectations - their priorities and needs.
- When cleaning bathrooms, cloakrooms and toilets (all high-risk areas) it is essential to understand the importance of cleanliness, and its significance in reducing the risks of cross contamination.
  Compile an inspection form for a cloakroom and toilet area.
  Draw a plan of the area and mark in red on your plan all the areas where bacteria could be passed from one person to another.
- Discuss the connection between cross-infection in a hospital ward and the cleaning schedule being used by the cleaning staff.
- If possible arrange a visit to, or a short period of work experience in a hospital or an old people's home, then write a report on your findings to include the following:
  the cleaning schedule being used,
  the types of cleaning agents being used,
  your impressions of the cleanliness of the premises.

CHAPTER 10 Maintenance

PREVENTATIVE MAINTENANCE

Preventative maintenance refers to taking regular care of fixtures, fittings and premises, both internally and externally, so that as soon as a potential fault is identified it may be dealth with in the appropriate manner. Preventative maintenance is carried out by all members of staff in all types of establishments. Everyone should:

- look for any change in colour, shape, texture, performance,
- listen for a change in sound of any mechanisms,
- check for any damage or inability to achieve the standard required.

In the housekeeping department, likely areas of preventative maintenance are:

- plumbing - dripping taps, drains slow to clear and blocked overflows,
- floors - frayed areas, stains, bare patches and uneven areas,
- windows - loose fitting frames, broken catches or sash cords and cracked glass,
- doors - loose fittings, ill-fitting knobs, faulty locks and the door catching the floor,
- furniture - chipped, stained, cracked, stuck castors and warped shelves,
- fittings - loose curtain rails, unsafe shelves and unsteady pictures,
- electrical - loose light fittings, flickering lights and unreliable clocks,
- equipment - unreliable, inefficient, noisy and frayed flexes,
- decorations - paintwork or wallpaper marked or chipped.

Any faults or potential faults must be reported to the duty housekeeper or supervisor who will then be responsible for the assessment of the fault and for reporting it to the maintenance department or contractor for repair.

REGULAR MAINTENANCE

Regular maintenance refers to the complete check that is made of every single item in a given area at regular intervals. The housekeeper instructs the supervisors to carry out maintenance checks in each room during less busy periods, and any faults are recorded on a check list. The housekeeper will then arrange for the necessary repairs and maintenance to be completed either by the in-house staff or by contractors. Maintenance check lists are designed to remind staff of all the furniture and fittings and items that should be checked.

Some establishments are closed at certain times of the year, such as hotels at seaside resorts, boarding schools and halls of residence. These types of establishments can complete their regular maintenance checks and work while they are closed from general trading. Redecoration is often completed at closed or less busy times of the trading year.

Maintenance is directly related to pest control. Table 1.23 on the following page lists the pests that can enter inadequately maintained premises.

ACTIVITIES

- Study Table 1.23 and then choose a specific area and compile a maintenace check list.

Table 1.23 Pests that could enter faulty premises

	Areas	Faults	Pests
Exterior	Brickwork	Cracked, faulty pointing	Rats, mice, ants
	Grease traps or drains	Broken grid, unclean	Rats
	Coal hole or shed	Broken doorway	Rats, cats, foxes, dogs, wasps
	Refuse bins	Unclean bins	Birds, flies, wasps, foxes
	Windows	Broken glass	Birds, mice, rats, bats
	Roof	Missing or slipped tiles	Birds, mice, rats, bats
Interior	Plasterwork	Damp	Wet rot, moulds, ants
	Plasterwork	Cracked, flaked	Cockroaches, silverfish, spiders
	Wallpaper	Damp, peeling	Woodlice, silverfish
	Sinks	Damp surrounding areas	Silverfish, spiders
	Woodwork	Damp	Dry rot, woodlice
	Wooden furniture	Old and uncared for	Woodworm, furniture beetles
	Carpet	Dirty, dusty	Carpet beetles, fleas
	Beds	Mattress and bedding dirty	Bed bugs, fleas, body lice
	Soft furnishings	Dusty and undisturbed	Moths, fleas, dust mites
	Refuse	Open, unclean bins	Flies, ants
	Ceramic tiled walls	Cracked, grout missing	Mould, woodlice, silverfish
	Houseplants	Not looked after properly, neglected	Ants, greenfly, earwigs
	Dry food stores	Unclean	Cockroaches, flies
	Vegetable stores	Unclean, damp	Silverfish, flies, wasps, ants

CHAPTER 11 Surfaces, finishes and linens

In this section the types of surfaces, finishes, furnishings and linens likely to be used in hotel and catering establishments will be considered. Note throughout how important it is to choose a material that will be suitable for the item or area in question. This involves analysing such factors as the use to which the item or area will be put, the type of wear it will receive and the overall impression that is trying to be achieved.

WASHROOMS, BATHROOMS AND TOILETS

The types of surfaces and finishes likely to be used in washrooms, bathrooms and toilets are shown in Table 1.24. It is important to note the characteristics of the materials used – some are favourable, others are not. However, it is more than likely that the advantages of the material greatly outweigh the disadvantages, which is why it is being used.

Table 1.24 Types of surfaces and finishes likely to be used in washrooms

Item	Material	Areas	Characteristics
Water closets and urinals	Enamelled steel	Residential and public areas or rooms	Hardwearing, but urinals will chip and scratch
	Porcelain	Luxury bathrooms	Not very strong, for light usage only
	Stainless steel	Public and industrial washrooms Work areas	Very strong, will stay looking good even with very hard use
Walls	Painted or sealed brick work or smooth concrete	Public and Industrial washrooms Work areas	Need regular maintenance Graffiti-proof seal or paint should be used
	Ceramic tiles	Industrial, residential and catering areas	Easily cleaned, an impervious surface
	Vinyl wall paper	Luxury residential areas	Short life, will tear and peel with water
Floors	Sealed concrete	Industrial and public areas	Uneven surface that will need regular resealing
	Quarry tiles	Industrial, public, and catering areas	Impervious surface, easily cleaned and very strong and hard
	Ceramic tiles	Residential areas	Attractive, impervious, strong surface Easily cleaned
	Thermoplastic tiles or sheets (sealed)	Residential and interior public areas	Need to be sealed regularly Not able to withstand constant dampness

Table 1.25 Types of floors used in hotels and catering establishments

Type	Area of use	Maintenance	Notes
Wood	Libraries, dance floors, gymnasia	Dust removal Buff as required	Always sealed Antislip
Cork	Libraries, hospitals (quiet areas)	Dust removal Avoid contact with water	Always sealed Antislip
Linoleum tiles or sheet	Heavy traffic areas, corridors	Dust removal Damp mop	Damaged by strong alkalis
Thermoplastic tiles	Schools, hospitals, shops, offices	Dust removal Damp mop	Always treated with water-based polish Buff if required
Polyvinyl chloride (sheet)	Hospitals (wards), theatres	Dust removal Damp mop	May be treated with water-based polish Buff when required
Polyvinyl chloride (studded)	Heavy traffic areas, corridors, arcades, airports	Dust removal Damp mop	Treated with water-based polish
Rubber tiles	Hospitals, entrances, corridors	Dust removal Damp mop	May be treated with a water-based polish
Terrazzo tiles	Entrance halls, shops	Dust removal Damp mop	May be treated with a water-based polish
Quarry tiles	Kitchens, store-rooms, toilets, laundries	Sweep Wet clean	

FLOOR SURFACES AND COVERINGS

The types of floor surfaces and covering used in hotel and catering establishments are shown in Tables 1.25 and 1.26. Note once again how the characteristics of the material are appropriate to the use to which the floor will be put.

CORIAN

Corian is a newly available material that is ideal for the hygiene-conscious establishment. It has the appearance of marble, with a smooth finish in seven different colours: white, beige, almond, grey, green, blue and dark grey.

Corian can be moulded for baths, basins and shower surrounds, and can be supplied in sheet form for wall cladding, work surfaces, dishes, WC partitions, bar tops, facias, display counters and service counters. Because it can be moulded, there are no joints or seams that can harbour harmful micro-organisms. It also has a far more attractive finish and appearance than its nearest rival – stainless steel.

The properties of Corian are of benefit to any residential establishment:

- it is nonporous and therefore resists staining,
- it will withstand tough treatment without cracking or crazing and is resistant to extremes of temperature,
- damage caused by knives, scratches and cigarette burns can all be removed by cleaning with an abrasive and a plastic scouring pad,
- it is easily cleared by using a natural detergent and is resistant to mild acids and alkalis,
- it will not fade or change colour in time – its natural finish remains near to perfect – and it is covered by a ten year warranty.

Table 1.26 Type of carpet used in residential establishments

Type	Uses	Maintenance	Notes
Woven *Wilton* *Axminster*	Heavy traffic areas, lounges, high comfort areas	Dust removal Spot clean as necessary	Warm, luxurious Wool or synthetic fibres Good underlay
Tufted	Medium to light traffic areas	Dust removal Spot clean as necessary	Tufted fibres on hessian or foam backing Good underlay
Corded	Light traffic areas	Dust removal	Spillage can cause a problem Foam backed
Bonded	Medium traffic areas	Dust removal Spot clean	Treated or minor 'wet' areas
Carpet tiles	Heavy traffic areas, offices, common rooms, bars	Dust removal Lift up and wash any soiled tiles then replace	May be rought to touch Very durable Can replace damaged tiles without relaying entire area

MATERIALS USED TO MAKE FURNITURE

Furniture is made from many different materials, each having its own set of characteristics:

1. Glass
   - plate glass with toughened, shatterproof qualities and bevelled edges is normally used for table tops.
2. Leather
   - hide is very expensive and must be kept supple and soft,
   - it is difficult to clean and may crack if not well cared for.
3. Metals
   - metals are used for frameworks and for complete items of furniture,
   - iron, steel, aluminium and chrome are often used with various finishes,
   - brass and copper are also used, mainly for handles and occasional table tops.
4. Paint
   - it may be a gloss, emulsion, matt finish or vinyl-based, and has a relatively short life as it is liable to crack or chip off the wooden surface,
   - it provides reasonable protection from heat and moisture.
5. Plastics
   - inexpensive, light weight and easily cleaned,
   - expanded polyurethane and rubber foam are used in upholstery,
   - bonded PVC or imitation leather are used for finishes,
   - they are used to coat metal frameworks,
   - reinforced, moulded plastic is used for chairs and tables.
6. Veneer
   - it is an economical and sound method of producing well-finished furniture,
   - hardwood sheets, or plastic sheets imitating woods, are glued to a base of soft wood, chipboard, blockboard, or plywood to create the desired finish with less cost and weight,
   - furniture is less liable to warp in heat than similar, solid wood types.
7. Wood
   - many types of wood are used to make furniture, including beech, birch, oak, deal, elm mahogany and teak,
   - it needs a protective finish when used in residential establishments:
     (i) a mix of beeswax and turpentine paste gives some protection against water and heat,
     (ii) melamine coating (a synthetic resin) gives an impervious coat, with good resistance to heat, water and abrasions,
     (iii) polyurethane varnish gives a high gloss finish with good protection from heat, water and scratches.

CARE AND CLEANING OF FURNITURE

The following points should be remembered when looking after furniture:

- avoid scratching and banging furniture,
- always remove soiling as soon as possible,
- rotate the position of the furniture in the room to avoid persistent wear, sun or heat damage,
- clean the furniture regularly and thoroughly, including the inside and back,
- maintain the furniture properly, checking the springs, legs, castors, handles and supports; worn parts should be replaced.

Table 1.27 show how furniture should be cared for and cleaned and what remedies can be employed if faults or damage occur.

UPHOLSTERY AND SOFT FURNISHING

The basis of upholstered furniture is a frame of wood, metal or plastic on to which padding is fixed. Traditionally, springs are attached to the frame, with hessian casing and wadding placed in layers to give softness. Some seats are constructed with horizontal tension springs or webbing fixed to the frame. Removable cushions are then placed on top of the springs, usually cut from latex or plastic foam blocks. If a plastic frame is used, the upholstery is placed directly onto the plastic and fixed with adhesive.

Fillings for upholstery include:

- animal hair which is expensive but is fairly strong and may be mixed with sisal fibres,
- flock (waste rags and wool which is cleaned before use),
- down feathers which are specially treated,
- polyester which is a soft, nonallergic filling,
- latex foam which is resilient and very durable,
- plastic foam which is very resilient and durable but highly inflammable and when it burns it gives off poisonous fumes.

Always check that upholstered furniture has passed the Upholstery Furniture (Safety) Regulations 1980 test. All safe items of furniture have a label attached to the frame indicating that they have passed the flame tests.

Table 1.27 Periodic care and cleaning of furniture

Material	Faults or damage	Remedy or care
Polished wood	Scratched surface Heat or water marks	Colour scratched area, repolish Rub mark with metal polish until it disappears and repolish
Varnished wood	Split or flaked varnish	Rub with fine sand paper and revarnish damaged area. It may be necessary to strip and revarnish entire item
Painted wood	Heat marks Resin stains Flaked or chipped paint	Rub with fine sand paper and repaint
Moulded plastic	General darkening Loosening of frame	Clean with neutral detergent solution and sponge Rebolt sections to the frame
Melamine coated	Heat marks Edging coming away	Use a nonscratch cleanser to remove the burn colour Use strong glue to refix the edging
Glass	Scratched or chipped at the edge	Send to glass specialist to have the edge rebevelled
Painted metal	Chipped, dented, bent Rubber 'feet' mising	Repaint and straighten Replace rubber ferrules
Copper or brass	Metal polish caught in moulding Tarnished	Brush mouldings, then clean with metal polish Buff up well
Upholstered fabric	Crumbs in crevices Darkened colour	Suction clean with nozzle, damp clean with shampoo and suction clean when dry

Criteria for the selection of furnishings

The following points should be considered prior to the selection of furnishings:

1. Effects
   - colour - it should blend with existing schemes and not dominate,
   - texture - decide upon the effect that the weave is to create,
   - pile - this can be used to create a luxurious effect,
   - comfort - it should be soft to the touch and never rough and coarse.
2. Heat
   - fading - it should not be damaged by sunlight or smoke,
   - insulation - some fabrics have heat retention properties,
   - flamability - it should be resistant to fire and excess heat.
3. Wear
   - strength - it should be strong enough for the purpose for which it is intended,
   - resistant - it should be resistant to insect infestation,
   - appearance - crease resistant properties are an asset,
   - fluffing - select a fabric which is not liable to fluffing,
   - weave - loose weaves will 'pull' easily, spoiling the effect.
4. Maintenance
   - cleaning - choose a fabric that can be easily cleaned,
   - stains - select a stain resistant fabric,
   - linings - these should be compatible with the main fabric,
   - resistant - select a fabric that is strong enough for the anticipated wear,
   - cost - choose a fabric that will give value for money.

CURTAINS AND BLINDS

All curtains should be properly lined, since this helps to reduce fading and draughts and makes them hang better. It also helps to insulate the building. If there is a radiator under the window sill, the curtains should be short enough to allow heat to come into the room.

The following are considered when hanging new curtains:

- one and a half times the width to be curtained is allowed for (including the wall at the sides of the area),
- an allowance for the heading and hems is added to the total length to be curtained,
- the tape is calculated by using the total width of the fabric,
- any design should be evenly plaed across or down the curtains (a small amount of fabric may therefore be wasted when cutting),
- various types of gathering tapes are now available - some give a high heading, some a regular pleating effect, and some are random gathering types. The type of tape to be used should suit the style being created.

Blinds are useful for keeping the strong glare of the sun from rooms, therefore preventing over heating in hot weather. Venetian blinds are:

- made from slats of aluminium or plastic, and come in a variety of colours and with nylon cords,
- they are hung vertically and can be adjusted to admit light and air at the same time as giving privacy to the occupants,
- frequently used in areas such as offices, shops, foyers, reception areas in hospitals,
- some are now being made of a fibreglass material with a woven finish, but these are expensive and are hung horizontally and, therefore, open from the side.

Any of the following methods may be used to clean venetian blinds:

- neutral detergent and sponge cloth to damp clean the slats,
- foam spray cleaner to remove grease marks,
- the entire blind may be removed, immersed in water and then rehung.

Roller blinds

- can be made-to-measure for any size of window,
- are made of glazed cotton or linen,
- roll up to the top of the window on a spring mechanism, making them liable to breakage if used frequently,
- create a pleasant effect as they may be any colour or design, but they are not very strong for industrial use.

HOUSEHOLD LINENS

Table cloths and napkins

- Cotton is still the most usual material used although man-made fibres with polyester or Terylene are sometimes used. Cotton makes the laundering easier if a commercial laundry is used.
- Always ensure that the style and pattern is in continuous supply before purchasing.
- Stains are easier to remove from 'neutral' fabrics.
- With table cloths, measure the table top and allow 55 cm all around for overhang and possible shrinkage.

Tea towels and glass cloths

- These are usually made of pure white linen often with a coloured strip at the side.
- They absorb water without leaving fluff on the glassware.
- Cotton cloths are not as absorbent or as strong and they tend to leave the glassware smeary.
- These cloths are used in the bars, still room washing up area, restaurants and are usually laundered on the premises.

Towels

- Towels are made of looped pile gripped by a woven backing. The wearing quality depends upon the backing being tightly woven, and the towels' absorbency depends upon the loop size and quantity.
- Linen towels are the most usual type used in residential establishments. They may be hot washed or boiled.
- Rayon towels are cheap but are not very strong or thick.

Sheets and pillow cases

- Linen sheets and pillowcases are the most expensive and are very hard wearing. However they are cold to the touch and are not used in residential establishments.
- Cotton is hard wearing, comfortable and easily laundered. It is the fabric used in most residential establishments.
- Twill or flannelette is reasonably hard wearing, warm and thicker than cotton. It is usually used as under-sheets and is easily laundered but heavy.
- Polyester cotton is reasonably comfortable and hard wearing. Since the sheets are made of thinner fabric than pure cotton, it is important to check that the laundry can clean this fabric before it is purchased. It needs a cooler wash than cotton and is suitable for small units with an on-premises laundry.
- Not used in commercial residential units.

The linen room

The linen room is the central store where all linen stocks are issued and received. It should be sited near to the back entrance to enable linen to be delivered easily. However, it should not be near the

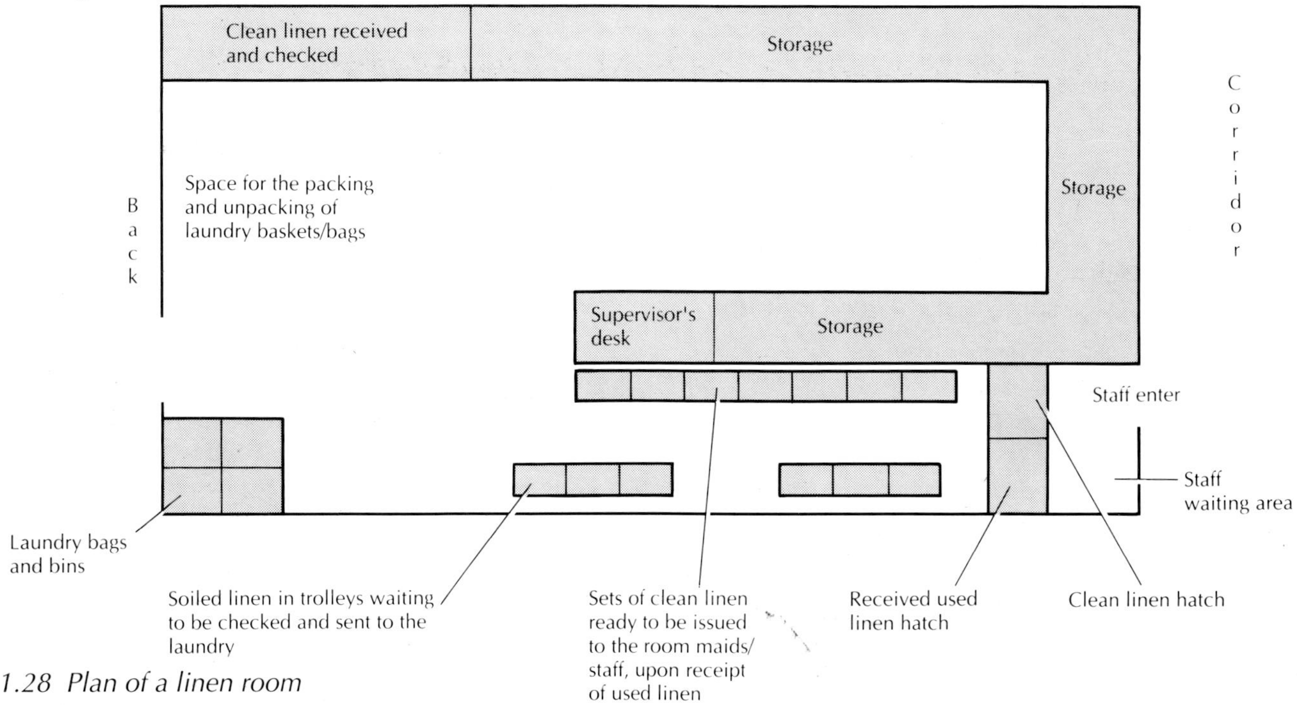

1.28 Plan of a linen room

kitchen because linen absorbs food odours very easily. The room must be well lit and well heated and central heating pipes around the walls will help to keep the linen aired.

Linen should be stored in cupboards with well-fitting sliding doors. Each shelf should be clearly labelled to indicate its contents as linen is difficult to identify when it is neatly stacked on shelves. The linen should be issued in strict rotation so that the entire stock receives equal wear. A counter or hatchway is essential so that the linen room staff can have proper control of the stock, and may issue clean sets of linen only upon receipt of used linen. An adequate waiting area should be provided for staff to prevent their assembling in the corridor.

ACTIVITIES

- Using Tables 1.24 and 1.25:

1. Identify six types of floor surface.
2. Compile a cleaning schedule for two of the surfaces.
3. Draw up a list of points to be considered when choosing a floor surface for:

- a kitchen floor,
- a restaurant floor.

PART 2

Food Preparation

This section will give you an understanding of the:

- methods used to prepare and cook foods in order to preserve their natural flavour, colour, texture and nutritional value,
- basic methods used in the preparation of sauces, soups, egg dishes, fish, meats, pastes and savouries,
- methods and effects of heat application upon foods,
- factors that influence the compilation of recipes and the planning of menus,
- implications of financial accountability and the necessity for financial control.

CHAPTER 12 Stocks and sauces

STOCKS

There are various flavoured and coloured stocks that are used in the kitchen in the making of sauces, soups, gravies, and braised and stewed dishes. A stock is an unthickened unseasoned liquid that contains some soluble nutrients. There are three basic types of stock:

Fish stock

The fish bones are rinsed under running water and sweated with other ingredients. They are added to water along with washed fish skins and onions and then simmered for 20 minutes. Once strained, the stock is ready for use when poaching fish or in the preparation of fish sauces.

Vegetable stock

For this type of stock, root vegetables, celery and onions are prepared and simmered to extract their flavour. The vegetables are kept whole in order to prevent over cooking and to produce a clear stock. This stock should be simmered for 2 hours at a very gentle pace before it is carefully strained ready for use in vegetarian dishes, vegetable braises, soups or sauces.

Meat stock

It is possible to make stock from various types of animal bones - beef, lamb, mutton, veal or chicken: all produce their own distinctive flavour. It is also possible to produce either a brown coloured stock, or a very pale coloured stock commonly known as white stock. Brown stocks are used to make gravies, meat based soups, brown stews and braises, and brown sauces such as demi-glace. White stocks are used in the making of white stews and braises, light coloured soups and velouté sauce. The bones and vegetables may be browned in a hot oven before being used for stock making; this causes the final stock to become brown in colour. If a white stock is required, the bones are blanched and then added with the uncooked vegetables to the water. All meat stocks are simmered for 6–8 hours, then carefully strained and cooled rapidly before being stored in clean containers in the refrigerator.

Convenience stock

These are readily available to the caterer in granular form. They are sold in various sized tins and are available in a variety of flavours such as brown beef, white chicken, vegetable, wine-based and fish. The advantages of these stock granules are that instead of lengthy preparation and cooking times, the granules can be easily reconstituted to produce a liquid stock by simply mixing with hot water. It is therefore feasible to mix up the exact amount of stock required as and when necessary instead of having to prepare large quantities and then having to waste some if it is not used.

SAUCES

Every sauce should be carefully and thoroughly cooked and blended. There should be adequate seasoning to complement the ingredients and the food that it is to accompany, and the finished sauce should be suitable in colour, flavour and texture. In a traditional kitchen, brigade sauces are prepared by the Saucier. There are three basic sauces from which it is possible to make a vast number of different tasting sauces. These are sauce béchamel, sauce velouté and sauce espagnol. Sauce béchamel is a creamy white sauce made from a white roux and milk, flavoured with onions and cloves. Sauce velouté is an ivory coloured sauce that is made from a blond roux and a white meat stock or a fish stock. Sauce Espagnol is made from a brown roux, browned mirepoix of vegetables and brown stock, and it has tomato purée added to enrich the flavour. Sauce Espagnol is used to make a demi-glace sauce, which is the basic sauce from which numerous other brown- based sauces are created.

Table 2.1 shows a comparison of the three methods used to create the three basic sauces.

Table 2.1 The three basic sauces

	Béchamel	**Velouté**	**Espagnol**
Thickening	White roux Equal quantities fat to flour	Blond roux Equal quantities fat to flour	Brown roux Slightly more flour than fat
Liquid	Milk	White stock	Brown stock
Flavour	Clouté onion	Determined by the type of stock used	Browned mirepoix vegetables and tomato purée and the flavour of the stock
Process	Make the roux, cool, add warm milk and bring to the boil; simmer, place in bain marie with cartouche and lid	Make the roux, cool, add hot stock, bring to the boil, simmer gently with the lid on the pan	Make the roux, cool, add tomato purée then add hot stock Bring to the boil and add browned vegetables Simmer, and skim
Cooking time	30 minutes if simmered; 1 hour in bain marie	40 minutes simmered	Simmered for 6–8 hours
Finishing	Remove the onion, pass through a chinois and reheat Check its consistency and flavour	Pass through a chinois, reheat Check its consistency and flavour	Pass through a chinois, reheat and skim Check its consistency and flavour Used to prepare sauce demi-glace

Table 2.2 How different sauces can be made from the three basic sauces

Sauce béchamel	
Sauce persil	Chopped parsley is added just before service; served with poached white fish and some boiled vegetables
Sauce mornay	Grated cheese and egg yolk are added, the sauce should not be allowed to reboil; served with poached white fish, pasta and cauliflower
Sauce crème	Cream is added to produce the consistency of double cream; it should not be reboiled; served with boiled root vegetables and poached fish
Sauce velouté	
Sauce supreme	Chicken velouté with mushroom trimmings, cream, egg yolk and lemon juice; served with cooked chicken dishes and vol-au-vents
Sauce aux champignons	Chicken velouté with sliced mushrooms, finished with egg yolk and cream; served with cooked chicken and sweetbreads
Sauce aurore	Sauce supreme with the addition of tomato purée making a pink-coloured sauce; served with cooked chicken and poached eggs
Sauce Espagnol	
Sauce demi-glace	Equal quantities of Espagnol sauce and a rich brown stock all boiled together and reduced by 50%. When strained, this produces *sauce demi-glace*
Sauce demi-glace	
Sauce chasseur	A reduction of white wine and shallots is added with sliced mushrooms and tomato concasse, chopped parsley and tarragon; served with chicken, chops or steak
Sauce diable	A reduction of shallot, white wine vinegar and black pepper and cayenne pepper is added to produce a spicy sauce; it is served with grilled or fried meats
Sauce Lyonnaise	Sweated, sliced onions and vinegar are added to the sauce and is served with fried liver and some rechauffé dishes

Sauce Espagnol

Espagnol sauce is a basic sauce that is seldom used in its own right. In order to achieve richness of flavour colour and gloss necessary to make a good sauce, the sauce has to be refined. Refining is achieved by using equal parts of sauce Espagnol and rich brown meat stock - estoufade. These two liquids are mixed together in a heavy based pan and then boiled vigorously (and skimmed frequently to remove impurities from the surface) to reduce the water content and therefore increase the strength of flavour and colour of the sauce. The more liquid that is evaporated from the pan the stronger the remaining sauce will become. Normally a 50% reduction is used.

The resulting sauce is called *sauce demi-glace.* This is a rich deep brown coloured sauce with a smooth texture and a natural gloss. Once the demi-glace has been made it can be used as a base for many other sauces.

THICKENING AGENTS

All the three basic sauces are thickened by an agent called a roux. A white or first stage roux is prepared for a sauce béchamel; a blond or second stage roux is prepared for a sauce velouté; and a brown or third stage roux is prepared for a sauce Espagnol.

White roux

Equal quantities of margarine or butter and flour are used (100 g of each to create 1 litre of sauce). The fat is melted in a thick-based pan, then the flour is added and allowed to cook gently for 2 to 3 minutes to allow the grains of starch in the flour to burst, thus allowing the starch to begin to cook. This process is known as 'cooking out the roux'. The roux should be allowed to cook without colouring and should be removed from the heat before the warmed milk is added. The milk should be added slowly to allow the starch to be evenly distributed and incorporated into the liquid, thus starting the thickening process. When all the liquid has been added and blended, the pan should be returned to the heat and constantly stirred while the sauce returns to the boil and thickens. The sauce is then finished.

Blond roux

Equal quantities of flour and lard or margarine are used (100 gm each to 1 litre of liquid). The fat is melted in a thick-based pan; then the flour is added and this time the roux is cooked out until a sandy brown colour and texture has been achieved. When this second stage roux has been prepared, it is removed from the heat and a warmed white stock is added gradually while constantly stirring the mixture to ensure that the roux and stock are well blended to a smooth consistency. This mixture has to be returned to the heat to thicken and is then finished.

Brown roux

For this roux, slightly more flour is required (125 gm flour and 100 gm of fat for 1 litre of stock). To make this sauce, lard or, better still, dripping is used and the stock should be a good brown colour and well flavoured. The reason for this increased quantity of flour is because when the roux is cooked out to the third stage, some of its thickening power is lost during the prolonged exposure to the heat. This third stage roux is prepared in a thick-based pan, by melting the fat and then mixing in the flour. This roux needs to be cooked out until the flour has turned a brown colour by the heat. The pan should then be removed from the heat and allowed to cool slightly before the other ingredients are added. As with the other sauces, it must be returned to the heat and boiled to allow the thickening and flavouring process to develop before it is finished.

Cornflour

This is pure corn starch that has thickening powers when mixed with boiling liquid. The cornflour is blended with a little cold liquid and then is poured into the remaining boiling liquid. When boiled, it produces a light, smooth sauce that may be flavoured or seasoned as required - the flavour of the liquid will determine the final flavour of the sauce, as cornflour has no flavour of its own. This method of thickening is often used for sweet sauces based on fruit juices, as a clear sauce can be achieved.

Arrowroot

This is a starch obtained from the tubers of the West Indian plant - maranta. It is used in the same way as cornflour, but it has the additional advantage of becoming transparent, and so it is often used for sweet glazes and sauces.

Beurre Manié

This is prepared by kneading butter and flour together to form a pliable paste. To use this form of thickening, the liquid should be brought to the boil and then allowed to simmer while small pellets of the kneaded butter are whisked in. The butter melts in the heat and the flour is evenly distributed into the liquid. This method is used when it is necessary to thicken sauces and soups at the end of their preparation process. For example, for petit pois à la française.

Modified starch

This is a prepared starch that can be whisked directly into simmering liquids to thicken them without forming any lumps. It is a convenient thickener that is colourless and flavourless, so that it can be used for both sweet and savoury items. This method can be used for a fruit pie filling, made from canned fruits and the juice thickened using the starch. Other uses are instant desserts and instant sauces to which boiling water is added. These starch preparations are very useful for the caterer to keep for emergency use because they are very quick to prepare, and keep well in air-tight containers.

Liaison

A mixture of cream and egg yolks at equal quantities is used for thickening and enriching some sauces, cream soups and white stews. The liquid should not be allowed to boil after the liaison has been added, as excessive heat will cause separation. For this reason, the liaison is added just before the sauce, soup or stew is to be served.

EGG-BASED SAUCES

Sauce Hollandaise

This is a rich egg and butter based sauce that is flavoured by the addition of a vinegar reduction and crushed pepper corns. It is served with hot fish dishes such as poached or grilled salmon or turbot, or with hot flower vegetables such as asparagus. The sauce is prepared by whisking the egg yolks with the vinegar reduction and peppercorns over a gentle heat so that the eggs begin to thicken. This thickened mixture is called a sabayon. At this stage of preparation, the creamy mixture is removed from the heat and the warmed butter is gradually added by whisking it in, creating a smooth, golden sauce that is then passed through a muslin to finish it before serving. Remember that this sauce is served warm. With the risk of *Salmonella* contamination in fresh eggs, it is common practice to use pasteurized egg yolks for the preparation of sauce Hollandaise.

Mayonnaise

This may also be prepared from pasteurized egg yolks to prevent the risk of *Salmonella* food poisoning. Mayonnaise is prepared by whisking egg yolks, vinegar and seasoning together, and then gradually adding oil a drop at a time to the mixture. The oil should be added slowly to produce a cold, thick-yellow sauce. Fresh mayonnaise should be treated carefully, as fresh egg yolks have not been subjected to heat and are therefore raw; it should be kept as cold as possible and used on the day of preparation to minimize the risk of *Salmonella* poisoning. The elderly and the young should not be allowed to eat fresh mayonnaise, because of the potential health risks involved. There are some good quality, commercially prepared, convenience types of mayonnaise now available should it be deemed inadvisable to serve fresh mayonnaise.

MISCELLANEOUS SAUCES

Tomato sauce

This is a rich tomato-flavoured and coloured sauce that is made with a blond roux, a white stock, and flavoured with a mirepoix of vegetables, bacon scraps and tomato purée. This sauce is simmered for approximately 1 hour to extract the full flavour from its ingredients. It is then passed through a chinois before being seasoned and served. The sauce is served hot as an accompaniment with cod à l'Orly, and some rechauffé and pasta dishes.

Curry sauce

This highly flavoured sauce is made with a blond roux to which curry powder and diced onion has been added, a white stock and, for flavouring, diced root vegetables and cooking apple, dessicated coconut, sweet chutney, tomato purée and sultanas. This has to be simmered gently for at least 1 hour, longer if possible, to extract the full flavour from all the ingredients. It has to be passed through a fine chinois before being seasoned for service. This sauce is served hot as an accompaniment to kedgeree and with some rice and egg dishes.

Bread sauce

Bread sauce is prepared by infusing milk with a clouté onion, then straining the milk and adding fresh, white breadcrumbs. The sauce is finished with butter and served to accompany roast chicken and game.

Apple sauce

Apple sauce is prepared with cooking apples that are cooked to a purée with very little extra water. This purée is then sieved, seasoned and finished with butter and served as an accompaniment with roast pork and roast duck, the tartness of the apple helps to counteract the fattiness of these types of meats.

Gravies

Gravies are prepared to accompany meat dishes, and should have the flavour of the meat. There are two types of gravy: jus-roti is a thin gravy that is prepared from the juices produced from a roast joint of meat. These juices are mixed with a stock, skimmed and then served to accompany the roast joint of meat. This gravy should be well seasoned and be a good brown colour. The skimming will remove all traces of fat that could be floating on the surface. The process of combining the meat sediment and juices with the brown stock is known as 'to swill or deglace the pan'.

Jus-lié is a thickened gravy that is prepared from a well-flavoured stock and thickened with arrowroot or cornflour. It is served to accompany rechauffé dishes and some meat pies and puddings. It should be a good brown colour and well seasoned, and be thin enough to pour easily from a sauce boat or jug.

Sweet sauces

These are used to accompany sweet pies, tarts and puddings, also some ice cream-based sweets that might be rather bland or dry to the taste without an appropriate sauce.

Custard sauce

A creamy yellow, milk-based sauce that is made from custard powder and sweetened to taste.

Sauce Anglaise

A delicate sauce made by combining milk and egg yolks and thickening over a gentle heat. It is flavoured with vanilla. If this sauce is overheated, the egg protein will harden and cause the mixture to separate and curdle.

Cornflour sauce

A white sauce made by thickening milk with cornflour. This sauce can be flavoured as required, for example with almond essence, vanilla essence, or peppermint flavouring; it may also be coloured if appropriate.

Jam sauce

For this sauce, jam is boiled with water to produce a liquid which is thickened with cornflour. Raspberry and apricot jams are frequently used for these sauces as they are not too sweet. Syrup sauce is also prepared in the same manner and is served with steamed puddings.

Safety when making sauces

It must be remembered that harmful bacteria will easily multiply in a warm sauce or gravy. It is therefore essential to keep sauces above 65°C. The easiest method of keeping sauces safely is to prepare them as near to service time as possible and then keep them in a double boiler or in a bain marie. To prevent a skin forming on the top, a greased paper cartouche should be placed over the surface. If a sauce has to be kept, it must be cooled rapidly and placed in a clean, covered container in the refrigerator. It must be used within 24 hours and then reboiled thoroughly before being served. As previously explained, egg-based cold and warm sauces are potentially dangerous and need special care, and they may not be reboiled as this will cause separation to take place.

Gelatinization

The process of gelatinization is important in all starch-thickening processes. It is the term used to refer to the action of heat upon grains of starch. When hot liquids come into contact with starch, the outer part of each grain of starch will gelatinize thus releasing its thickening properties into the liquid. It therefore follows that if the grains of starch are allowed to form small clusters or 'lumps', they will be unable to release their thickening properties and the liquid will remain thin in consistency. The gelatinization process may be affected by the addition of an acid to the liquid before the thickening process has taken place. For example, when preparing the lemon filling for a lemon meringue pie, it is advisable to thicken the water with the cornflour and then to add the lemon juice and rind. Sugar can also have an adverse effect upon the gelatinization process.

CONVENIENCE SAUCES

There are many types of convenience sauce preparation available to the caterer. These may be divided into those that are entirely ready for immediate use, and those that require some cooking before completion.

Canned or vacuum packed sauces

Condensed sauces are diluted 50/50 with water and are then ready for immediate use. It is possible to obtain a large variety of sauces that are packed in this manner. These sauces are expensive but can be useful in an emergency.

Bottled sauces

These are prepared to accompany foods that are offered to the customer as they are. Sauces that come in this category include tomato ketchup, brown sauce, spicy sauce, tartar sauce, mayonnaise, salad cream and mint sauce. They are also packed in individual sachets for fast-food catering units.

Dehydrated mixes

These are in powdered or grandular form and require blending with water and then simmering for approximately 20 minutes before being ready to serve. These sauces are generally not as acceptable to the customer as the 'real' sauce as they lack the depth of flavour that can be obtained from fresh ingredients.

Instant granules

These can be reconstituted by simply pouring boiling water onto the granules. They are made with a modified starch base so that it thickens as soon as it comes into contact with the water.

ACTIVITIES

- What effect does the fat have when a roux is prepared to thicken a sauce?
- Why is cornflour blended with a little liquid before being used to thicken a liquid?
- Why should sauces be stirred constantly while being brought to the boil after liquid has been incorporated into the roux?
- How many sauces can you find that can be devised from the three basic sauces - sauce béchamel, sauce velouté and sauce Espagnol. Give an example in each case of a suitable dish or food that the sauce could be served with.
- How many safety points can you think of that could relate to the following: the temperature of the sauce, cross-contamination, keeping the sauce hot, and the ingredients used.
- Look around the shops, at trade price lists and trade journals and make a list of different types of flavoured convenience sauces and stocks (look for savoury, sweet and cold sauces).
- For each of the sauces given, answer the five questions listed (sauce tomato, sauce anglaise, and sauce persil):
  1. List the ingredients required to make 1 litre of sauce.
  2. What are the significant temperatures in the preparation of the sauce?
  3. Describe how the sauce may be kept hot successfully.
  4. List some likely faults that could occur during its preparation and during the keep hot period.
  5. Give two examples of suitable foods with which to serve each of the sauces.

CHAPTER 14 Soups

In a traditional meal, soups are served between the hors d'oeuvre course and the fish course. All good soups should be served very hot (at 90°C) or, very occasionally, chilled (at 3°C) and they should complement the rest of the meal and provide a contrast in flavour, texture and colour. There are numerous types of soups, prepared by the potager in a large, traditionally organized kitchen brigade. Soups are classified according to their method of preparation - some may be clear and light in substance, such as a consommé, while others may almost be a meal in themselves, such as minestrone soup or French onion soup. It is necessary to take care when preparing soups that are not puréed or passed - the vegetables should be cut to even sizes so that the presentation is not spoilt. For example, Scotch broth uses a brunoise cut of vegetables and potage paysanne uses a paysanne cut of vegetables (see page 114).

BROTHS

These consist of a good flavoured stock containing diced meat or vegetables. They are thickened by the starch from either pearl barley or rice that is cooked with the other ingredients in the stock. As this soup is not passed in any way, it is essential to have a neat brunoise cut of vegetables and finely diced meat. The stock should be well flavoured and be of the same type as the diced meat. Broths are a substantial soup because of the cereal, meat and vegetable content, and are normally garnished with freshly chopped parsley.

POTAGE SOUPS

These soups involve the precise cutting of vegetables to a paysanne cut. The vegetables are gently sweated in butter in a heavy based pan until they are just beginning to soften. The remaining ingredients and the stock are then added, and the soup is gently simmered until the vegetables are just cooked. Overcooking will cause the vegetables to lose their shape.

Some potages are passed through a fine chinois at this stage and then finished as appropriate, for example, potage bonne femme is finished with butter and cream. Other potages are served unpassed - the ones that have been made using paysanne vegetable cuts. These soups do not have any further garnish or accompaniment, except for minestrone soup where finely grated Parmesan cheese is offered separately. Like broths, these potages are substantial soups because of their high vegetable content.

PURÉE SOUPS

Purée soups are another type of substantial soup because they are based upon vegetables, and are served with croutons as their accompaniment. The soups are made from one main vegetable, such as lentils for a purée of lentil soup. Other vegetables are included in smaller proportions for flavouring purposes, these usually include onion, celery, leek, carrot, swede or a combination of some of these vegetables. A purée soup is prepared by cooking all the vegetables previously diced in a good flavoured white stock. When the vegetables are cooked, the soup is passed through a sieve or liquidized, then passed through a conical strainer. The resulting soup should be smooth and thick enough to coat the back of a spoon, while still being able to pour from a ladle. After reheating and seasoning, the purée soup is served with croutons as an accompaniment.

VELOUTÉ SOUP

Velouté soup is a lighter richer soup than broth, potages and purée soups. These soups are made using a blond roux and white stock with a few base vegetables for flavouring. The base vegetables are diced and sweated in fat until softened, but not coloured. The flour is then added and a blond roux is made. The well flavoured white stock is added and, after bringing to the boil, the soup is simmered for 40–50 minutes.

The type of stock used will depend upon the type of soup that is being prepared. For example, if a chicken velouté is being prepared, then chicken stock is used and diced chicken will be added for garnish just before service.

After being simmered, the soup is passed through a fine strainer, reheated, and then checked for consistency and flavour. To finish the velouté soup, cream or milk is often added just before service and an appropriate garnish is added.

CREAM SOUPS

A cream soup is a smooth, rich soup that has a definite main ingredient and flavour. It is a soup that has been made from another base soup. For example, it can be made from any of the three following methods, a purée base that is finished with milk to create a creamy texture; a purée base that is combined with a thin, béchamel sauce to create a creamy based soup of a lighter texture than full purée soup; and a veloute base that is finished with the addition of a cream and egg yolk liaison to create a creamy, rich soup (care must be taken when adding the liaison, see page 81).

CONSOMMÉ

This is a clear soup made from a well-flavoured stock that is cleared by the action of egg white protein (albumen) and meat protein which rise to the surface during cooking, bringing insoluble particles with them. When cool, the coagulated protein is carefully strained off and the resulting stock should be crystal clear. Consommé may be garnished in numerous ways and should be served very hot or chilled. The stocks used are either chicken, beef or game. They should be free from fat globules and thoroughly strained before being used in the preparation of consommé.

ACTIVITIES

- Research the answers to the following:
  What is meant by the term 'liaison' in soup making?
  Describe how liaison should be made to a velouté soup.
  What category does the soup vichyssoise come into?
  Describe how to prepare and serve soupe l'oignon.
  Describe how to incorporate beurre manié into a soup and explain its function.
- Which types of convenience soups are available to the caterer. List examples under the following categories: canned, dehydrated, condensed and instant soups.
- Discuss merits of each of the categories listed in the previous question for the following type of consumer:
  1. a student in a self-catering hall of residence,
  2. a caterer in a large general hospital,
  3. a chef in a medium-sized city hotel with table d'hôte menu.
- Collect advertisements and other promotional material that you find for convenience soups and discuss which type of consumer they are aimed at.

CHAPTER 15 Cereals and farinaceous foods

Cereals are cultivated grasses whose edible seeds are highly nutritious. The principal grains used are wheat, which is used for the production of flour; maize, which is mainly refined to produce cornflour; barley, which is grown to produce malt extract; and oats, which are refined to make rolled oats, and rice. The word farinaceous is often used to describe cereal-based products which have a high starch content. All cereal products are capable of absorbing moisture, and so it is necessary to store them in airtight containers.

WHEAT

Wheat is cultivated in many areas of the world. Some varieties of wheat contain a higher proportion of protein than others, and these are used to manufacture strong flour that is essential for bread and pasta making. Wheat with a lower protein content is used to make soft flour, which is used for cakes and general baking purposes.

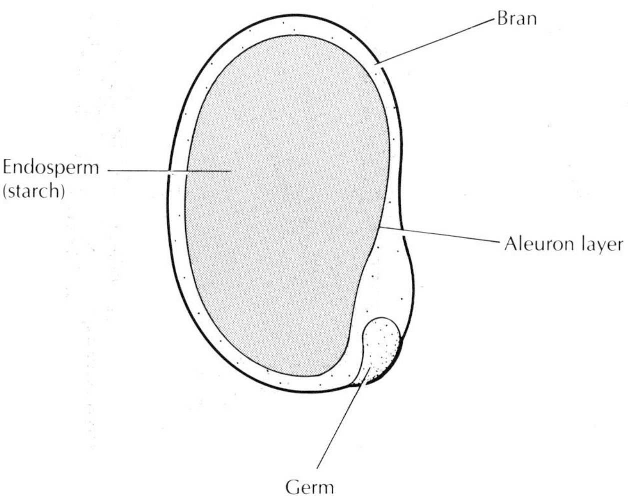

2.1 Cross section of a grain of wheat.

The outer skin of wheat grains are removed before the 'real' flour production can begin. The three remaining parts of the grain are the endosperm, 85 per cent of the grain; the bran, 13 per cent of the grain; and the germ (the seed) 2 per cent of the grain (see figure 2.1).

Flour production

The grains of wheat are cleaned, crushed or milled, and then sieved to separate the bran, germ and endosperm. This enables the endosperm to be used on its own for white flour production, or with varying amounts of the bran to produce brown flours. The removal of the bran when making white flour results in the loss of dietary fibre, protein and vitamin B. In the UK, it is essential to replace these losses. Each 100 g of flour must contain 1.24 mg of thiamin, 1.60 mg of niacin, 1065 mg of iron, and 240 mg of calcium. These are all added after the refining process. Flour improvers – ascorbic acid, ammonium chloride and potassium – are sometimes added to improve the baking quality of the flour. Brown flour without whole wheat grain has what is known as a 100 per cent extraction rate. If 30 per cent of the grain is used, then the extraction rate is 70 per cent.

Common types of flour

- Wholemeal (100 per cent extraction rate) is a brown flour that has its own distinctive flavour due to the presence of the bran and germ of the wheat. It may have been milled or stoneground and is used to make breads.
- Wheatmeal (80–85 per cent extraction rate) is commonly known as brown flour. A light brown flour with a mild flavour, this flour is used primarily for bread making but also for scones and pastry.
- Wheatgerm (70 per cent extraction rate) is a brown flour containing 10 per cent processed wheatgerm, making it rich in vitamin B and giving it a distinctive flavour. It is used for bread making.

- Strong white flour (72–73 per cent extraction rate) has a protein (gluten) content of 12–15 per cent. Strong flour is ideal for all yeast cookery, breads bun doughs and for puff pastry and batters. The gluten produces the light, open structure required for breads and puff pastry.
- Soft white flour (72–72 per cent extraction rate) may be bleached or unbleached and has a protein content of only 7–10 per cent. It is the ideal flour for short and sweet pastries, cakes and sponges.
- Self raising (70–72 per cent extraction rate) is a soft white flour which has a chemical raising agent added (see page 121) to create the raising power necessary to bake successful cakes and puddings.
- Semolina is the milled endosperm of the grain of wheat. The best quality is produced from the wheat that makes strong flour, for example, durum wheat, which is grown in Canada and the USA. Semolina is used to make milk puddings and has a high starch content.

SOME OTHER CEREALS

- Rye is used in large quantities in Eastern Europe, to produce a dark-coloured flour that is used to make rye bread and rye biscuits. It has a lower protein content than wheat and produces bread in a close texture. It is also used to make crispbreads.
- Maize is largely produced in North and South America. Cornflour is made from maize, and as 'corn on the cob' is eaten widely as a vegetable. Pop corn and cornflakes are made from maize.
- Oats are cultivated in cool climates – the north of England and Scotland produce good crops. The grain contains fat which, in time, will go rancid, causing it to have a relatively short shelf life. Rolled oats are oats with the bran removed and the grain flattened to produce oval-shaped, flat grains, which are used to make porridge, biscuits and flapjacks. They are used in many popular breakfast cereals and muesli preparations. For oatmeal, the grains are ground to produce coarse flour that is used for porridge, oatcakes, parkin and haggis.
- Barley is a cereal that is low in protein and is grown to produce malted products for the brewing industry. Malt flour, used in malt bread, is also made from barley.
- Pearl barley is the endosperm of barley grain and is used to thicken stews and broths.

Starches that are used in food preparation and that are not, by definition, cereals include the well-known sago and tapioca, which are used to make milk puddings, and arrowroot, which is used to thicken sauces and gravies and to make clear fruit glazes for flans.

Examples of cereals can be seen below.

RICE

Rice is harvested from an annual grass which is grown in South East Asia, Africa, Australia, Italy, Spain and America from where most of the rice imported into the UK comes. The bran and husks are removed from the grains of rice, and this is

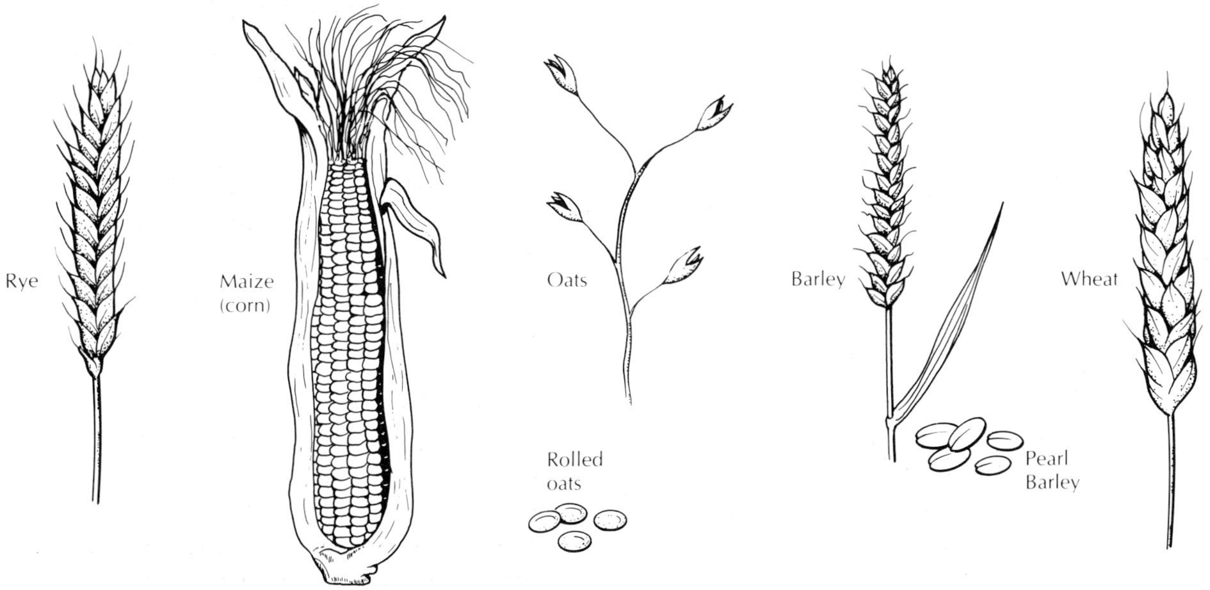

2.2 Different types of cereals.

normally done before the rice is exported. There are several types of rice:

- Patna or long grain rice has a narrow pointed grain that is used for boiled rice served with curries, etc.
- Carolina rice has a short, rounded grain that is used for making milk puddings.
- Medium grain rice is medium sized and is used for braised rice dishes.
- Brown grain rice is the term used to describe rice that is unpolished, has a brown colour and more roughage.
- Ground rice is a fine, white, granular powder made from rice grains, and is used to thicken soups.
- Rice paper is a thin, edible paper used as a base for sweets and macaroons.

Nutritional value of rice

The protein content of rice is extremely low. However, the amino acids present in the rice are a valuable complement to other foods, such as chicken or fish. Here the combined nutritive value of the amino acids in the rice and the other food is greater than would be obtained if the two foods were to be eaten separately. Rice may be used in low-protein and low-fat diets. It needs only a little skill and imagination to produce a large variety of nutritious dishes using rice as the base commodity.

FARINACEOUS FOODS

Farinaceous is the term used to group together foods or items on menus that contain a high level of starch. These are mainly pasta dishes, and although gnocchi (dumplings made from pieces of semolina pasta) and rice also contain high levels of starch, they are strictly not included in this group. Farinaceous foods are traditionally served hot with an appropriate sauce, either as an alternative to an hors d'oeuvre or as a separate course offered immediately before the fish course in a full dinner menu. In recent years, pasta dishes have increased in popularity and they are now served in considerable quantity both as part of a traditional menu and as a light meal or snack. Pasta may be freshly made on the premises for use in a day or two, or it may be commercially prepared and sold in its dried form. Dried pasta is made from white or wholemeal durum wheat flour, combined with salt, fat, water and, sometimes, egg. The nutritional value of pasta depends upon the type of flour used and whether egg is used in the recipe. Durum wheat flour is used because the gluten content is higher than other flavours, which enables the pasta to hold its shape. Green pasta, known as pasta verdi, is produced using spinach in the manufacturing process to give it its green colour.

Cooking pasta

Pasta should be placed in a large pan of boiling salted water, and then simmered gently until the pasta is just softened right through – al denté is the term used to refer to pasta that is still firm after cooking. Sufficient water must be used when cooking pasta; this is especially important when cooking lasagne or canelloni as the pasta is inclined to stick together unless care it taken. A few drops of cooking oil may be added to the boiling water to help prevent the pasta pieces from sticking to each other, and this also helps when rinsing the starch from the pasta after cooking. As soon as the pasta is cooked, it must be drained and rinsed under running water to prevent

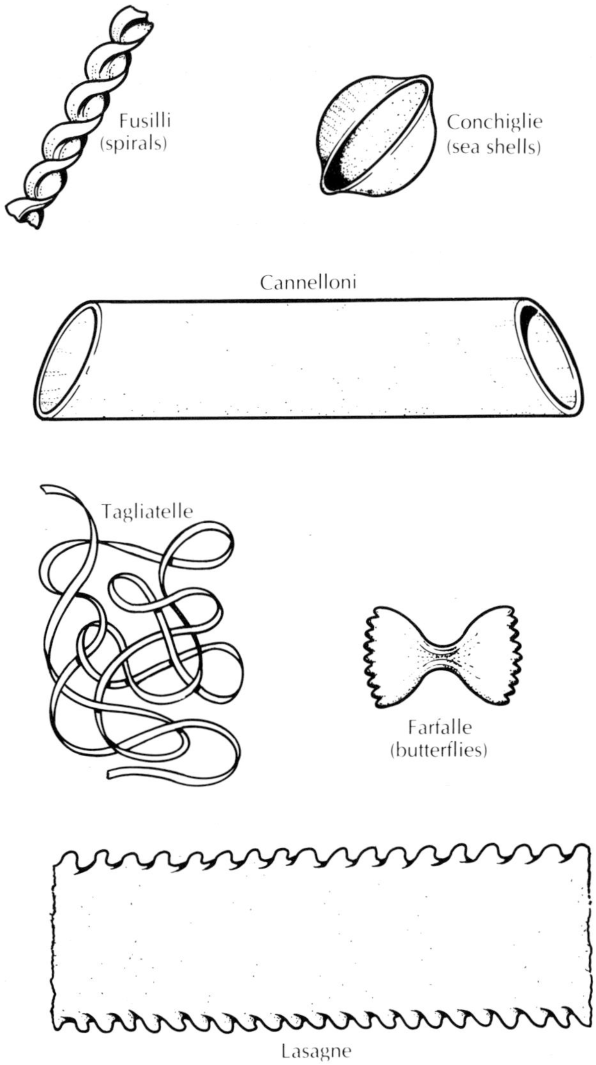

2.3 Types of pasta.

over cooking and to rinse the excess starch away; this rinsing process is called 'refreshing' the pasta. If the refreshing process is not carried out, the pasta will be sticky and difficult to handle. Cooked pasta may be held in ice cold water until it is required and then reheated by either plunging it into boiling water for a minute, tossing it in melted butter over a fierce heat, or by combining it with an appropriate sauce and then finishing it off in a hot oven, or under a salamander, or in a microwave oven. Pasta is normally served in shallow earthenware dishes placed on a dish paper and a silver flat or other suitable under dish. Figure 2.3 on the previous page illustrates different varieties of pasta.

Pasta is popular because:

- it is economically priced,
- it has a long shelf life if kept in a cool, dry atmosphere,
- it is easily obtainable from wholesalers and cash and carry outlets and can be used in times of emergency,
- the cooking time is relatively short and it does not need to be prepared,
- it may be used in a wide variety of dishes with a wide range of different flavoured sauces and ingredients,
- it is a sustaining food,
- it creates a nutritional dish when sauces are added,
- it makes a tasty, snack-type meal.

ACTIVITIES

- Look at a selection of breakfast cereals and find out from which cereal each has been made.
- List examples of biscuits and snack products that have a high cereal content.
- Give one example where a cereal is used in the following ways:
  cereal as a garnish,
  a cereal as a main filler in a dish,
  a cereal as a thickening agent,
  a cereal to create the main flavour of the dish,
  a cereal to create a sweet dish,
  a cereal used in a confectionary product,
  a cereal used to make a savoury snack item.
- Make a display of different pasta shapes and give a recipe example for each shape.
- List types of convenience foods that are based on pastas.
- Give a short menu description of the following pasta dishes and in each case describe how the finished dish should be served:
  spaghetti Italienne,
  spaghetti Napolitaine,
  spaghetti Milanaise,
  ravioli,
  canneloni,
  spaghetti bolognaise.

CHAPTER 15 Dairy products, vegetable fats and oils

The term dairy products is used to cover foods that traditionally come from the dairy, that is milk, cream, cheese, fats and eggs. There are recommendations in the NACNE and COMA reports for reductions in the consumption of fats (see page 168). This has resulted in increased use of skimmed and semi-skimmed milks, dried skimmed milk, low-fat artificial cream, low-fat cheese, cottage cheese, vegetable fats and spreads as alternatives to butter. If these products are to be used in cooking, it is advisable to check their effects upon the quality of the finished dish, as some recipes will need to be adjusted to create the desired effect with less fat in the ingredients.

MILK AND CREAM

Milk and cream are sold in many types of bottles and cartons from litre size to small cartons for individual servings of tea and coffee. The cartons are used extensively in fast-food establishments (see page 177).

Types of milk treatment

Pasteurized milk

Pasteurization is the process of heating milk to a temperature of 72°C for 15 seconds and then cooling it to 4°C in order to destroy harmful micro-organisms. There is a definite cream line visible in each bottle of pasteurized milk. It is suitable for normal use and has a 48 hour shelf life if stored in a refrigerator. It has a minimum fat content of 3.8 per cent.

Homogenized milk

Here the fat globules are evenly distributed throughout the milk, which is then pasteurized. The cream is evenly spread in the milk making this type of milk ideal for catering uses. It will keep for 48 hours if stored in a refrigerator and has a minimum fat content of 3.8 per cent.

Ultra-heat-treated (UHT) milk

This milk is homogenized, then heated to 130°C for 1 second, packed into sterile cartons and rapidly cooled. The unopened cartons will keep for six months. Because of the high temperature used in this process, there is a definite cooked flavour to the milk. It has a minimum fat content of 3.8 per cent. Much of this type of milk is packed into small cartons for use in fast-food outlets for individual servings of tea and coffee.

Sterilized milk

Here the milk is homogenized, bottled and then heated to 104°C and held at this high temperature for 30 minutes. It is then rapidly cooked. This treatment destroys bacteria and other micro-organisms more completely than pasteurization, but the high temperature causes slight caramelization of the lactose in the milk, resulting in a definite cooked taste. For this reason, it is seldom used in the catering industry. The bottles of sterilized milk are hermetically sealed and the milk will keep fresh for up to three months if left unopened. It has a minimum fat content of 3.8 per cent.

Channel Island milk

This milk comes from a Channel Island breed of cattle, which produce milk that has a minimum fat content of 4.0 per cent and, therefore, the milk is creamy and richer than the ordinary milk. It is homogenized and is expensive as it is in short supply. It is not used in catering establishments as it is far too rich for most recipes and there is a limited supply.

Processed milk

Milk powder is produced by evaporating the water content from the milk. There are two methods of evaporation: roller drying, where the milk is spread on revolving drums, allowing the water to evaporate, leaving a thin powder of solids; and spray drying, where the milk is pumped as a fine spray into a hot air chamber where the water evaporates leaving a fine powder of solids. The temperature is carefully regulated so that the milk proteins do not coagulate and the powder does not take on a cooked flavour.

Skimmed milk

Skimmed milk is the liquid left after the cream has been skimmed off or separated. This process results in a light, nutritious milk with a low fat content.

Evaporated and condensed milk

The milk is homogenized and then heated to 80°C for 15 minutes. To complete the process, the milk is boiled to evaporate the excess water from the milk. This has the effect of producing a concentrated milk with a cooked, creamy flavour. Evaporated milk is twice the concentration and condensed milk two and a half times the concentration of ordinary milk.

Nutritional value of milk

In fact, 87 per cent of milk is water. However, milk is a very valuable food as it contains proteins, lactose (milk sugar) and mineral salts. The proteins in milk are of high value as they contain all the essential amino acids required by the human body. The main proteins are casein, lactalbumin and lactoglobulin.

Table 2.3 The nutritional value of milk (one pint)

	Pasteurized	*Skimmed*
Protein	19.3 g	19.9 g
Fat	22.2 g	0.6 g
Carbohydrate	27.5 g	29.3 g

Types of cream

Creams vary in thickness because they contain varying amounts of fat. Coffee, single and double creams are pasteurized which increases their safety and shelf life:

- Coffee cream has a thin consistency and a fat content of 15 per cent.
- Single cream is slightly thicker (it has a fat content of 25 per cent), can be whipped, and is used for enriching sauces and soups.
- Double cream is thicker (it may be whisked 'to a peak') and has a fat content of 59 per cent. It is used to decorate gateaux and pastries, and as an accompaniment to fresh fruits and compotes.
- Clotted cream is produced from milk from Devon and Cornwall by a gentle process of heating milk to 82°C and then skimming off the resulting thick cream from the surface. This cream is rich with a fat content of 65 per cent) and is not smooth enough to pipe. It is traditionally served to accompany fruit pies, compotes and scones for afternoon tea.
- Synthetic cream comes in the form of powder or granules, to which water is added. It is then whipped to a thick cream that may be piped. This type of cream is very susceptible to bacterial contamination and should, therefore, be used as soon as it has been reconstituted. It is unsafe to keep in this reconstituted state, so small quantities should be whipped as required.

CHEESE

It is believed that the first form of cheese was made accidentally by nomads who carried milk in bags made from the stomachs of sheep. The rennet present in the stomach lining would have helped to turn the milk into curds. Today, cheese is made throughout the world from a variety of different milks: cow, sheep and goat. It takes approximately 5 litres of milk to produce 1/2 kg of cheese. The composition of cheese will vary according to the type of milk and the treatment used to make the cheese.

Cheese contains protein, fat, carbohydrate, minerals and water. Table 2.4 compares the contents of Cheddar cheese with cottage cheese. From the figures one can tell that cottage cheese will be the choice of cheese for those on a low-fat diet. Cottage cheese is a curd cheese that has a delicate flavour and is used in a variety of dishes and snacks, for example, in cheese cakes, with fruits, in salads, and in filled rolls and sandwiches.

The quality of a cheese is important - stale cheese will change in flavour and texture. There should be no signs of any mould or mildew, and no dry cracks or fat globules on the surface. The cheese should smell pleasant, and soft varieties should not be runny in the centre.

Popular English cheeses

Hard cheeses include the following well-known varieties:

- Cheddar cheese may be yellow or red in colour. It has a mellow, buttery flavour and is used in many cooked or raw dishes.

- Cheshire cheese has a mellow flavour, and is slightly salty in taste. It may be white or red in colour and has a loose, crumbly texture, but it is too mild for successful cooking.
- Gloucester cheese is a close, crumbly cheese with a medium flavour. It is a pale orange colour and is best raw in salads or on a cheese board.
- Derby is a mild cheese with a close texture and pale colour. It may have sage added to produce a cheese with a distinctive flavour and colour.
- Leicester is a deep orange-coloured cheese with a mild, mellow flavour, and a loose and flaky texture.

Lightly pressed cheeses include the following varieties:

- Caerphilly is a white-coloured cheese with a very mild flavour. It is used on cheese boards.
- Lancashire is a deep orange-coloured cheese with mild, mellow flavour and a loose and flaky texture.
- Wensleydale is a white, soft, creamy cheese with a mild, yet distinctive flavour, and is usually served on cheese boards. It complements fruits.

Blue-veined cheese include the following:

- Stilton cheese has a thick, protective crust and a distinctly flavoured, blue-veined interior. It is served with port at the end of a meal. It is occasionally used in cooking for soup or flavouring sauces.
- Limeswold is a flat-shaped cheese with a thin crust and a soft, blue-veined interior. It has a short shelf life.
- Wensleydale is a blue-veined, soft, mild cheese. It is not very widely available.
- Cheshire is a mild, blue-veined cheese.

Table 2.4 The nutritional value of Cheddar and cottage cheese

Nutrient	*Cheddar (100 g)*	*Cottage (100 g)*
Protein	26.0	13.6
Fat	33.5	4.0
Carbohydrate	trace	1.4
Minerals	3.4	1.4
Water	37.0	78.8
Kilocalories	406	96

FRESH MILK

↓

Heated to destroy harmful micro-organisms

↓

Pumped into vats and tested for acidity

↓

Ripened – started with lactic acid; the enzyme, rennin is added to cause clotting

↓

Curd produced and cut

↓

Curd is scalded using steam at for 30 minutes

↓

Curd settles and the whey drawn off

↓

Curd is cut and the blocks stacked up

↓

Salt is added to preserve the curd and bring out the flavour

↓

Curd is packed into muslin-lined moulds

↓

Curd is compressed in the moulds

↓

Curd is removed from moulds and stacked to mature at 4.4°C to 10°C

2.4 The manufacture of cheddar cheese.

Using cheese in the kitchen

When using cheese in dishes one must consider the overall flavour to be achieved, the texture to be created and the finished appearance of the dish. Generally, hard, well-flavoured cheese is recommended; Cheddar is frequently used as is Swiss Gruyère and the well-flavoured Italian Parmesan cheese. The advantage of using these hard cheeses is that they do not create a fatty or greasy finish to the dishes as their fat content will not readily melt in the heat applied during the cooking process. Another point to remember is that due to the high fat content, cheese will become increasingly difficult to digest as the heat will eventually melt the fat content, and the protein content will coagulate and become indigestible. When producing dishes that have to be gratinated, a very hot salamander should be used to ensure the desired browning effect without making the cheese indigestible. When adding cheese to sauces, it should be finely grated and mixed into the very hot sauce; no further boiling should take place as this will make the mixture very difficult to digest.

FATS

Fats are used in many ways in the catering industry, and they come in many forms. They can be categorized into those of animal origin and those of vegetable origin. Each type of fat has its own characteristics, and is therefore best suited to a particular range of used.

Animal fats

- Lard is produced from rendered down pig fat to create a smooth, soft, white fat that is used for roasting, frying and baking.
- Suet is prepared from the hard fat found around the kidneys of beef cattle. It is usually sold in grated form ready for use in suet puddings, pastes and dumplings.
- Dripping is the name of the fat that is produced from the rendered down fat from beef and pork carcasses. It is a light, sandy-coloured fat that tastes of the original animal meat. It is used for roasting, shallow frying meat products, in the preparation of sauce Espagnol (see page 80), and in meat-based soups.
- Butter is the most commonly used animal fat and is produced by churning fresh milk to separate the fat from the liquids. It is a rich yellow colour and readily available either from a particular region, such as Devon or Wales, or as a blended butter, which is produced from milks that have come from a variety of regions. Butter has a unique flavour and is used in cooking to impart its flavour into the dish being prepared, for example, plaice meunière, omelette or sauce Hollandaise. Butter has also been used to glaze vegetables, but since the NACNE and COMA reports, this practice is less common (see page 168).

Vegetable fats

Vegetable fats are manufactured from varying quantities of whole milk, cream and vegetable oils. Vitamin A and D and a small amount of colouring agent are added to margarine to increase its acceptability as an alternative to butter and other animal fats.

- Block margarine is a hard or a semi-hard fat for use in a white roux, sauces,
- Table margarine is a soft fat sold in tubs or boxes. It is used as a substitute for butter in sandwiches, 'butter icings' and some creamed cake mixtures.
- Pastry margarine is a semi-hard fat blended especially for pastry making.
- Vegetable fat is a white, smooth, semi-hard fat used in most cooking processes as a substitute for lard.

EDIBLE OILS

Various edible oils are used in the catering industry. The oil most often used because of its distinctive flavour is olive oil. In Mediterrean countries, where olive trees are cultivated, olive oil is used extensively. However, in the UK its flavour is often considered to be too strong for cooking, and so its main use is in salad dressings, such as vinaigrette.

There are many oils, both pure and blended, which have little flavour. Most are used for deep or shallow frying and important factors in their use is their smoke and flash points. The smoke point is the temperature at which the oil will begin to smoke. The flash point is the temperature at which the oil will burst into flames. With the finish frying temperature for potato chips at 190°C, it is obvious that the caterer requires an oil that will not smoke or burst into flames up to that temperature. For example, olive oil has a smoke point of 148°C and a flash point of 270°C; corn or maize oil has a smoke point of 232°C and a flash point of 360°C.

Oils should be stored carefully, away from direct heat, to prevent it overheating. If oils become too warm, they will go rancid and will result in a strong, unpleasant flavour and the oil being wasted. If oil is allowed to become too cold, it will solidify. Fortunately, on warming it will liquify once more, but the risk of it becoming rancid will increase.

It should be remembered that used oil must never be poured down the waste drains or waste disposal units as it will clog up the waste pipes; this also applies to liquid fats, as these will solidify and block them completely.

EGGS

Eggs are a very versatile food. They have the ability to retain air in their foam-like structure when whisked. They will also coagulate at relatively low temperatures: whites start to set at 60°C and yolks at 65°C. Because of this, eggs should be cooked quickly and to order, since excessive heat, or even keeping them warm once cooked, will over-harden them, making them indigestible and unpalatable.

The weight is significant when cooking in large quantities, here the total weight of eggs is often quoted rather than the actual number of eggs. This is also the case if pasteurized, liquid eggs are to be used.

Eggs may be cooked in numerous ways: in a traditional kitchen brigade they are prepared by the garde-manager (larder chef) if they are cold egg dishes; the entremettier (vegetable chef) prepares hot egg dishes to be served for lunch or dinner; while at breakfast time they will be prepared by the breakfast chef.

Table 2.5 The grading of eggs by size and weight

Grade	Size	Weight
Large	1	70 g
	2	65 g
	3	60 g
Standard	4	55 g
	5	50 g
Medium	6	45 g
Small	7	40 g

Methods of cooking

Boiling

The whole egg is carefully immersed into a pan of boiling water. The heat is reduced to maintain the water at simmering point, which gently coagulates the egg protein. To obtain a soft-boiled egg, the egg should be removed from the water after 3 1/2 minutes. If a hard-boiled egg is required, it should be simmered for 7–10 minutes. To keep the hard-boiled egg a perfect colour, it should be cooled rapidly in cold water before being shelled. Obviously, soft- boiled eggs must be served immediately, in an egg cup.

Examples of hard-boiled egg (oeufs dur) dishes:

- oeufs dur Indienne are halved eggs placed on a bed of boiled rice and coated with a curry sauce,
- oeufs chimay are halved eggs stuffed with duxelle, then coated with mornay sauce and finally gratinated,
- oeufs à la tripe are halved eggs coated with a white onion sauce and garnished with parsley.

Table 2.6 The uses of eggs in cookery

Uses	Description	Examples
Thickening	Care must be taken to keep the mixture below 65°C otherwise hardening of the protein and curdling will result.	Sauce Anglaise Crème Caramel Sauce Hollandaise Custard tartlet
To bind foods	Beaten eggs will coagulate and hold the ingredients together.	Stuffings for meats and fish Duchesse potato mixture
To coat foods	Beaten eggs are used in order to protect and to add a crisp outer coat to the food. The egg coagulates.	Pané process for foods such as fish Potato croquettes Rissoles
To glaze foods	Beaten eggs are brushed over uncooked pastry products to give a gloss and a coloured finish.	Puff pastry items Vol-au-vents Pastries
To enrich foods	Whole eggs or egg yolks may be added to foods. This will increase the nutritional value and enrich the flavour.	Mornay sauce Duchesse potato
To aerate foods	Whisking makes a foam-like structure that lightens the mixture. The air cells expand and coagulate in the heat and this aerates the mix.	Genoise sponge Fatless sponge Meringues Souffle
To emulsify foods	Beaten whole egg or egg yolks act as an emulsifier to hold ingredients in suspension, thereby preventing separation.	Mayonnaise

Poaching

The egg is carefully broken into water that is just below simmering point. As soon as the white of the egg comes into contact with the heat, it begins to coagulate, thus preventing the egg from spreading out in the water (see figure 2.5). The egg coagulates quickly and should be lifted out of the water with a draining spoon as soon as the yolk begins to thicken (after 2 to 3 minutes). It is common practice to add a couple of drops of vinegar to the water to help coagulation and to keep the albumen white.

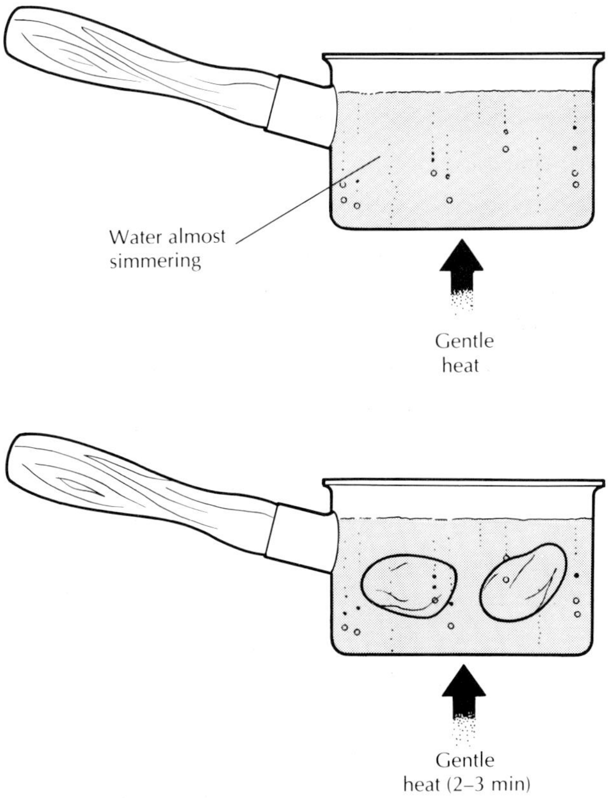

2.5 Poached eggs.

Steaming eggs

Eggs can be steamed by breaking them into cocotte dishes, which are placed into a pan of simmering water before putting on the lid and allowing them to cook (see figure 2.6).

Examples of steamed eggs (oeufs en cocotte) dishes:

- oeufs en cocotte à la crème – the cooked egg is finished with a thin cordon of double cream just before being served,
- oeufs en cocotte jus lie – the cooked egg is finished with a thin cordon of jus lie just before service.
- oeufs en cocotte Portugaise – tomato concasse is added to the cocotte dishes and then the eggs are added and cooked. The eggs are finished with a cordon of tomato sauce just before service.

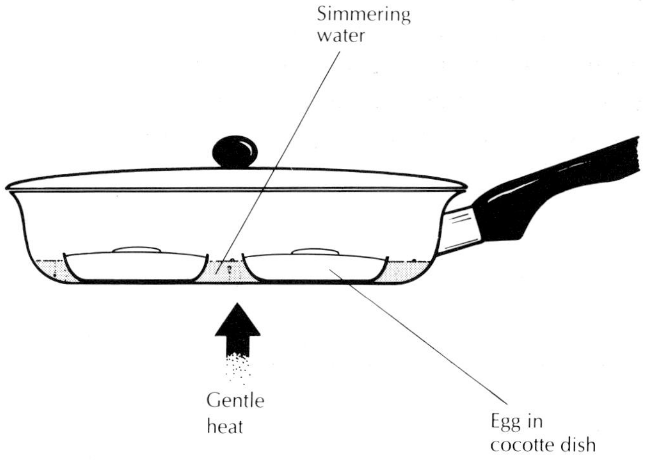

2.6 Egg in cocotte.

Simmering

Two eggs are broken and beaten together with a tablespoon of cold milk, salt and pepper. A small knob of butter is melted in a heavy based pan, and the egg mixture is added and stirred over a gentle heat until it coagulates and forms a soft mixture. Scrambled eggs are usually served upon hot buttered toast; they

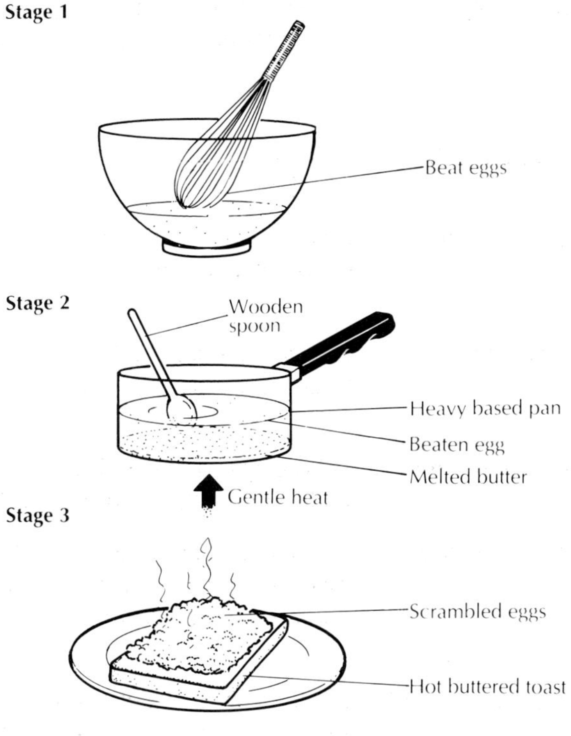

2.7 Method of making scrambled eggs.

will not keep hot for very long and so they are served immediately they are cooked. Over heating the mixture will cause separation of the eggs.

Examples of scrambled eggs (oeufs brouillés) dishes:

- oeufs brouillés aux champignons are scrambled eggs with cooked, sliced mushrooms forked in just before service,
- oeufs brouillés Portugaise are scrambled eggs with tomato concassé forked in just before service,
- oeufs brouillés aux croutons are scrambled eggs garnished with croutons and chopped parsley just before service.

Frying

There are two methods of frying eggs:

- Shallow fried eggs – the shells are broken and the eggs are cooked gently in shallow fat so that the whites are coagulated and the yolks are just becoming thick. The white of the egg that covers the yolk should be just set and opaque.
- Deep fried eggs – the eggs are broken into hot, deep oil then the white is 'moulded' around the yolks as they coagulate to form a round shape (a long-handled spoon is used). When lightly coloured and set, they are removed carefully with a perforated spoon.

Oeufs sur le plat

This is the method used to cook and serve fried eggs in shallow, individual dishes. The shell is broken and the egg placed onto a heated, greased sur le plat dish (see figure 2.8). The cooking is started with base heat from a solid- top range then completed with radiant heat from a salamander. When cooked, the white should be just set, and the yolk just beginning to thicken.

- oeuf sur le plat au lard – a grilled rasher of bacon is placed beneath the egg in the dish.
- oeuf sur le plat Bercy – the egg is served with a small grilled chipolata and a cordon of hot tomato sauce.
- oeuf sur le plat à la crême – when almost cooked, a spoonful of cream is poured over the top of the egg.

Sauté

Omelettes are made using this method of cooking. To make a successful omelette, it is important to have a thick-based pan with sloping sides 2–3.5 cm high, that is used exclusively for this purpose. The surface will be damaged if it is used for general purpose frying, thus causing omelettes to stick and break up. This pan must be cleaned prior to cooking the omelettes by heating it over a fierce heat with its base covered in cooking salt; when this layer of salt gets hot it will begin to colour, and when the pan is wiped out with a clean, dry cloth, the salt will remove all traces of dirt. The pan will then be 'proved' ready for use for cooking an omelette.

Two or three eggs are beaten with a fork and then poured into the pan containing foaming butter. They are then cooked rapidly and when the mixture is lightly set, the pan should be shaken continuously to ensure the egg coagulates rapidly. To serve, the omelette is half folded in the pan and placed onto a dish. It is essential that omelettes are served immediately, as they will loose height and become leathery in texture if kept warm even for a few minutes. Examples of omelette dishes:

- omelette aux fines herbes – a pinch of parsley, chervil and chives are added to the mixture,
- omelette aux champignons – sauté sliced mushrooms, then cook the omelette in the same pan,
- omelette au fromage – grated cheese is added just before the omelette is folded in half in the pan.

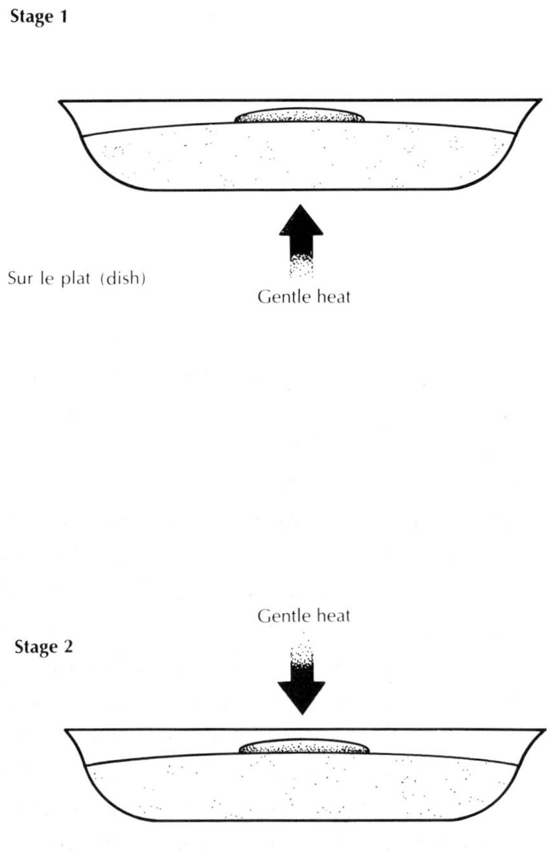

2.8 Oeufs sur le plat.

Flat method

These omelettes are cooked by the same method as basic omelettes, but when the mixture has just set in the pan, they are tossed as a pancake and then turned onto a heated flat dish and served immediately. It is sometimes found easier to 'flash' the omelette under a hot salamander instead of tossing, this has the same effect of just sealing the top surface.

Examples:

- omelette paysanne - diced, fried potato, onion and bacon are placed into the pan, and then the omelette is cooked to create a thick, fairly substantial omelette,
- omelette fermière - chopped parsley is added before the egg mixture is cooked, to create a lightly flavoured and light textured omelette.

Fresh egg storage

It is essential to store fresh eggs at a temperature between 2 and 5°C where they will be cool enough to prevent the multiplication of *Salmonella* bacteria. This, in effect, means that they should be stored in a refrigerator, and kept there until ready to be used. Eggs will deteriorate if stored long periods, as the water will evaporate from the albumen through the porous shell, making the air sack inside the egg larger. The albumin becomes thinner and so does the egg yolk, and eventually the smell and flavour of the egg changes, bacteria multiply and the egg becomes unfit to eat.

Egg products

Because of the risk of *Salmonella* poisoning in fresh eggs, many establishments now use pasteurized egg products. Welfare catering units and those establishments that cater for the elderly, the young and the sick, are more likely to use pasteurized eggs as their customers are less able to survive food poisoning than fit adults are.

The pasteurization process used for shelled eggs involves a process of heating the bulk egg mixture to a temperature of 64.4°C and holding it at this precise temperature for 2 minutes 30 seconds, before cooling it rapidly to 3.3°C if it is to be used within 24 hours. Otherwise, the temperature is reduced still further to below freezing point, at which time the egg can be safely stored without any risk of bacterial growth.

Frozen pasteurized egg can be purchased for use in catering establishments in cartons from 1 pint size, which is equivalent to 1 dozen eggs, upwards. However, as soon as the pasteurized egg is thawed, it has a limited shelf life and should be treated as a potentially dangerous, perishable food commodity. For this reason, the smaller cartons are obviously safer to use than the larger volume containers.

To thaw these cartons of pasteurized frozen egg, the carton should be placed in a refrigerator or immersed in a sink full of cold water at 15°C overnight; it will then slowly thaw without becoming too warm, ready for use in the morning. Any thawed pasteurized egg should be kept in a refrigerator at 3°C until it is required for cooking.

There are various types of pasteurized egg product now available to the caterer:

- Whole egg - the egg is shelled, beaten together, then pasteurized and frozen; it may be used in the place of fresh, beaten eggs for cooking processes,
- Egg white - the white is separated, pasteurized and then frozen; it may be used in place of fresh egg white for cooking or baking.
- Egg yolks, which are separated, pasteurized and then frozen.
- Sweetened whole eggs are eggs that have been pasteurized and filtered, before sugar is added and they are sealed in packs (0.5 kg size or larger). These sweetened eggs are suitable for use in the preparation of cakes, puddings and sweet sauces, etc.
- Salted egg yolks are pasteurized eggs that have salt added before they are sealed in packs. These egg yolks may be used for savoury dishes. (These and sweetened whole eggs may be stored at a cool temperature for up to 12 months if left unopened).
- Dried eggs are eggs that have been dried to a powder, and which can be reconstituted by whisking with cold water. The resulting mixture may be used as a substitute for fresh egg, but in baking, a chemical raising agent will be required as the drying process reduces the eggs aeration properties. Once reconstituted, the egg mixture should be used immediately.
- Substitute egg white may be reconstituted and used to prepare meringue or royal icing. Because of its manufactured process, it is unsuitable for aerating products such as souffle´.

Dried and synthetic egg products will become perishable as their sealed packets are opened. They should therefore be treated carefully and only stored in small quantities in sealed containers.

ACTIVITIES

- Dried milk is used extensively in catering establishments. Discuss the reasons why this statement is true, taking into account considerations of cost, convenience, nutritional value, uses and taste.
- What type of establishments and catering operations are likely to use individual cartons of UHT milk and cream? Why do customers like them? Why does the caterer like to use them?
- Compile a list with descriptions of some commonly used and served foreign cheeses.
- Describe and provide an illustration of a well-balanced cheese board to serve in a restaurant at the end of a meal.
- List some dishes that have grated cheese offered as an accompaniment.
- Describe the advantages of using cheese in snack foods, referring to storage, availability, cost, acceptability and versatility.
- Describe some common uses of cheese in cooking when grilling, baking, sauce making and frying for flavouring.
- Discuss the use of pasteurized eggs in relation to the safety of foods to be served to the general public.
- Imagine that you are a manufacturer of pasteurized egg products. Compile a fact sheet to send to your customers giving them information about the advantages of using pasteurized egg products.

CHAPTER 16 Fish and shellfish

There are many types of fish available to the caterer. They are sold either fresh (wet) or frozen (so that they may be available throughout the year). Fish is a versatile food that is high in protein, and is often eaten as an alternative to meat.

Fish responds favourably to a variety of cooking methods and can be served in a multitude of different ways, and for these reasons fish is often put on the menu. Some oily fish are preserved by canning, including salmon, tuna, pilchards and sardines. Some fish are preserved by smoking, for example, haddock, salmon and trout. Smoked herrings are known as kippers.

White fish contains only a trace of fat, but its flesh contains protein, calcium and potassium. It also contains a significant amount of water. Oily fish has fat distributed throughout the flesh, making it a more sustaining food than white fish, but it is also less easily digested than white fish. Its flesh has significant quantities of vitamin A and D.

FISH CLASSIFICATION

White round fish

Cod

A deep water fish whose colour varies from a brown to pale green, with yellow and green speckles. The white, curved line running from the gills to the tail is known as the lateral line. Cod's liver gives an oil rich in vitamins A and D. Cod roe is often served fried. Darnes may be cut from cod (see page 102). Cod is a versatile fish that may be cooked in a great number of different styles, for example, deep fried in batter or steamed with parsley sauce as a basis for fish pies and cakes.

Haddock

A fish that is never more than 1 metre in length and often much smaller. It has 'thumb marks' of a dark colour just behind the head, and a black lateral line from head to tail. The back is a grey or brown colour which gradually fades to produce a silver or white coloured underside. Haddock has a firm, white flesh that may be cooked and used in the same types of dishes as cod. Some haddock is smoked after being gutted and trimmed: this produces a bright yellow flesh, which has a distinctive flavour. Smoked haddock is usually poached and served with melted butter, or it may be used to create other fish-based dishes, such as kedgeree and fish mousse.

Whiting

A smaller fish than haddock, weighing approximately 700 g. It is silvery white in colour with a grey back, a pale lateral line, and a black spot behind the gills. The flesh is white, fine grained and easily digestible. Whiting is often poached or deep fried (see figure 2.9).

White flat fish

Plaice

A plentiful fish that weighs approximately 500 g in weight. In has a white underside and a black back with distinctive orange spots. The flesh is white and has a fine texture. Plaice is frequently cooked by shallow frying or poaching. It is a fish that cooks very quickly, particularly when filleted, or when cut into goujons and deep fried (see figure 2.10).

Sole

A darkish, brown-grey coloured fish whose average weight is around 500 g. It is more expensive than plaice, but it has slightly thicker flesh which is white and firm when cooked. This makes it a versatile fish. Sole is a much used fish: it is ideal for filleting, poaching and then finishing with a variety of sauces.

Turbot

A fish with a brown, mottled skin and large, boney lumps beneath the skin's surface. It is a large fish that can be 1 m in length and weigh anything from 11 to 20 kg. It is wide and flat, producing a delicately flavoured, firm, white flesh that is either poached or

White round fish

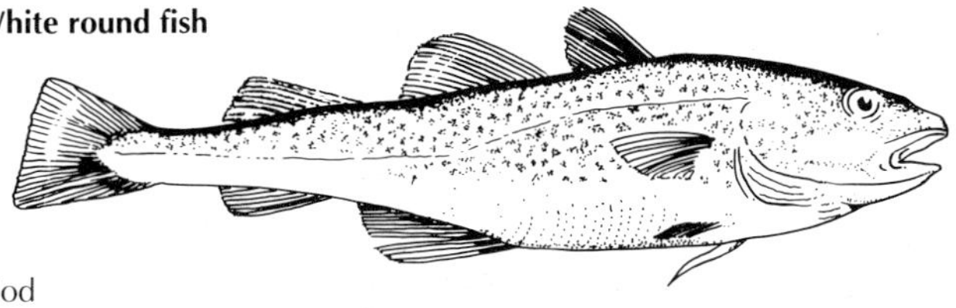

Cod

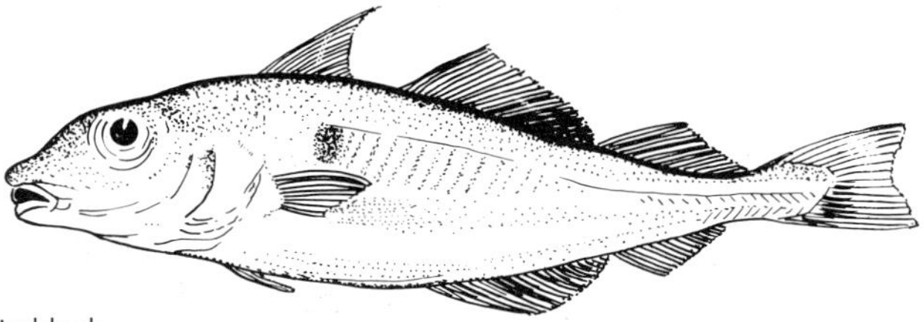

Haddock

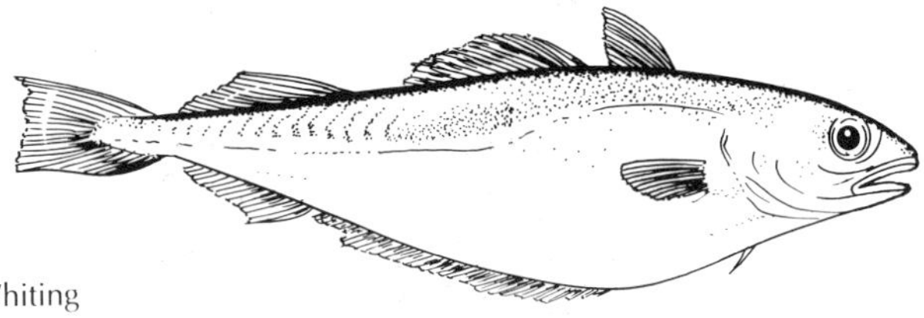

Whiting

2.9 Some types of white round fish.

White flat fish

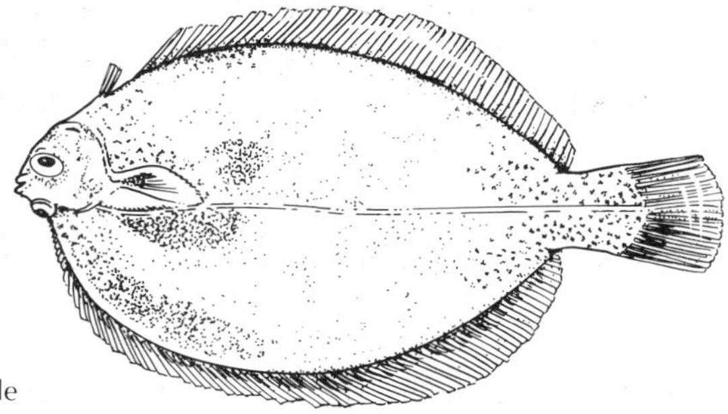

Sole

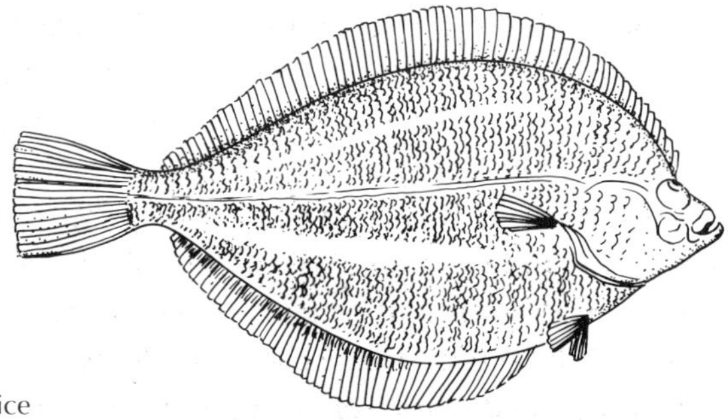

Plaice

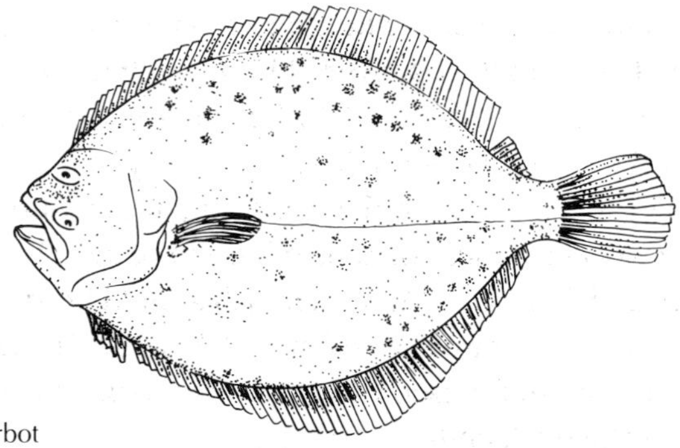

Turbot

2.10 Some types of white flat fish.

grilled. Suprèmes and tronçons may be cut from turbot (see page 102).

Oily fish

Herring

A small fish that is silvery grey in colour, with a pale-coloured flesh. It is often served whole and cooked by grilling, either plain or in oatmeal served with mustard sauce. It may be smoked whole to produce bloaters, or split and smoked to produce kippers. It is also possible to fillet and pickle herrings to produce rollmops or Bismark herring, or to cook and smoke the herring simultaneously to produce beckling.

Mackerel

A fish with a very high fat content, which makes it rather indigestible. It has distinctive vertical zig-zag stripes of a blue or green colour. Mackerel is usually grilled for service – it is too oily to be fried.

Salmon

A blue or grey coloured fish with a pink-coloured flesh. It has a firm flesh that is nutritious, sustaining and full flavoured when cooked. These properties

Oily fish

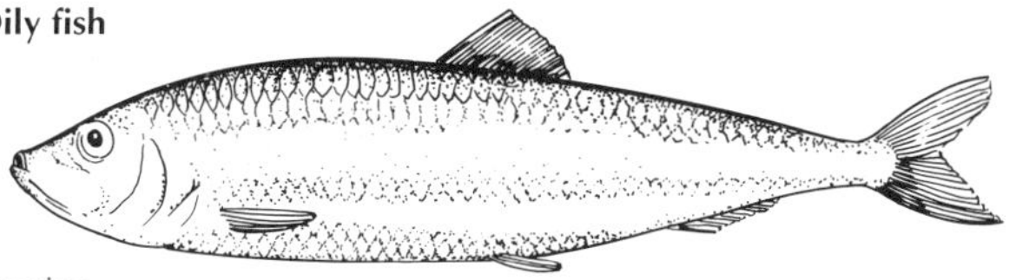

Herring

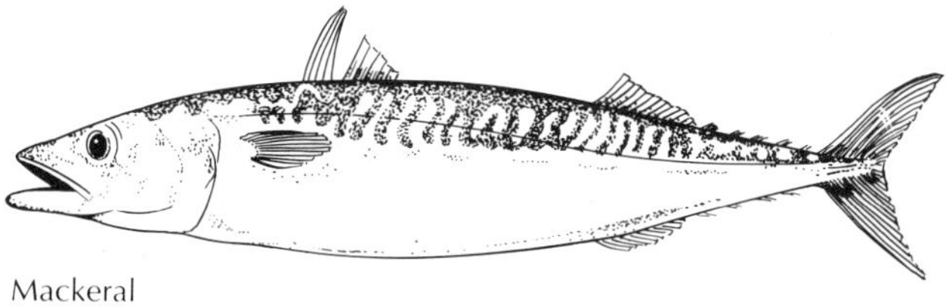

Mackeral

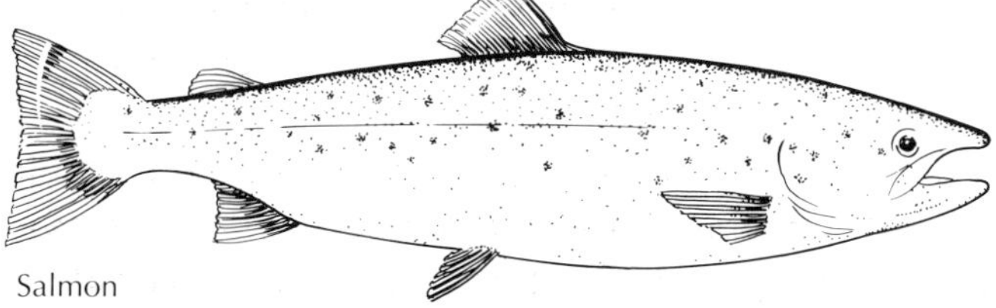

Salmon

2.11 Some types of oily fish.

Molluscs

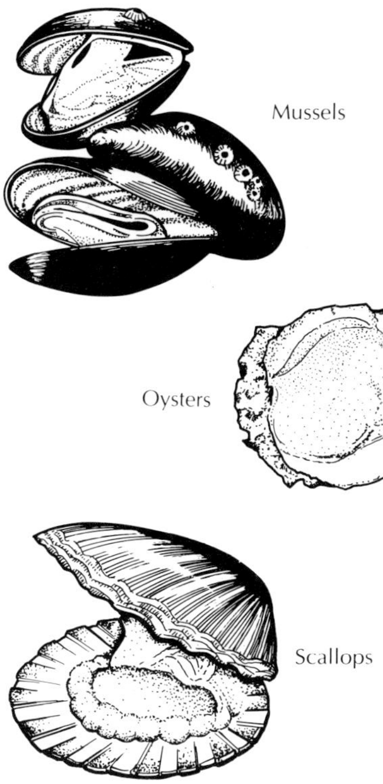

2.12 Some types of mollusc.

make salmon a versatile fish that may be eaten either hot or cold; it can also be boiled, poached, steamed or grilled. Darnes can be prepared from salmon, which may be served hot with hollandaise sauce, or cold with mayonnaise. Smoked salmon is prepared, sliced thinly and served with black pepper and lemon wedges, and accompanied by fingers of brown bread and butter as an hors d'oeuvres. Salmon is also sold in large quantities, canned and made into pastes and spreads (see figure 2.11).

Shellfish

Shellfish are easily infected by bacterial organisms if they come into contact with sewage-polluted water or poisonous algae. Therefore, care should be taken to purchase shellfish only from known and registered suppliers. Many shellfish, particularly prawns, are supplied frozen as they are imported from Norway where their breeding grounds are free from infection. The freezing also helps to reduce the risk of contamination caused by undue handling or by the shellfish being stale. All shellfish have a very short shelf life and can easily be the cause of viral gastroenteritis. The flesh of shellfish has a high protein content, but is very firm and compact, which makes it difficult to digest. Some shellfish are enclosed inside a hard shell and are called molluscs, for example, clams, cockles, mussels, oysters, scallops, welks and winkles

Crustaceans

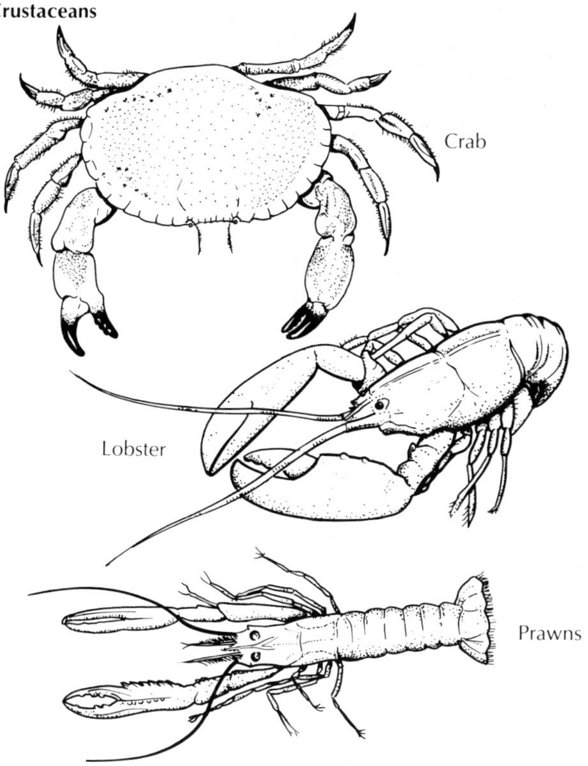

2.13 Some types of crustacean.

(see figure 2.12). Others have a thick, hard shell that covers the main part of their body. These are called crustaceans, such as crabs, crayfish, lobsters, prawns, scampi and shrimps (see figure 2.13).

THE PURCHASE AND STORAGE OF FISH

Prepared fish should be purchased by weight and by the number of fillets or cut, for example, 50 plaice fillets at 75 g per fillet. This ensures that all the fillets will be the same size, thus giving equal-sized portions and equal profit margins.

All fresh fish should have bright eyes and red gills, a pleasant smell and flesh that is moist and firm. Fresh fish should be delivered in crates between layers of ice to keep the fish chilled. These trays should be immediately placed in the refrigerator away from other foods until the fish are to be prepared for cooking. Frozen fish should be delivered in sealed bags or boxes, each clearly labelled, indicating the type and size of the contents. Some frozen fish may be purchased ready prepared for cooking, for example, fillets of plaice pane, portions of cod in batter, fish fingers and fish cakes. All frozen fish should be stacked in deep freeze cabinets as soon as they are delivered or, if they are to be used that day, placed on trays to thaw in the refrigerator.

CUTS OF FISH

There are seven classic cuts of fish (see below):

- *Fillets* are the flesh that has been removed from the back bone - one fillet from either side of the fish. Filleted fish, such as plaice, haddock, sole and herring, may be poached, grilled or fried either as they are, or with a coating of butter, egg and breadcrumbs, seasoned flour or oatmeal.
- *Délice* cuts are quarters or small fillets of plaice, sole or haddock that may be folded, poached, and them napped with a sauce.
- *Goujons* are cut from the fillet of the fish, and are strips of flesh 0.5 to 6 cm in size. They are flavoured by covering the strips with egg and breadcrumbs, and deep fried to order.

2.14 (a) Cuts of fish: (back row) fillet; (middle row) suprême, délice, paupiette; (front row) goujons, goujonettes

- *Suprème* cuts are fillets of large fish, such as turbot, cut on the slant, and then poached or grilled (b).
- *Paupiette* cuts are fillets of fish that are skinned, spread with forcemeat, and then rolled up and gently baked in a moderate oven or poached.

(b) Suprême of salmon

- *Tronçon* cuts are slices from a flat fish, such as turbot, cut across the back bone. They may be grilled or poached (c).

(c) Tronçon of turbot

- *Darnes* are slices from a round fish, such as salmon, cut across the back bone. Often called a 'cutlet', they may be cooked by grilling or poaching (d).

(d) Darne of salmon.

THE COOKING OF FISH

If fish is to be boiled or poached, the liquid used is often a court bouillon. The content of this is determined by the type of fish to be cooked. For white fish, a mix of water, milk, lemon juice, salt and a bay leaf is used. For oily fish and shellfish, a mix of water, vinegar, carrot, onion, bayleaf, parsley, thyme, white peppercorns and salt is used. In both cases, the ingredients are brought to the boil and then allowed to simmer for 30 minutes, then strained and reboiled before being used.

Boiling

A fish kettle has a perforated rack for the fish to rest on and allows a whole fish to be cooked successfully. The fish should be placed in cold liquid and brought slowly to the boil and then simmered gently until cooked. Rapid boiling will cause damage to the flesh of the fish. If the fish is to be served cold, it is usually cooled in the liquid to enable it to stay moist and retain its full flavour.

Poaching

The fish is just covered with the liquid, well covered with a cartouche of greaseproof paper to prevent evaporation, and placed in the oven to cook. If the liquid is allowed to boil, the flesh will be damaged.

Grilling

The prepared fish is often coated with seasoned flour before grilling. It is then placed on a baking sheet and lightly brushed with oil before being cooked. The centre of the fish should be cooked when the outer skin is just coloured and turning crisp. The fish should be turned over only once to prevent damage to the skin.

Shallow frying

The fish may be pane before cooking or simply coated with seasoned flour before being fried. When shallow frying, care needs to be taken to place the presentation side downwards first as the fish is turned only once. This means that the presentation side is cooked in the hot, clean pan first.

Deep frying

Fish is coated either in flour, egg and breadcrumbs, or in flour and batter. The fat should be at the correct temperature before the fish is lowered carefully into the fryer, one piece at a time. A frying basket is not used as the fish would stick to the mesh. The fish will sink when it first goes into the fat, but will rise to the surface in a couple of seconds and then a further piece of fish may be added if required. A 'spider' (a long-handled draining implement) is used to turn the fish over and to remove it from the fryer. Fish cooked by this method should be well drained and served immediately, as the coating will lose its crispness if it is kept hot for any length of time. This method of cooking will increase the fat content of the fish.

ACTIVITIES

- Draw a diagram of a deep fat fryer and label the safety points. Why do you think safety is important when using a deep fat fryer?
- Find out into which category the following fish fit and write a brief description of them to include their size and an appropriate method of cooking: brill, skate, halibut, mullet, whitebait, and trout.
- There are many rechauffe dishes for fish. Find recipes for three such dishes, one using pastry, one using potato and one using a pané coating. Write a brief description of your three examples and suggest an establishment that might use each dish on its menu.

CHAPTER 17 Meat, offal and poultry

COMPOSITION OF MEAT

Meat is made up of bundles of muscle fibres held together by connective tissues called collagen (muscles contains about 7 per cent moisture). Some fat is dispersed between the muscle fibres, which is known as marbling (fat contains about 15 per cent moisture). The presence of fat increases the flavour and moistness of meat, and so very lean meats are sometimes larded to increase their flavour and juiciness. This is done by threading thinly cut strips of fat through the lean meat. If larding is not possible, then layers of fat may be wrapped around the piece of meat before it is cooked. Different animals produce meats and fats of different colours and textures and, in general, the older the animal is, the darker the flesh and the yellower the fat.

- Beef has red flesh and semi-hard, cream-coloured fat.
- Lamb has red-blue flesh and brittle, white fat.
- Mutton has dark red flesh and brittle, cream-coloured fat.
- Pork has pink flesh and soft, white fat.
- Veal has pink flesh and very little fat.

The longer and thicker the muscle fibres, the tougher the meat. Long fibres should be cut across during meat preparation to increase tenderness. The collagen turns to gelatin when cooked. The muscle fibres are made of thin tubes containing water and extractives (mineral salts, and proteins, albumin, globulin and myosin), which give the meat its flavour. The walls of the tubular fibres are made of elastin, which is also connective tissue. Elastin is inedible and is visible in older animals in the form of rubbery, connective tissue found running parallel to the back bone.

TENDERIZING MEAT

As an animal ages, the muscle fibres lengthen (the muscles that are used most frequently being the longest). The tenderness of beef may be increased by hanging the carcass after slaughter (this is not necessary for lamb, veal or pork). Game birds should always be tenderized by hanging. Hanging will increase the tenderness, flavour and moisture content of the meat if the carcass is hung in a fairly humid atmosphere at a constant temperature of 1–2°C. Another method of tenderizing meats is to marinade the joint or cut of meat in an acid solution for 24 hours prior to cooking. Solutions used to marinade meats usually contain either wine or vinegar to produce the acidity and flavour, which may be enhanced by the use of herbs and spices. Small cuts of meat, such as steaks, can be tenderized by pounding or beating the muscle fibres. This breaks down the connective tissue between the fibres and shortens the fibres, so increasing tenderness. A commercial method of tenderizing meats is to inject proteolytic enzymes into the animal prior to slaughter in order to break down the connective tissue between the muscle fibres, but this is unacceptable to many health conscious consumers.

THE COOKING OF MEATS

Some methods are more suitable for the short-fibred, tender cuts of meat, and some are more suitable for the longer-fibred, tougher cuts. As a general rule, the tougher cuts should be cooked by moist, slow methods such as boiling, steaming, stewing or braising. The more tender cuts of meat can be successfully prepared by shorter and, sometimes, drier methods of cooking; these include roasting, frying, grilling and poaching, shallow frying and sauté.

Meat is an expensive commodity, and so the most appropriate method of cooking should be the one that yields the greatest number of portions per kilo of meat, and that enhances the natural flavour and properties of the meat.

The traditional joints cut from a carcass are shown in figures 2.15 to 2.18.

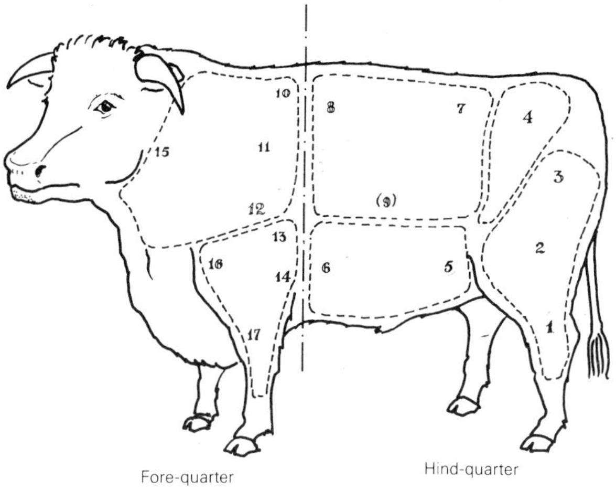

2.15 *Cuts of beef.*

Table 2.13a Cuts of beef

	Cuts	Cooking method
Hind-quarter	1 Shin	Mince; soup
	2 Silverside	Salted, boiled; braise
	3 Topside	Second-class roast; braise
	4 Rump	Roast; steaks – fried
	5 Thick flank	Braise
	6 Thin flank	Stew; sausages; mince
	7 Sirloin	Roast; steaks – fried
	8 Wing rib	Roast
	9 Fillet	Roast; steaks – grilled
Fore-quarter	10 Fore rib	Second-class roast; braise
	11 Middle rib	Braise; pies; puddings
	12 Chuck rib	Stew; pies; puddings
	13 Plate	Mince; sausages
	14 Brisket	Salted, boiled, pressed
	15 Sticking	Mince; sausages
	16 Leg of mutton	Stewed
	17 Shank	Soup; mince; beef tea

Table 2.13b and figure 2.16 Cuts of lamb

Cut	Use	Service	Accompaniments
1 Scrag end	Broth	In soup bowl	Bread roll
2 Middle neck	Braised	In shallow dish on a silver flat	Turned root vegetables, and chopped parsley
3 Best end	Roast – crown	On flat dish with forcemeat stuffing	Roast gravy
	Cutlets, fried	With tomatoes and mushrooms on flat dish	Mint jelly or sauce
4 Loin	Roast	On flat dish with château potatoes	Roast gravy, mint sauce and glazed root vegetables
	Loin chops, fried	On silver flat with watercress	Straw potatoes, tomatoes and mushrooms
	Saddle, roast	On large flat dish	Roast potatoes and gravy
5 Leg	Roast	Sliced onto flat dish with roast potatoes	Roast gravy and mint sauce
	Chump chop	On silver flat with watercress	Mint jelly and glazed root vegetables
6 Breast	Roast	Sliced onto shallow dish with forcemeat stuffing	Thickened gravy and roast potatoes
7 Shoulder	Roast	Sliced onto flat dish with turned root vegetables	Roast potatoes, mint sauce and roast gravy
	Pies	Pie dish cleaned, placed on plate with pie frill	Thickened gravy

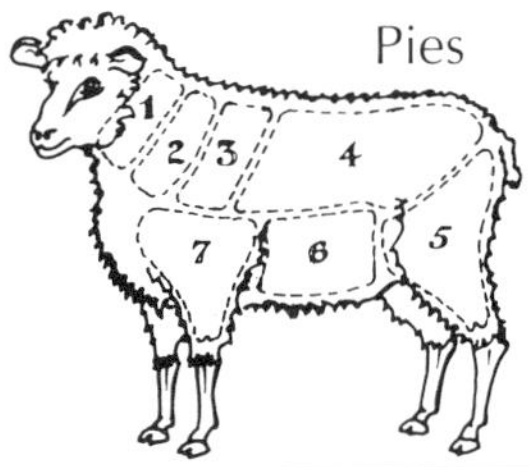

2.16 *Cuts of lamb.*

Table 2.13c Pork cuts

Cuts	Uses	Service	Accompaniments
1 Leg	Roast	Sliced on a flat dish with roast potatoes	Roast gravy; apple sauce
2 Loin	Chops	In shallow dish with apple rings	Sauté potatoes; gravy
3 Spare rib	Roast Pies	Sliced on silver flat with thickened gravy Cold with salads	Sage and onion stuffing Mayonnaise
4 Belly	Fried	In shallow dish	Jacket potatoes; tomatoes
5 Shoulder	Sausages	In shallow dish with bacon rashers	Sauté potatoes; tomatoes

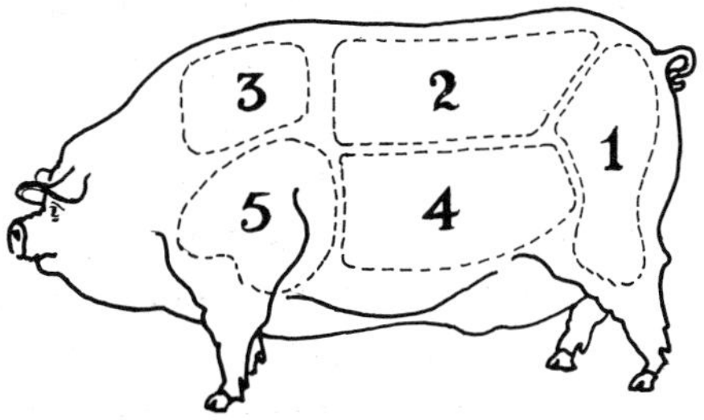

2.17 Pork cuts.

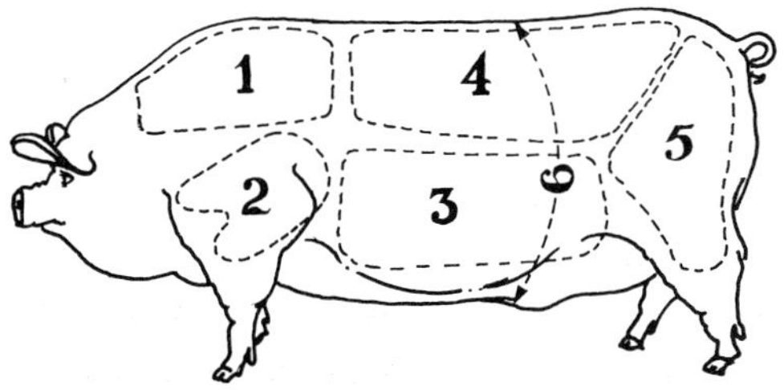

2.18 Baconer pig cuts.

Table 2.13d Bacon pig cuts

Cuts	Method of cooking	Uses/accompaniments
1 Collar	Boil; rashers – fried	Boiled bacon and pease pudding
2 Hock	Boil; diced in pies	Savoury pies or flans
3 Streaky	Rashers; bacon rolls	Savoury flans
4 Back	Rashers – grilled or fried	With eggs, sausage, tomato
5 Gammon	Boil; sliced – grilled	Grilled gammon and pineapple
6 Through-cut	Rashers – cut using the whole side of bacon, streaky and back	Grilled or fried

Roasting

The joint of meat should be sealed in a hot oven and then the temperature reduced to 180°C to cook the joint through. To prevent the joint from frying on the base of the roasting tin, it is placed on a trivet of vegetables and bones. To reduce shrinkage and drying out, the meat is basted frequently with the fat from the base of the roasting tin. To increase portion yield, meat should be rested for 30 minutes before being carved to prevent loss of juices. Joints are placed in a holding oven for this resting period. Roasting may be successfully done in a combi-oven to reduce the problem of the meat shinking.

Cooking times

These are calculated according to the weight of the joint to be roasted.

- Beef: 15 to 20 minutes per 1 kg plus 15/20 minutes.
- Lamb: 20 minutes per 1⁄2 kg plus 20 minutes.
- Pork: 30 minutes per 1⁄3 kg plus 30 minutes.
- Poultry: 25 minutes per 1⁄2 kg plus 25 minutes.

Accompaniments

- Beef: Yorkshire pudding, roast gravy, horseradish sauce and mustard.
- Lamb: mint sauce and roast gravy

- Mutton: leg – red currant jelly; shoulder – onion sauce and roast gravy.
- Pork: sage and onion stuffing, roast gravy and apple sauce.
- Chicken: bread sauce, roast gravy, stuffing, game chips, watercress and bacon rolls.
- Duck: sage and onion stuffing, apple sauce, roast gravy and watercress.
- Turkey: cranberry sauce, roast gravy, bread sauce, stuffing, game chips, watercress and chipolatas.

Braising

For this method (half stewing and half roasting), the meat must firstly be sealed by either frying it quickly or by placing it in a very hot oven for a few minutes. This process will prevent undue shrinkage and loss of juices from the meat into the sauce. A base of rough cut vegetables – a mirepoix of carrots, celery and onions – is prepared and placed in the deep braising tin. Then the sealed meat is placed on the mirepoix of vegetables and a stock or appropriate sauce is added, sufficient to just half cover the meat. The dish is then covered with a well-fitting lid, brought to the boil and finally placed in a moderate oven to allow gentle cooking of the meat. The meat is self-basted by the condensation that builds up inside the container, and falls from the lid onto the meat below. This results in a slow, moist method of cooking that is ideal for joints and smaller cuts of meat that are too long fibred to be cooked by quicker methods. When the meat is cooked through, it is removed from the tin, kept hot while the vegetables and the liquid from the braising tin are strained and finished to produce an appropriate sauce to accompany the meat.

Cooking times

Generally 25 or 30 minutes per 1/2 kg plus 25 minutes.

Garnishes

A few of the many suitable garnishes that may be used to complete a braised meat dish:

- Allemande: buttered noodles, rosettes of mashed potatoes and jus lié.
- Dubarry: cauliflower florettes, mornay and chateau potatoes and jus lié.
- Jardinière: root vegetables and jus lié.
- Maçedoine: vegetables and jus lié.

Stewing

There are three different methods used to create a stewed meat dish: natural thickening, white stews and brown stews. In each method, the meat is trimmed and diced, and then immersed in a gently simmering liquid. To prevent shrinkage and loss of juices, the diced meat is sealed at the beginning of the cooking process.

Natural thickening

Water containing the diced meat is brought to the boil and the surface blood and impurities are removed and the meat is refreshed. The process of blanching will seal the diced meat without colouring it. The meat is then stewed gently; it is simmered together with all the vegetables and an appropriately flavoured stock until the meat is tender and the potatoes have caused the stock to thicken slightly by losing their starch into the liquid during the cooking process. An example of this method is Irish Stew.

White stews

There are two methods of preparing white stews. In the first method, the diced meat is blanched in water, refreshed and simmered in an appropriate stock. Once the meat is cooked it is removed from the stock. This stock is then strained and used to prepare a velouté sauce. To complete the dish, the cooked meat is added to the velouté sauce, liaison is added and seasoned for service. An example of this method is blanquette of lamb.

In the second method, the diced meat is fried to seal it quickly without colouring it. A velouté sauce is prepared using an appropriate white stock and the meat is gently stewed in this sauce until it is tender. To complete the dish, the sauce is skimmed and seasoned, and may be finished with a liaison of cream and egg yolks. To give it a richer taste and gloss. Care should be taken not to allow the sauce to reboil after the liaison has been added, as this will result in curdling. An example of this dish is fricassée of veal.

Brown stews

For this method, diced meat and mirepoix are browned by frying. The addition of flour and cooking oil produces a brown, flavoured roux. Brown stock is then added to produce a well coloured, rich flavoured sauce. The meat and the vegetables are simmered in this sauce until the meat is tender. To complete this method, the meat is placed in an entrée dish and kept hot while the sauce is finished, by straining, skimming, seasoning and adjusting the consistency. The completed sauce is finally strained over the meat. Examples of this method are ragout of beef and navarine of lamb. These dishes may be garnished (see Garnishes on page 129). Curries and goulash are also prepared by the brown stewing method of cooking.

Cooking times

The tender cuts of meat with short fibres, such as veal and chicken, will take no more than an hour. Those with long, tough fibres, such as poorer cuts of mutton and beef, may take up to 3 hours to cook. It is always advisable to taste a piece of the meat to determine if it is cooked. Stewing is a slow method of cooking that cannot be hurried.

Steaming

This method of cooking is used for large joints of meat that are unsuitable for dry cooking methods such as a leg of mutton, a joint of silverside, a whole gammon or boiling fowl. It is usual to steam such joints in an atmospheric steamer or combi-steamer until tender. Joints cooked in this way suffer very little shrinkage and should be rested for 30 minutes before being carved. The cooking time is calculated by allowing 30 minutes per 1/2 kg plus 30 minutes. Welfare catering establishments use this method for large-scale cooking. As a general rule, any joint or cut of meat that is suitable for boiling may be successfully steamed; for example a leg of mutton, a joint of silverside, a whole gammon or boiling fowl. There are traditional accompaniments for boiled and steamed meats.

Accompaniments

- Leg of mutton: caper sauce and boiled potatoes.
- silverside: dumplings, boiled whole carrots and onions and cooking liquor.
- gammon: pease pudding, boiled potatoes and cooking liquor.

Boiling

Boiling is used for tougher joints of meat, such as those that are steamed. It is a slow method of cooking and the meats are served with the accompaniments listed under steaming. A large capacity pan is used to allow the joint to move freely while being cooked. The joint is placed in cold water and when the water reaches boiling point, the heat is reduced to maintain a simmering temperature. The water should be skimmed regularly, and a lid should be used to prevent evaporation. This method causes the joint to shrink more than steaming, and it also results in a loss of nutrients into the boiling liquor. This liquor should be used for stock and not discarded.

Grilling

Small, prime cuts of meat can be successfully grilled. They should be trimmed, marinated if required or seasoned, and then sealed by placing under a hot salamander, turning once to seal the second surface. Once the sealing has taken place, the temperature should be reduced to allow the meat to cook through without overcooking and blackening of the surface. To keep the cut of meat moist, it may be basted with melted fat or oil. It may be necessary to place the meat on a shallow tray before grilling. To allow the meat to cook through, it is sometimes necessary to lower the complete grill shelf, thus reducing the intensity of the heat. Meat may also be successfully grilled by using a char grill where the heat source is below the meat. Meat that may be grilled include chops (lamb and pork), steak (fillet and sirloin), sausages, gammon slices and bacon rashers, beefburgers and portions of chicken.

Garnishes

There are some standard garnishes that are used for grilled meats, particularly lamb and steak grills.

- Vert-pré: pommes paille, watercress and parsley butter.
- Aux champignons: mushrooms, watercress and pommes paillé.
- Henry IV: pommes pont neuf and watercress.

Frying

There are two methods of frying that may be used successfully for small, thin, prime cuts of good quality meats and commercially prepared meat products, such as burgers, sausages, pané slices and tenderized portions.

Shallow or sauté frying

When shallow frying meats, it is important to remember that the surface should be sealed quickly before the actual cooking takes place using a reduced heat. Cooking times will vary according to the type and thickness of the food. For example, a thin rasher of bacon will need only seconds to fry, whereas a thick sausage may need up to 6 minutes to cook.

With all shallow frying, the meat will tend to absorb some of the fat or oil used in the pan and, for this reason, it is not a very popular method of cooking.

Deep frying

This method of cooking is seldom used as it is difficult not to overcook the outside of the meat. However, some meat products, such as Scotch eggs, are successfully cooked by this method: by panéing the meat and by reducing the frying temperature, overcooking may be avoided. Here again the oil will be absorbed by the outer coating of the meat product, making it an unpopular method of cooking.

STORAGE OF MEATS

Meats should be stored carefully in any catering establishment: to reduce the potential risk from contamination, and because meat is one of the most expensive commodities on the menu.

- Carcasses and large joints should be hung in a refrigerator. Care must be taken to prevent blood from dripping onto any other food or surface.
- Small cuts and joints should be placed on separate, clean trays in the refrigerator and lightly covered to prevent evaporation of their natural moisture.
- Frozen meats should be kept in a deep freeze unit at below -18°C. They should be placed on a deep-sided tray in the refrigerator to defrost before being cooked; and they should be kept separate from raw meats and all other food commodities. The liquid should be collected in a tray and carefully disposed of as it will be contaminated.
- Sliced or cooked meats must be placed on trays and covered to keep fresh. They must be kept in the refrigerator at all times, away from raw meats.
- Cooked meats in sauce must be cooled rapidly and placed in a clean, covered container in the refrigerator. They must be kept very cold (3°C).

OFFAL

The term 'offal' refers to all edible parts of the animal carcass that are not actually joints of meat (muscle fibre). Because of the risks of contamination and its short shelf life, offal is usually purchased frozen.

Kidneys

All kidneys contain vitamin A and iron and are, therefore, a nutritious food, useful in a balanced diet.

Kidneys vary in size and use, according to which animal they come from:

- Ox kidney is the largest type of kidney and is widely used diced and stewed or braised, often to make steak and kidney pies and puddings.
- Lamb's kidney has a delicate flavour and is much smaller; it is usually halved and then sauteéed or grilled.
- Pig's kidney may be either stewed or sautéed; it has a stronger flavour than lamb's kidney.

Liver

- Ox liver has the strongest flavour and is dark in colour and large in size. It is normally skinned, then sliced and braised with onions, and served with a thickened sauce.
- Lamb's liver is usually shallow fried and served with bacon rashers; it has an acceptable, mild flavour and takes only a few minutes to cook.
- Pig's liver is a very strong in flavour and is often used to make pâté.
- Calf's liver has a distinctive flavour, a soft texture and is very expensive; it is served sautéed or made into pâté.

Tongue

Lamb's tongues are small and have a fine texture and delicate flavour; ox tongues are larger. Tongues may be served boiled as a hot meat dish, but are usually served pressed and cold. The cold, boiled tongue slices well and is served with salads, in filled rolls and sandwiches, and on cold buffets.

Tripe

This is the lining of the ox stomach and is creamy white in colour. Smooth tripe comes from the first part and rippled honeycomb tripe from the second part of the stomach. Honeycomb tripe is the more expensive of the two. Both types of tripe are prepared by being blanched and then stewed with onions to make a white, creamy stewed dish.

Sweetbreads

These come either from the thymus in the throat or from the pancreas in the abdomen. They are creamy, smooth glands with a delicate flavour. Those from the pancreas are of superior quality to those from the throat. Only sweetbreads from lambs or calves are eaten. They are firstly cleaned by blanching, and then stewed or panéd and deep fried.

Oxtail

These are normally chopped into pieces 2 cm thick and, after cleaning, are browned and either braised or stewed. Soup may be successfully prepared using oxtail flesh.

POULTRY

The term poultry refers to birds that have been bred specifically for human consumption, such as chicken, turkey and duck. The meat from poultry is of a similar composition to that of meat taken from cattle or sheep. It is made up of moist, lean muscle fibres, the majority of which are fine in texture and short in length, and so it is easily and quickly cooked and digested. But as with red cattle meat, the older the fibres and the more they have been used, the longer they will become and therefore the tougher they will be unless they are cooked by a slow, moist method. These frequently used fibres are found in the thighs, legs and wings of poultry, and are a darker colour than the lesser-used white muscle fibres found around the plump breast area of the bird.

Frozen poultry

Owing to extensive poultry farming, it is possible to purchase frozen poultry which vary in size. The chef may, therefore, specify which size and weight of poultry is required, in order to eliminate over-ordering and wastage. It is essential that frozen poultry be thawed out thoroughly before being used (see page 109).

Chicken is an easily digested meat with a high protein, low fat content. Fresh and frozen chickens are readily available all the year round.

Turkeys are considerably larger than chickens: they can weigh from 5 kg to 20 kg. Approximately 90 per cent of turkey meat is served in the month of December, and a large majority of the meat sold in this period is from frozen birds. There are a few self-basting frozen birds available, but most need frequent basting during cooking as turkey meat tends to be dry. For greater portion yield, it is often considered advantageous to bone out the bird prior to roasting; this also makes more even cooking possible.

Duck is a rich, dark meat containing more fat than other types of poultry. It should be cooked in a way that eliminates excess fat in the final dish; for this reason it is often served braised with fruit, such as oranges or apples or cherries, to aid its digestability.

Ducklings are popular and are frequently served roasted (one bird will serve two people).

Table 2.7 is a summary of the commonly used poultry and its uses. The edible parts from inside the bird are called the giblets. These consist of the neck, heart and liver. The giblets are used for making stock for use when making roast gravy and for soups. The livers are also used to make pâté, omelette fillings and some savouries.

If the poultry is fresh:

- the top of the breast bone will be soft and pliable,
- the feet will be smooth and the scales smooth and flat,
- the skin will be unbroken and white with a blueish tinge,
- the breast will be plump and have little fat.

With frozen poultry:

- the wrapping should be clean and unbroken, with properly listed details including the total weight and date of packaging,
- the giblets should be packed separately and found in the body cavity,
- there should be no lumps or chalky white areas visible, or any ice inside the packing.

Table 2.7 Poultry and their uses

Poultry	Size	Uses
Chicken	1.5–2 kg	Roast, sauté, many convenience items are available
Boiling fowl	2–2.5 kg	Soup, fricasee, pies
Capon	2.5–3 kg	Roast
Spring chicken	1–1.5 kg	Grill, roast
Pousin	1 per portion	Grill, roast
Turkey	5 kg upwards	Roast
Duck	2–3 kg	Roast, braise

ACTIVITIES

- Compile a list of prices of all the joints and cuts of meat that can be obtained from a carcass of lamb.
- Give menu examples for each cut and joint for the following types of establishments:
  (a) a boarding school,
  (b) an industrial canteen,
  (c) a luxury restaurant.
  Explain what considerations you have taken into account for each type of establishment, such as costs, customer preferences, equipment availability, service methods, staff availability and time.
- In your own words give a brief description of the following terms which are connected with meat cookery:
  collagen, elastin, spinal cord, extractives, marbling, marinade, tenderizing, basting, mirepoix.

CHAPTER 19 Vegetables

STORAGE OF VEGETABLES

All vegetables should be stored in a cool, airy room in their trays, crates or sacks, placed on well-spaced slatted racks out of direct sunlight and lightly covered to keep as clean as possible. Green leaf vegetables, flower, fruit, blanched stems, legums and fungi all have short shelf lives and should, therefore, be purchased as required.

COOKING OF VEGETABLES

In a traditional kitchen brigade, the cooking and preparation of vegetables, other than those that are deep fried, is the responsibility of the entremettier. The rotisseur looks after the deep frying of foods. It is essential to preserve as much of the nutritional value as possible when preparing and cooking vegetables. When preparing vegetables, a sharp knife should be used and the vegetables peeled as thinly as possible to prevent undue cell damage. All vegetables should be prepared as near to their cooking time as possible, and never left soaking in water as this enables the soluble vitamins B and C to be lost in the water. For the same reason, the minimum quantity of water should always be used, and steamed vegetables will retain more vitamin B and C than boiled vegetables. Care should be taken never to over cook vegetables as this will not only destroy some of their nutritional value but also cause loss of colour, texture and, eventually shape. All cooked vegetables should have a slight crispness to them when served. This is referred to as serving vegetables al dente.

When vegetables are cooked, heat causes their cellulose to soften and they lose moisture. In vegetables with a high starch content, heat will cause the starch grains to gelatanize, which means that the grains swell up in the water that they have absorbed. When they are swollen, they are tender and the starch is more easily digested. For example, chips are firstly blanched at 158–166°C to enable the starch to gelatinize; they are then immersed into hot fat to creat a crisp, golden-coloured outer coating.

Table 2.8 Vegetables can be placed in specific categories

Group	Examples
Roots	Carrot, parsnip, beetroot, swede, turnip
Tuber	Potatoes, old and new. Jerusalem artichoke
Bulb	Onion, leek, shallot, garlic
Leaf	Cabbage, lettuce, spinach, spring greens, brussel sprouts
Flower	Cauliflower, broccoli, Globe artichoke
Fruit	Tomato, cucumber, marrow, courgette, aubergine, peppers
Legumes	Peas, broad beans, runner beans, mangetout
Blanched stem	Celery, asparagus, chicory
Fungi	Mushrooms

To allow for even cooking, vegetables should be trimmed or cut to appropriate sizes. Overcooking of vegetables should be avoided at all times to avoid loss of colour, shape and crispness. All vegetables should still have a slight crispness to them when they are served – never so soft that they offer no resistance when bitten. This crispness is referred to as "serving vegetables al dente".

Boiling

The vegetables that grow below ground are placed in cold salted water and brought to the boil in a deep, lidded pan. The heat is then reduced and the vegetables are simmered until just softened. Those vegetables that grow above ground are placed in boiling water, and then simmered until just softened (new potatoes are also placed into boiling water). Once cooked, the vegetables should be rapidly cooled in cold water and put onto trays; they may be immersed in simmering water for 1 minute before being drained and placed in their service dish. Tubers, roots, green leaves, flowers and pulses can all be boiled.

Atmospheric steaming

The vegetables are placed on steamer trays, sprinkled with salt and steamed for 30–60 minutes depending upon the size and type of vegetables being steamed. This method is used in large scale vegetable cookery where all the portions are required cooked with a very short service time, for example, in the school meals service, industrial canteens and other welfare establishments.

Pressure steaming

As this is a quick method of cooking, small quantities may be cooked almost to order. The advantage of this method is that the vegetables may be placed directly in their service dish, steamed for a few minutes, and then served direct to the customer. A mixture of vegetables can be steamed in the same dish at the same time, as long as they have the same density and size. Tubers, roots, flowers and pulses can all be pressure steamed.

Baking

This is a method that is used for old potatoes only. The potatoes are scrubbed clean, pricked all over with a fine skewer, and then placed on a baking sheet and sprinkled with salt. They are cooked in a moderate oven until the potatoes are soft inside. To serve, the potatoes are slit open and a knob of butter is added. Sometimes the baked potatoes may be scooped out, mixed with a filling such as cheese and onion, and then re-filled before being gratinated and served.

Roasting

For this method, the potatoes are cut to even sized and shaped pieces, and then browned quickly on top of the range, before being roasted in the oven. This initial browning may be completed by placing the roasting tray with the fat and potatoes already in, on the hot, solid top range hob, where the outer surfaces of the potatoes will quickly colour. Once this colouring has been achieved, the tray is placed in a medium oven to allow the potatoes to cook through; basting is required to achieve a good, crisp finish to the potatoes. Because of the fat used for this method, the potatoes have a high fat content when cooked. Parsnips and onions may also be roasted using this same method.

Braising

For this method, the vegetables are blanched and then cooked whole or after being stuffed with a suitable filling or forcemeat. A lidded dish, tray or casserole must be used, and into this a mirepoix of root vegetables is placed in the base of the dish. The prepared vegetables are placed on top and sufficient stock or appropriate sauce is added to just cover the mirepoix. The lid is put on and the dish placed in a moderate oven to allow the liquid to gently simmer while the braising is in progress. As the liquid simmers, the lid collects the condensation which gradually falls onto the vegetables being braised, keeping them moist and helping to impart the flavour of the mirepoix while the oven heat softens them. To serve, the braised vegetables are placed in their service dish and napped with a jus lié or appropriate sauce, often made from the braising liquor itself. Celery, stuffed onions and cabbage may be braised.

Deep frying

The surface of any vegetable to be deep fried must be dry as moisture will cause the hot oil to spit. The vegetables are cut and then blanched in the oil at 170°C. They are then immersed in hot oil at 190°C to brown and crispen the outer surface. Except for potatoes, types of vegetables are either paned or coated in seasoned flour or batter before being deep fried. Obviously, this method results in the vegetables having a high fat content. Examples of vegetables that may be cooked successfully by this method include: potatoes, croquettes, crisps and chips, mushrooms, cauliflower, aubergines, courgettes and onions.

Sauté

This method is used as a quick method to create a crisp, golden surface to potatoes that have usually been partially cooked, and for vegetables that cook very quickly, such as mushrooms, onions and tomatoes. To cook potatoes by this method, it is necessary to boil or steam the whole potato in its skin first. Once softened, it is skinned, cut into slices 1 cm thick, and then tossed in a hot sauté pan using a minimum quantity of fat or oil to achieve a golden crisp finish to the slices. Onions are sautéed by slicing and then finishing the resulting rings in a hot sauté pan. Combining the potatoes and onion will result in pommes Lyonnaise, a very popular sauté potato dish.

Stir fry

This method of cooking vegetables originated in the Orient. It requires a deep pan with sloping sides and a high, fierce flame to create sufficient heat. The pan is oiled and then finely chopped raw vegetables are added and tossed until they are glazed, slightly softened and just beginning to colour. At this point they are removed and served immediately. Peppers, mangetout, onions, carrots, Chinese lettuce, mushrooms, courgettes, bean sprouts and parsnips can all be cooked using this method.

VEGETABLE CUTS

There are five traditional vegetable cuts that are frequently used either to create a vegetable garnish or to prepare vegetables for cooking (see below):

- *julienne*: vegetables that are cut to 3 cm X 2.5 mm thin strips and used for garnishing,
- *brunoise:* vegetables cut to 2.5 mm dice and used for broths and garnishing,
- *macédoine:* vegetables cut to 5 mm dice and used as mixed vegetables,
- *jardinère:* vegetables cut to 5 X 5 X 25mm batons and used with mixed vegetables,
- *paysanne*: vegetables cut to thin shapes - a mixture of circles, triangles, rectangles and edged rounds - used in soups where the vegetables are not passed.

VEGETABLE PREPARATIONS

There are some traditional vegetable preparations that are frequently used in the kitchen:

- *Duxelle:* finely chopped onions or shallots are sweated in butter until they are soft. Chopped mushroom stalks and peelings are added and the mixture is cooked for 6–10 minutes until the moisture has evaporated. Sometimes, fresh breadcrumbs and chopped parsley are added when duxelle is required for a stuffing.
- *Tomato concassé:* whole, firm tomatoes are peeled then quartered and the pips removed. The remaining tomato flesh is coarsely chopped; often it is combined with chopped, sweated onions. It is used in some sauces and in pommes marquise.
- *Clouté onion:* small, peeled onion that is studded with cloves and then used to flavour liquids by infusion. It is used in the flavouring of a béchamel sauce.
- *Mirepoix:* diced carrots, celery, onions, parsley and thyme used as a base for food to be braised. It can also be used as a flavouring for basic sauce preparations such as Espagnol.
- *Diced vegetables:* vegetables cut to even-sized, neat dice or cubes.
- *Rough chopped:* vegetables chopped into small pieces, used in soup and sauce making where the vegetables will eventually be passed through a

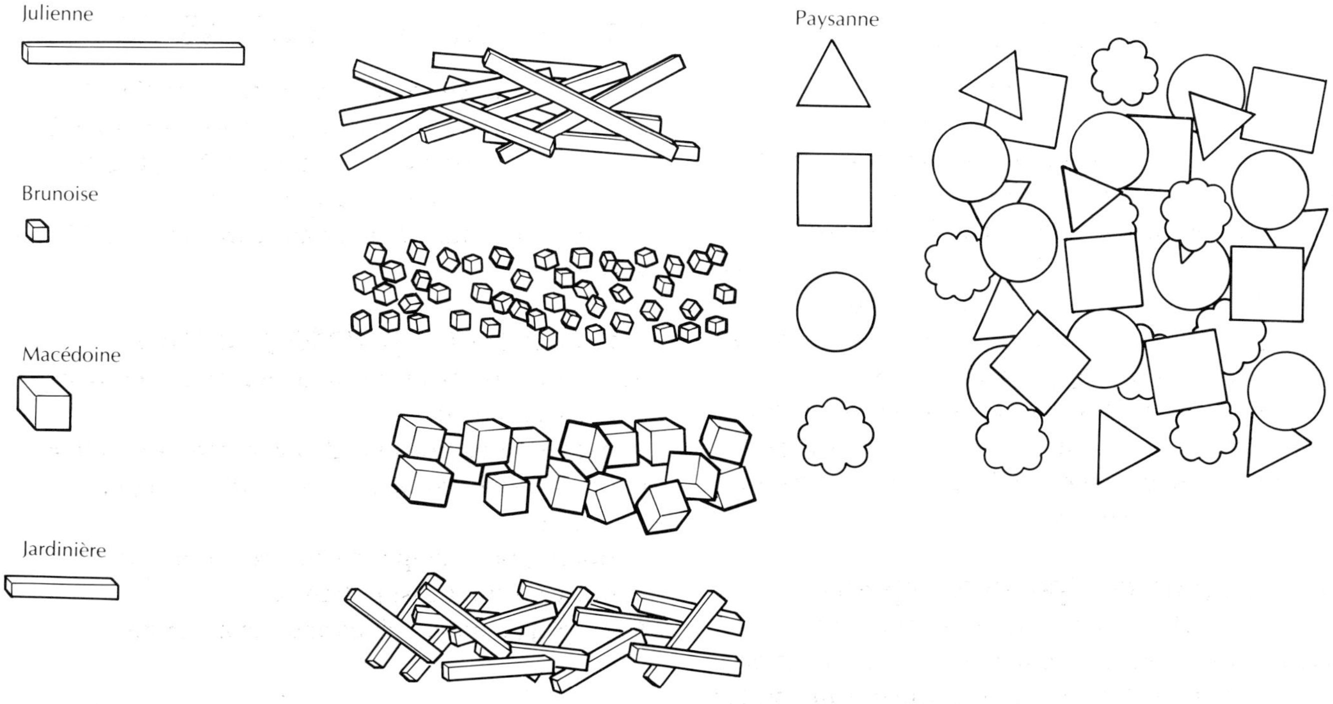

2.19 Standard vegetable cuts.

sieve making their shape unimportant. However, their density should be the same to allow for even cooking time.

- *Tourné vegetable:* tuber or root vegetables shaped using a small knife to create eight-sided barrel shapes 3 cm in length. These are cooked and used for garnishing ragouts and similar dishes. This technique is also used for the preparation of pommes fondants and pommes chateau.
- *Bouquet garni*: a bundle of herbs that is used for flavouring foods during their cooking process, and then removed prior to service. Bay leaves, thyme, parsley stalks and celery are wrapped in an outer leek leaf.

POTATOES

There are many varieties of potato available to the caterer, some of which are suitable for frying, while others are at their best when boiled or baked.

- Majestic is a main crop potato with a white skin and fine texture that is suitable for frying.
- Arran Pilot is an early variety of potato that is generally used for boiling and baking.
- Epicure is an early variety of potato with a pink skin and white flesh that is suitable for boiling and banking.
- Golden Wonder is the potato that is most suited for deep frying.

For boiling, steaming, roasting and deep frying, old potatoes are most satisfactory. But when shallow frying potatoes or making potato salad, new potatoes will give better results as the young potato has a very solid, waxy texture when cooked. Potatoes can be prepared and cooked in a variety of ways; the following are some of the standard methods:

Boiled potatoes (pommes natures)

New potatoes (pommes nouvelles) are served with melted butter and chopped mint leaves.
Duchesse potatoes (pommes duchesse) are cooked potatoes that have been riced, had beaten egg yolk and butter added, and piped into a coiled heap and glazed and browned in a hot oven.

Steamed potatoes (pommes vapour)

Even sized jacket potatoes (pommes en robe de chambre) are scrubbed and cooked in the steamer and served unpeeled with a slit in the top and butter inserted.

Shallow fried potatoes (pommes sauté)

Parmentine potatoes (pommes parmentier) are peeled potatoes that have been cut to 1 cm cubes, rinsed and dried, and then cooked in hot, shallow fat until golden brown. They are then drained, seasoned and served with chopped parsley sprinkled over.

Deep fried potatoes (pommes frites)

Matchstick potatoes (pommes allumettes) are peeled potatoes that have been cut into 5 cm × 3 cm × 3 mm strips, rinsed, dried and deep fried until golden brown and crisp. They are then drained and seasoned with salt.

Baked jacket potatoes (pommes au four)

Baked jacket potato with cheese (pommes gratinée) are baked potatoes that have been cut in half, removed from the skin, mashed with butter and seasoning, and then returned into the skin. They are then sprinkled with grated cheese, brushed with melted butter and baked until golden brown.

Roast potatoes (pommes roti)

Chateau potatoes (pommes chateau) are small potatoes that have been turned into eight-sided barrel shapes, rinsed, dried and placed in a roasting tray containing fat or oil. They are cooked in the oven and basted frequently until golden brown and then drained and served.

CONVENIENCE VEGETABLES

Convenience vegetables are prepared by the manufacturer so that they require little attention by the caterer. Examples of such vegetables, together with the treatment they have received and preparation required afterwards, are give in Table 2.9. (overleaf)

Convenience potato products

There are now many convenience potato products available, such as:

- canned potato salad and new potatoes in water,
- dried potato flakes, potato powder and potato 'pearls',
- frozen potato chips, potato croquettes, saute potatoes and Roast potatoes,
- deep fried potato chips and potato crisps.

Table 2.9

Treatment	Examples	Preparation
Accelerated freeze dried	Peas, carrots, shredded onions and potatoes	Soak for 20 minutes in warm water and then simmer until soft. Mix, add boiling water and butter and then serve.
Frozen	Roots, legumes, flowers and tubers	Cook as appropriate from the frozen state
Canned	Legumes in water or sauce, roots in water and pulses in sauce	Heat through as these are cooked during the canning process

The advantages of convenience vegetables

There are many advantages of convenience vegetables, the most important point is that they can be used to supplement fresh vegetables, for example, a few frozen peas will add interest and colour to a dish of fresh carrots served in the middle of winter. Other points to be remembered include their ability to add colour and variety of shape to meals, for example, the use of baby sweetcorn. Dried pulses will add nutritional value to meals in the form of protein and carbohydrate. Frozen and canned vegetables need no preparation time while dehydrated vegetables need only a very small storage area.

PULSES

Pulses are the edible seeds of plants that produce pods. The pulses usually come dried. They are an inexpensive source of vegetable protein and carbohydrate, and are a valuable contribution to the diet, especially in vegetarian dishes. Before being used, hard, dried pulses have to be soaked in cold water for 24 hours to allow them to regain moisture and their original size. Pulses should be cooked by simmering gently in fresh, salted water until soft. They are frequently used in soups, for example, purée of lentil soup (Purée St. Germain).

Table 2.10 Some well-known and frequently used pulses

Type	Characteristics
Butter beans	Large, oval and off-white in colour
Haricot beans	Small, oval and off-white in colour
Green peas	Whole or split and dull green in colour
Yellow peas	Whole or split and dull yellow in colour
Lentils	Split and orange coral in colour
Soya beans	Oval, off-white in colour and high in protein

Textured vegetable protein (TVP)

This is extracted from soya beans (and other vegetable sources such as wheat). In making TVP, particular care is taken to reproduce the protein, vitamin and mineral content of meat and poultry. Once extracted from the soya, the protein is mixed to a paste with water and colour, vitamins and minerals are added. It is then partly cooked, extruded and cut into pieces of the required size and either dried or frozen. TVP may be spun in a similar manner to textiles to give it the texture of meat before being partly cooked out.

Meat extenders

These are TVP chunks or 'mince' that has been dried. Once rehydrated in stock, they are added to fresh meat or poultry mixes to make the 'go further'. Generally, they take on the flavour of the food to which they have been added. They are used to supply extra protein to a dish and to reduce the costs involved if using 100 per cent fresh meat. Extenders are used in stews, curries, goulashes, pasta dishes, meat pies, meat puddings, pasties, croquettes and rissoles.

Meat substitutes

These may be either frozen or dried, and are intended to replace the total meat portion in any meal; they are therefore useful when creating dishes suitable for vegetarians. They are available in chunks, pieces or 'mince'. Most substitutes are particularly suited to dishes with a strong, definite flavour, such as curries, carbonnades, goulashes and food such as pasties that contain onion and herbs. Care must be exercised when using meat substitutes, as the flavouring will require some thought - they do not have the natural flavouring that meats and poultry have, and this makes seasoning more difficult.

SALADS

In recent years, salads have become very popular - no longer does the word 'salad' conjure up the image of flabby lettuce with a slice of tomato and cucumber. Nowadays, salads are frequently offered as an alternative to cooked vegetables, and some are offered as a complete meal. Salad bars are now found in the high street, shopping precincts and many welfare and industrial catering establishments.

There is no rule that states exactly what a salad should contain, however, there are salad classifications that are still followed in a traditional kitchen where the garde manager is responsible for the preparation of salads and hors d'oeuvres. There are numerous combinations of vegetables and other foods that can be prepared to create attractive salads. Whatever combination of foods are to be used, it is essential to use fresh and clean products that are firm, crisp and well coloured: no sign of bruising or insect infestation should be apparent.

Salads should always contain at least one or two items that are crisp and that give the customer the need to bite and chew their food. To keep salads crisp, they should be prepared as near to service time as possible, kept in a chilled cabinet or refrigerated area, covered to eliminate contamination, and made from top-quality, fresh ingredients.

Since most salad vegetables are grown on the surface of the soil, they are likely to contain particles of soil around the leaves and roots. Careful washing is therefore essential to remove the soil that contains bacteria such as *Clostridium perfringens*, and minute insects and slugs that feed on the leaves. Careful washing must be followed by a thorough draining of all surplus moisture as this can cause the leaves to lose their crispness. Another cause of limp salads is the addition of too much dressing too far in advance of the service time. Raw vegetables contain many nutritious ingredients, including water soluble vitamins B and C and soluble minerals. They should therefore be prepared in the most appropriate manner so as to preserve these nutrients.

Simple salads

Green salads

These contain green salad vegetables such as lettuce, cucumber, celery, watercress, mustard and cress, parsley and green pepper. These salads are often served with a vinaigrette-based dressing, offered separately or sprinkled over just before service. They are often offered to accompany the entrée course such as grilled meats.

Combination salads

These contain green salad vegetables and coloured vegetables such as tomato, red and yellow peppers, carrot and onion. Fruit, such as apple, pear or grape, may also be included, as may nuts and pasta. Combination salads are usually served with protein-based foods to form a main course. All combination salads are served in shallow dishes and should be carefully garnished to reflect the main ingredient.

Fruit or vegetable salads

These are based on a particular fruit or vegetable, for example, Florida salad has segments of grapefruit and orange dressed on a bed of lettuce and with acidulated cream offered separately. In a beetroot salad, the cooked beetroot is diced and mixed with onion, and has vinaigrette sprinkled over it and is dressed in a ravier garnished with parsley.

Compound salads

These are all made from a combination of ingredients, such as vegetables, meats, fish, eggs and poultry. The ingredients are prepared and then marinated in an appropriate dressing, either vinaigrette based, mayonnaise based or with an acidulated cream dressing. Compound salads may be served on their own as a single hors d'oeuvre, as part of an hors d'oeuvre various or as part of a cold buffet presentation; or they may be served in larger dishes or in bowls as appropriate.

Salad dressings

All salad dressings should be used to compliment the salad ingredients, never to smother the natural tastes and textures. Dressings may be created by adding different ingredients to either a basic mayonnaise or vinaigrette, although acidulated cream and natural yoghurt are also popular bases. Salad dressings include:

- sauce andolouse: 3 parts mayonnaise to 1 part tomato ketchup with pimento added,
- sauce verte: mayonnaise with finely chopped spinach, tarragon, chervil and chives,
- vinaigrette: 4 parts olive oil to 1 part vinegar dressing,
- acidulated cream: 4 parts cream to 1 part lemon juice, with seasoning,
- mayonnaise (see page 81).

ACTIVITIES

- Compile a list of rules to be followed by persons working in the vegetable preparation and cooking areas of a kitchen, taking into account such points as cross-contamination, storage, cooking, reheating and nutritional retention.
- Compile a chart to show the methods of cooking vegetables and some examples of vegetable dishes.
- Compile a list of dishes that involved the use of the five traditional vegetable cuts.
- Find some further examples of potato dishes that come under the categories summarized on page 115.
- List the main disadvantages of using convenience vegetable products.
- List examples of convenience potato products and try to find illustrations of each. Give the finishing methods necessary for each example, and the package weight and retail price. Which do you think are the best value for money?

CHAPTER 19 Fruits and nuts

FRUITS

Fruits may be used in many different ways and parts of the menu, for example, they are frequently used in hors d'oeuvre or fruit cocktails and juices; they are used to garnish foods such as poultry and fish dishes; and they are also made into sauces to accompany meat and poultry, their tartness complementing the richness of the meat.

Table 2.11 Classification of fruits

Group	Fruits	Convenience products
Soft	Strawberries	Jam, syrup, frozen, canned, jellies
	Raspberries	Jam, frozen, canned, jellies
	Currants	Jam, jellies, frozen, canned, pie filling
	Gooseberries	Frozen, canned, pie filling
Hard	Cooking apples	Solid pack canned, frozen, dried rings, flakes
	Dessert apples	Juice
	Pears	Juice, canned, dried
Stone	Plums	Canned solid pack (prunes)
	Damsons	Canned solid pack, pie filling, jam
	Peaches	Canned, dried
	Cherries	Canned, pie filling, glace, cocktail, jam
	Apricots	Jam, canned, pie filling, dried
Citrus	Oranges	Marmalade, juice, canned, candied peel, jellies
	Grapefruits	Canned, juice
	Lemons	Juice, segments (for bar), candied peel, jellies
	Mandarins	Canned, juice
Tropical	Pineapples	Canned, juice, jam, jellies
	Melons	Frozen
	Bananas	Dehydrated slices, dried whole
	Dates	Dried, dried sugared

In sweet desserts, fruits are used in hot and cold pastry and sponge-based dishes. Their versatility and variety of colour, flavour and texture makes them invaluable to the caterer. Fruits are used to increase the nutritional value of the meal as they have a high vitamin C content, particularly citrus fruits, which are usually eaten raw. Other fruits high in vitamin C include strawberries, blackcurrants and redcurrants. Fruits are a valuable source of fibre. Fruits are classified into groups, and all are available in preserved form to increase their usage (see Table 2.11).

Tropical fruits are readily available, and so no longer should the traditional fruit salad simply comprise of apple, orange, pear and banana.

Fresh fruits should be used when they are just ripe - under-ripe fruit may be difficult to digest if eaten raw. There is a large quantity of water in fruit - up to 85 per cent which is why fruits are so refreshing. Cooking increases the digestability of soft, stone and hard fruits by softening the fibres, but it destroys most of the vitamin C. Examples of cooking methods commonly used for fruits are detailed in Table 2.12 (overleaf).

Dried fruits

The dried fruits that are used in the pastry section of a kitchen are usually either currants, sultanas or raisins. These dried fruits are produced by the drying of different sized and types of grape, which will have a shelf life of at least a few months if they are stored in airtight containers. They will require a thorough wash before use, and for many recipes this will have to be followed by adequate drying to prevent moisture spoiling the resulting mix, for example when making a Dundee or other fruit cake.

Table 2.12 Cooking methods for fruits

Method	Fruits	Summary
Poached	Soft fruits Other fruits	Hot syrup is gently poured over the prepared fruits and allowed to cool Prepared fruits are gently poached in syrup until they are just soft, but not broken, and cooled
Baked	Cooking apple	Apple is cored whole, the centre filled with dried fruit and demerara sugar, and baked in a moderate oven until the flesh is soft
Steamed	Apple Pear Rhubarb	The fruits are prepared, placed in a shallow tray with a small quantity of water, and placed in the steamer to cook until soft and unbroken. This method is useful when preparing pies and puddings as the flavour is not lost. Sweeten as required
Deep fried	Apple Banana Pineapple	The fruits are prepared, flavoured and dipped into yeast based frying batter and then fried in a medium heat until the batter is crisp and golden. Drain well and serve immediately

NUTS

Nuts are the edible seeds produced from a tree or plant. Nuts contain a high percentage of carbohydrate in the form of starch, and fat in the form of oils and protein, which makes them a valuable food, particularly for vegetarians. Nuts are often added to breakfast cereals and so are eaten by many people who would otherwise probably not benefit from their protein. Some nuts are prepared by dry roasting and are then tossed in seasonings and salt. These are then sealed in foil sachets or tubs and retailed for snacks and to accompany alcoholic drinks. Peanuts and cashew nuts are frequently treated in this way. Mixed nuts are prepared and sold in packets or tubs with raisins, or used in many different types of snack biscuits and chololate-based bars of confectionery. Nuts are used in the kitchen in the preparation of sweet and savoury dishes. They may be purchased in any one of the following forms: ground, nibbed, flaked, halved, whole, dessicated and in essence form. All nuts should be stored in air tight containers to retain their crispness, and in a cool atmosphere to prevent their fat content from turning rancid. Nuts commonly used include the following: almonds, hazelnuts, walnuts, pistachio nuts, coconut and chestnuts.

ACTIVITIES

- Compile a list of fruits under the following headings: hors d'oeuvre, sauces, garnishes, pies, puddings, cakes and gateaux.
- Compile a list of fruits and their prices (visit your local supplier and look in the trade press). Do you know how all these products can be used and, more importantly, what they taste like?
- Compile a list of three dishes in each of the following categories:
  1. nuts used for flavouring,
  2. nuts used as the main ingredient to provide protein,
  3. nuts used as a garnish for a savoury dish,
  4. nuts used to decorate a sweet dish.

CHAPTER 20 Pastry work

In a traditional kitchen brigade, all the sweets, pastries and bakery items are made by the pâtissier (unless there is a baker who makes the bread items). The pastry section is always situated furthest from the stoves and main ovens, in a relatively cool area where there is plenty of space for pastry preparation and as little humidity as possible so that the pastry can be prepared successfully.

Hygiene is an important consideration in this section, as a large quantity of cream, fresh and synthetic, fresh and pasteurized eggs and fresh and dried milk is used. All these items are susceptible to bacterial contamination, especially when used in sauces, creams and mousses. Other food items that are susceptible to bacteria and small insect infestation include dessicated coconut, flour, meringue and cake and pastry mixes. It is also essential to keep all fats cool in a refrigerator to prevent them from turning rancid. Modern pastry sections now use induction hobs as these help to keep the temperature at a more acceptable level. Thorough cleaning of this section is important, as is a good supply of refrigeration units and well organized storage areas for food commodities and for clean equipment.

RAISING AGENTS

Mixtures may be made light by the use of the following actions or additions – the recipe will indicate if such a process is necessary. Cakes and gâteaux certainly require a raising agent of some kind.

- Sifting is where the flour and other dry ingredients are sifted together to incorporate air, for example, in the making of shortcrust pastry, cakes and shortbread.
- Rubbing in is a method used to incorporate the fat into the flour, for example, in the making of shortcrust pastry.
- Aerating is achieved by incorporating air into eggs by whisking either a whole egg or just the white of an egg. Utensils must be free from grease and slightly warm to increase the stability of the eggs. This is used in the making of meringues and genoese sponge.
- Beating is used when mixing and creaming the fats and sugar together to incorporate air, for example, when making Victoria sponge and sponge puddings.
- Chemical raising agents in the form of baking powder, when added to the mixture, leads to the giving off of carbon dioxide, which causes the mixture to rise and bubble. Used in the making of Victoria sponge, suet paste, rubbed in mixes and scones.
- Biological agents in the form of yeasts, when added to mixtures, will react and cause carbon dioxide to be given off, and cause the mixture to rise. Yeast is used in the making of sweet bun dough and breads.
- Layers of fat in pastry, produces steam when they melt, which causes the layers of paste to rise, for example, in Scotch and full puff pastry.

Baking powder

This chemical raising agent is usually made from a mix of one part sodium bicarbonate to two parts cream of tartar. The sodium bicarbonate produces carbon dioxide when mixed with an acid (the cream of tartar) in the presence of moisture and heat. The carbon dioxide causes the mixture to rise, but excessive amounts lead to the mixture collapsing in the centre, and to a strong, unpleasant taste. Baking powder should always be sifted with the flour so that it is evenly distributed throughout the mixture. It should be stored in airtight containers and will only be effective for two months.

Since it is the acid that causes sodium bicarbonate to give off carbon dioxide, it is sometimes possible to substitute cream of tartar with another acid, such as vinegar (ascetic acid), lemon juice (citric acid), sour milk (lactic acid), or the acid found in jam, treacle or golden syrup. One well-known example of the latter is the use of treacle and bicarbonate of soda in the making of gingerbread.

Yeast

Fresh yeast is a form of plant life and should be used while it is sandy-coloured, moist, crumbly and fresh smelling. It may be stored for a short time 2–3 days by being kept well wrapped in a refrigerator. Dried yeast is in a concentrated form and is approximately twice as strong as fresh yeast. The optimum temperature for yeast activity is 21–27°C; yeast is destroyed by temperatures above 56°C, which occurs in the baking process once the dough as risen.

Salt gives the yeast dough its flavour, but it should be used carefully as to much salt will kill the live yeast. Sugar provides food for the yeast and is added to the flour to start the action of the yeast. Liquid should always be added at blood heat 36-37°C in order to enhance the yeast's action upon the flour.

BREAD MAKING

Quantities will vary slightly from recipe to recipe but, in general, 1.5kg of strong white flour will require 25g of fresh yeast and approximately 750ml of tepid liquid, 25g of lard, and 20g of salt and half a teaspoonful of castor sugar to produce a good standard dough. Eggs and/or dried fruits are sometimes added to make an enriched dough suitable for fruit loaves and other types of enriched mixtures such as savarins and babas. In some of these mixes, the quantity of sugar is also increased.

To produce a good dough it is necessary to understand what happens to the ingredients once they are mixed together. The method of incorporating the ingredients together to form a dough depends upon the volume of dough being mixed and on whether fresh or dried yeast is being used. There are five stages to successful bread making:

1. Mixing will produce an elastic, smooth, pliable dough.
2. Kneading ensures that the gluten is strengthened and obtains a good rise and an even textured dough by allowing the protein and the starch to be well incorporated into the flour.
3. Fermentation or proving allows an enzyme (diatose) in the flour to convert the starch to maltose, which can then be absorbed by the yeast cells. This is done by the yeast enzyme changing the sucrose to maltose and glucose (this process is called hydrolysis). The yeast cells absorb the glucose which is then converted into carbon dioxide and alcohol. The alcohol is evaporated by heat and the carbon dioxide causes the dough to rise. The dough is allowed to ferment until it has doubled its size.
4. Proving is when the dough is left to rest in a humid atmosphere at a temperature of 30°C after it has been 'knocked back', cut and shaped. Thus the dough continues to rise until, again, it is double its size after shaping. This is the reason why loaf tins should be only half filled before proving.
5. The proved, well-risen dough is next baked in a hot oven at 200°C, during which all enzyme activity ceases as the heat kills the yeast, the moisture in the dough is given off in the form of steam, the flour starches gelatinize, and the gluten coagulates thus 'setting' the shape and size of the dough. During the baking process, the starch cells near the surface of the dough turn to dextrin to form a sweet crust, and browning occurs. The bread is tapped sharply with a knuckle and if it sounds hollow, it is cooked. Commercial breads are usually baked in a steam oven to produce the characteristic soft crust found on all sliced loaves of bread.

PASTRY MAKING

The secret of success when making a pastry for a particular use or dish is to use the correct ingredients in the correct proportions. During preparation, all pastry with the exception of choux paste and hot water crust paste, should be kept as cool as possible. The flour should be soft for short pastries and strong for layered pastries, such as rough puff, and full puff pastry. Wholemeal flour may be used with white flour to produce short pastries of an acceptable flavour and texture, but the bran content makes it unsuitable for puff pastry. The fats used should be cold, firm and of good quality. Margarine and butter give good colour, texture and flavour to pastry. White lard gives a good, crisp texture, but lacks the colour and flavour. To create the best results, it is often best to mix the types of fat. All fats can retain air during mixing and produce steam during cooking, which aerates and lightens the pastry. Special pastry margarine is used for layered pastry. This has a higher melting point than ordinary fats, making it easier to handle, but it does not give the same flavour as butter. A pastry margarine must be manipulated to the same consistency as the paste to give the best results.

Shortcrust pastry

This pastry is prepared by rubbing the fat (½kg) into the flour (1kg) in order to coat each particle of flour with fat. This breadcrumb-like mixture is then mixed with sufficient ice cold water to form a short dough that can be handled without feeling sticky. To use shortcrust paste, it is necessary to roll it out lightly, using just enough flour on the pin to prevent it sticking. One rolling is all that is necessary, as this pastry does not produce the layered effect that puff pastries have. Any trimmings can be used again. When the pastry is placed in a hot oven, the water evaporates and an open, short texture to the finished pastry results. Shortcrust pastry is used for jam tarts, treacle tarts, quiches, tartlet cases and cornish pasties.

Scotch puff or rough puff pastry

To make this pastry, it is necessary to use two-thirds fat to flour. The fat is cut into small knobs the size of walnut and then roughly mixed into the flour. Sufficient chilled water is added to create a softish, lumpy mixture. The fat is gradually incorporated into the mix by a process of rolling and folding the dough until a smooth, layered dough has been achieved. Each time the dough is rolled out, and then folded into three, the fat is incorporated and the air is trapped between the folded layers. Air bubbles should become visible as the rolling and folding progresses. The paste should be 'rested' after each third rolling to keep it cool. A minimum of six rollings should be completed and all the fat should be incorporated before the paste is given its final rolling ready for cutting and using. The pastry is baked in a hot oven where the trapped air will expand, causing the pastry to rise. Over handling and uneven rolling and folding can easily ruin the baked result. Scotch puff pastry is used for meat pie, crusts, jam puffs, eccles cakes and suasage rolls.

Full puff pastry

Equal quantities of butter and strong flour are used to make puff pastry. Twenty-five per cent of the fat is rubbed into the flour and then this is mixed to a soft, pliable dough after cold water and a few drops of lemon juice (to strengthen the gluten and improve elasticity) have been added. To incorporate the remaining fat, the dough is first rolled out into a circle, and then the sides are further rolled to create a star-like shape (see figure 2.20). The remaining fat is

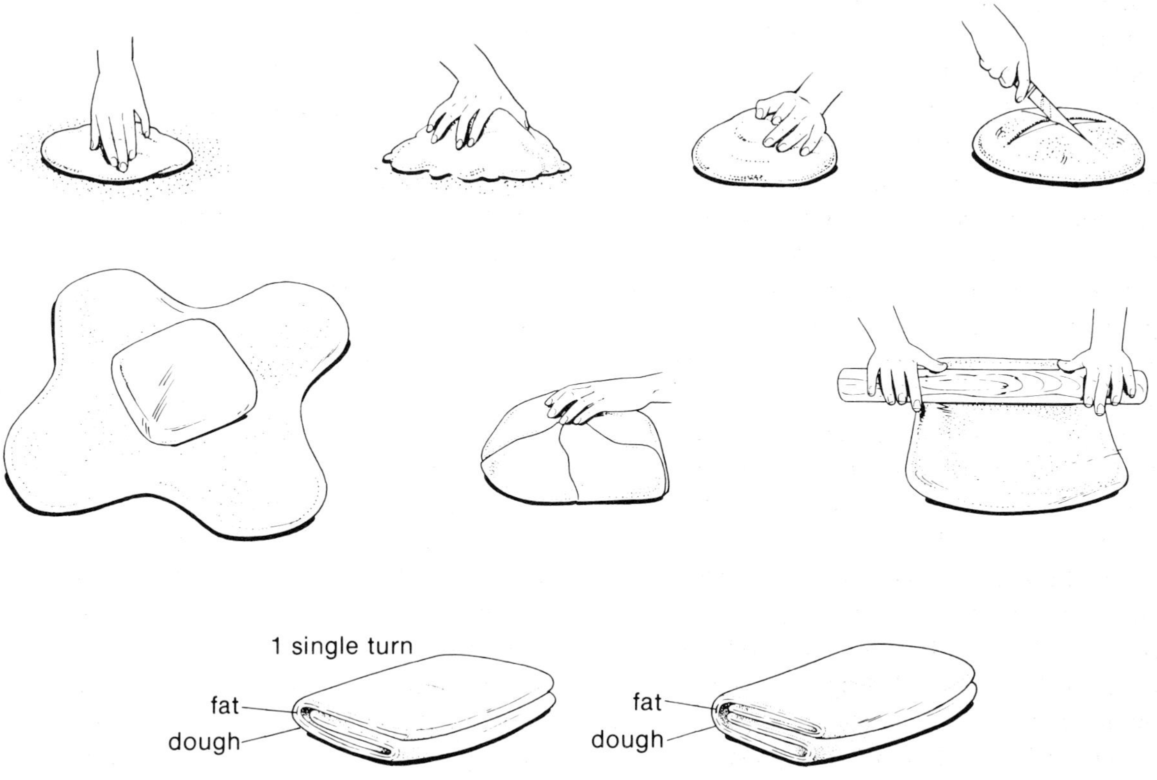

2.20 Method of making puff paste.

added to the centre of the 'star' and the four 'flaps' are folded over to enclose the fat in the centre. This is rolled and folded carefully to incorporate the fat into the dough, to trap as much air as possible in between the layers (the layering of fat and dough is referred to as laminating). As the pastry is rolled and folded, the number of layers of fat and dough increases. When the pastry is placed in a hot oven, the water is driven off in the form of steam, which forces the layers of dough apart, causing the layers to rise. The fat will melt and be absorbed into the dough as it cooks. If the oven is not hot enough, the fat will melt and ooze out making the pastry dough greasy and unrisen. Over handling and uneven rolling and folding can also easily ruin the final result. Puff pastry is used for meat pie, crusts, cream horns, jam puffs, jalousie and band aux fruits.

Sugar pastry

The ingredients used to make this pastry are 200g soft flour, 50g caster suagar, 125g butter and one size 3 egg. The flour and sugar are mixed together and the butter is incorporated by 'rubbing in' to create a mixture that resembles the texture of fine breadcrumbs. The fat will have separated each grain of flour and sugar. The egg is beaten and gradually added to create a soft, but not a sticky, dough. Before handling, this paste needs to cool and rest for at least 30 minutes in a refrigerator. One rolling is all this is required for this pastry. Any trimmings may be carefully saved and used again, although over handling will cause the final pastry to be tough. Because of the egg content, this pastry does not normally require glazing before it is baked (some recipes recommend glazing during the final few minutes of baking). The high sugar content will cause this pastry to colour easily, so a moderate oven temperature should be used. Sugar pastry is used for sweet flan cases, apple flan, fruit tartlets and date and apple slice.

Suet pastry

Here the prepared suet is mixed into the flour, half suet to flour. Baking powder is added to provide the raising power required for a light finish. The ingredients are mixed to a paste with chilled water and should then be cooked immediately, in order to gain maximum lift from the baking powder. This mix may be baked or steamed. Suet pastry is used for apple dumplings, baked jam roll, steamed puddings and rolls, spotted dick and dumplings.

Choux paste

With choux paste, the mixture is prepared as a thick roux (panada) using strong flour, water and butter. Initially, it is cooked in a saucepan allowing the flour to gelatinize, and then allowed to cool slightly before the eggs are added by beating them into the mixture gradually, until it is smooth, shiny and pliable. To cook, the paste is piped to the required shape – domes for profiteroles and oblongs for éclairs – placed on greased baking sheets, and then baked in an oven at 200°C. Choux paste is a soft, pliable mixture that is raised by the action of the eggs and the steam created when the mixture is cooked. A poor texture will result if the water is allowed to evaporate when preparing the paste, or if the panada is not cooked thoroughly before the eggs are added. Strong flour provides the gluten necessary to make the elastic dough so that it can rise when baked. Some choux recipes require the paste to be deep fried, either as it is, or with additions. Choux paste is used for chocolate or savoury eclairs, profiteroles, and deep fried, dauphine potatoes and cheese fritters.

Hot water paste

This paste is used when preparing raised pies. It is prepared with 1/4 fat (lard) to flour with salt added and mixed with water (315ml to 500g of flour). It is called hot water paste because it is made by melting the lard in the boiling water and then mixing this liquid into the flour and salt to create a hot, pliable, kneadable dough. It must be used while hot, so it is necessary to have the pie filling and pie mould ready before making the paste. Once moulded and filled, the pie should be baked in a moderate oven to ensure that the filling is cooked thoroughly while not overcooking the paste. When cooked, the pie is usually removed from the mould, glazed and then placed in a hot oven to give the paste an acceptable colour and gloss. The pie should be served chilled so it should be topped up with jellied stock and then cooled rapidly and refrigerated. Fillings may be any combination of meats, such as veal and ham.

CAKES

Large quantities of cake mixture can be mixed by machine more successfully than by hand, (always check the maximum machine load before commencing, for example a US 12qt mixer takes a 6 lb load). Ensure that the correct mixing attachment is fixed to the central spindle: the balloon-shaped whisk is for whisking; the flat beater is for general mixing and creaming; and the dough hook is for heavier mixing and kneading doughs. The speed of mixing is regulated by the machine gears. Too high a speed will damage the finished texture of the cake because it will damage the starch granules in the flour. There are several different methods used to mix a cake mixture, coupled with different combinations of ingredients, to create the various textured cakes when baked.

Rubbing in

The fat is incorporated into the flour using the minimum speed, so that each grain of flour is coated with fat, producing a texture like fine breadcrumbs. Any flavourings are added at the minimum speed, followed by the liquids. When the mixture is baked in a hot oven, the moisture is driven off between the fat particles. This creates an open, short-textured result. Scones, rock cakes, raspberry buns and fruit cakes are made using this method.

Creaming

The fat and castor sugar are mixed or 'creamed' together at high speed until the fat becomes a lighter colour and the mixture is soft and fluffy. Half the beaten eggs are then gradually beaten in at medium speed. Frequent scraping down of the bowl is essential to ensure thorough mixing. The sifted flour, raising agent and any flavouring is carefully folded in alternately with the remaining eggs until all the ingredients have been incorporated into a smooth, light, soft mixture. Creamed mixtures should be baked in lined cake tins at a medium temperature. While in the oven, the heat expands the air which has been trapped in the mixture, causing the cake to rise; the gluten in the flour and the egg albumen coagulate. The finished texture should be soft and light with an even rise created by the air expansion. Note that if this method is being used to make a rich fruit cake, the baking temperature will be lower to allow the cake to be cooked through to the centre. Victoria sponge cakes, small sponge cakes, buns and rich fruit cakes are made using this method.

Whisking

A whisked mixture may be made without the addition of butter, which will produce a dry sponge that does not have a very pronounced taste and that does not keep fresh for more than a few hours. Most whisked sponges are made with butter (a Genoese sponge). If a Genoese is to be successful, it is essential to warm the butter so that it is able to pour from a spoon but not be actually liquid. To make a Genoese the eggs and castor sugar are whisked until the mixture becomes a light colour and thick enough to hold its shape. This whisking begins to heat the eggs, making the mixture slightly thicker. The sieved flour is then carefully folded into this foam-like mixture and, finally, the butter is folded in. The mixture is baked in a greased and floured or sugared tin at a temperature of 190°C. During baking, the heat expands the air bubbles which cause the mixture to rise. At the same time, the protein in the egg albumen and the gluten in the flour coagulates. The resulting cake is a very light textured sponge due to a large quantity of air that has been incorporated during the whisking process. Swiss rolls and flan cases are made from fatless sponges. Genoese sponges are used in the making of various types of gâteaux.

Convenience pastry and cake mixes

There are a wide variety of convenience products available to the caterer. The most commonly used varieties are in a ready mixed, dried form, and simply require the addition of water; in the case of cake mixes, eggs are also required. Scone mix, bread roll mix, sponge mix, shortcrust pastry mix, suet paste mix and puff pastry mix are all available.

It is essential to read and follow the manufacturer's preparation instructions carefully in order to achieve satisfactory results. Although these dried products are not able to reproduce the flavour and texture of a freshly-made pastry or cake, in many respects they are of a very good quality, and they make cost control easy, reduce preparation times, and give consistent results.

Frozen pastries and gâteaux are also available. Puff pastry requires skill and time to make, but frozen puff pastry comes either in a block or ready rolled and cut ready for baking. These products are frequently used in the modern catering unit, especially if there is not a separate specialist pastry section. Frozen gâteaux and pies, tarts, cheese cakes, éclairs and flans are used widely in either a traditional menu or a fast-food type operations. Consistent results, portion and cost control are easily achieved by the use of these ready to serve frozen sweets, which also applied to all sorbets and ice creams.

ACTIVITIES

- Write a short definition for each of the following terms:
  1. lamination,
  2. rubbing in,
  3. creaming,
  4. folding in,
  5. whisking,
  6. rolling and folding.
- Make a collection of convenience cake and pastry mix information for shortcrust pastry, suet pastry, puff pastry and sponge cake. Compare:
  1. the cost against the fresh ingredients used,
  2. the temperatures recommended for cooking,
  3. the ingredients themselves,
  4. the estimated preparation time.
- A traditional Victoria sponge mix may be used to prepare a large variety of different flavours and types of products. Using the following headings, create a chart to show how this basic mix may be used to produce six completely different products.
  1. the name of the dish,
  2. the additional ingredients,
  3. the cooking method and time,
  4. the finishing method.

CHAPTER 21

Canapés, savouries, sandwiches, hot snacks and Hors d'oeuvre

CANAPÉS

Canapés are appetizers that are placed in a neat and tidy way on a large oval or circular tray and offered to guests as they sip their aperitifs before a meal. They are neat slices of toasted white or brown bread that have been cut small enough to be eaten in one mouthful. Selected slivers of meat and fish are arranged on each toasted base, which has been spread with a complementary flavoured butter, for example, butter flavoured with horseradish for thin slivers of roast beef. The canapés are then garnished delicately with vegetables or herbs and thinly sliced aspic.

Substitutes for canapés

In many catering establishments, canapés are not served. The preparation time is long and the types of service offered is not always compatible with such elaborate foods. Potato crisps, potato sticks, savoury shaped biscuits, salted nuts, nuts and raisins and vegetables, such as silverskin onions, are often served instead. Crudites - fingers of prepared, raw vegetables arranged around a central bowl filled with a savoury dip - are often seen on lounge bars and on occasional tables.

SAVOURIES

These are small, tasty items that may be eaten in a few mouthfuls, and are traditionally served after the sweet course on a dinner menu. They should be neatly garnished and have an attractive, appetizing appearance, and are designed to remove the sweetness left from the dessert course and add variety to the menu. Scotch woodcock, anchovy fillets, devilled kidneys and Welsh rarebit are served on toast. Shrimps or mixed shell fish and chicken and ham in sauce supreme are served in pastry bouchées. Devilled soft roes and prawns in sauce crème are served in pastry barquettes.

SANDWICHES

Sandwiches are served in many different types of outlets and at various times of day. They are used as part of a traditional afternoon tea (miniature sized) served as part of a finger buffet, as part of a pub lunchtime menu, at railways and motorway service areas, from vending machines and from sandwich bars and shops. Filled rolls are also served in increasing quantities and their advantages to the caterer are numerous. They need minimum preparation time and cause no wastage; they allow for good quality control and portion control; there is a large variety of shapes, sizes and flavours of bread rolls available; and there is often a larger profit margin than on sandwiches as rolls provide more bulk and take less filling. Both sandwiches and filled rolls are normally filled and packed in a clear, plastic box or triangular packet. This is sealed and labelled, then placed on a cold shelf to ensure that the contents remain at 3°C until sold to the customer. If large quantities of sandwiches are to be prepared, care must be exercised to ensure that all the ingredients remain below 3°C at all times - making them in a cold room is advisable. Cross contamination must be avoided by keeping all meats and other ingredients on separate trays and by using clean knives and boards every time the ingredients are changed. A small amount of salad will enhance any filled roll or sandwich - they will not be bought unless they look appetizing.

There are many ways that rolls and sandwiches may be filled and seasoned, either by using just one main ingredient or a combination of ingredients to create more flavour and a variety of colours and textures. Different types of breads will also increase the flavours.

Toasted sandwiches are made by either inserting a hot filling between two slices of hot buttered toast, or by making the sandwich and then toasting it in an electric sandwich toaster. Double-decker sandwiches are made from three slices of bread, two brown and one white, and two types of filling. A three-tiered sandwich may be created and then sliced to make thin fingers. Open sandwiches originate from Scandinavia, and each slice of bread is buttered and topped with an attractive selection of meats and salads, and is garnished. The finished effect is a base piled high with attractively arranged items.

HOT SNACKS

These are served in many types of catering outlets and are not restricted to a particular time of day. The foods are often based upon the traditional breakfast dishes with modern and regional variations. Chipped potatoes or jacket potatoes are often available. Pasta based dishes are often available as snack items and, with the growth of healthy eating habits, salads are increasingly popular.

Other items may be served as a savoury, as a hot snack or as an incidental meal with salad or a bread roll, and may be made individually or portioned as required.

- A soufflé is a panada that has been flavoured with cheese and eggs, baked in a hot oven until well risen and served immediately. It is cooked to order.
- Fritters with a yeast batter to ensure a crisp coating and flavoured with cheese or mushrooms are popular. They will keep hot for only a few minutes.
- Flans with an egg custard base in either a short crust, puff paste or choux paste case. They may be flavoured with onions, ham, cheese, mixed herbs, fish, tomato or any combination to create an acceptable flavour and colour filling. Flans must be baked carefully to ensure that the pastry is properly cooked and that the egg-based filling is set without allowing it to overheat and curdle.
- Croutés are thick slices of white or brown bread with the centre scooped out, deep fried and filled with a tasty filling such as cheese or tuna fish.
- Baked jacket potatoes with their centres scooped out and mixed with a tasty savoury filling such as cheese and onion, then refilled and flashed in a hot oven to reheat and gratinate the top.

HORS D'OEUVRE

The hors d'oeuvre course of a meal may consist of either a single item or a variety of items, served as an appetizing start to the meal. It is designed to stimulate the gastric juices and therefore aid digestion of the meal to follow. Hors d'oeurvre consist of a well-balanced selection, well-flavoured, colourful foods served either singularly or from a saladiere or special hors d'oeuvre trolley.

Hors d'oeuvre variés is a selection of items served on a trolley or a saladiere. It should include a meat dish, a fish dish, a simple salad, a compound salad, an egg dish and a fruit dish.

Many vegetables are used in hors d'oeuvre. These are sometimes prepared à la Greque or à la Portugaise. These two methods of presentation greatly increase the variety of items available for inclusion in the hors d'oeuvres varies.

Hors d'oeuvre à la Greque are vegetables that have been cooked in a liquid prepared from water, olive oil, garlic, tomatoes, parsley, thyme, bay leaf, salt and pepper, to produce a tomato and garlic flavoured vegetable.

ACTIVITIES

- Look around your nearest shopping centre and make a list of all the establishments that sell snack type foods to be eaten on the premises, or as a takeaway facility. Compare the items available, their cost and the facilities offered by each establishment. Which to you think is offering the best value for money? Are any of the establishments aiming their products at a particular type or age of customer?
- Look at the packaging of sandwiches and filled rolls. Which type is the most successful? Why should the packages have the contents and the sell by date written on? List the hygiene points that must be followed when preparing sandwiches to order in a sandwich bar.

CHAPTER 22

Food presentation, menus and adapting recipes

However good a meal tastes, it is vital to ensure that it looks appealing, if not irresistible, to the customer. The entire eating experience begins when the consumer first sees food or detects its aroma. To look appetizing, food should be presented in a recognizable and well-balanced manner. One part of the food should never be allowed to predominate over the other parts of food on the plate or dish. For example, a sauce should not be so thick and in such volume that it is impossible to determine what it is covering up. Likewise, the garnish should never be permitted to predominate the food that it is designed to compliment. For example, in a ragout de boeuf, there should be one small dumpling per portion and three or four tiny turned root vegetables, so that the ragout is definitely the main component of the dish.

GARNISHES

Garnishes are used for two reasons: to make the food look more appetizing and colourful and to compliment the flavours and textures of the food without overpowering its main flavour (see table 2.13 overleaf).

Standard garnishes

In traditional menus there are numerous classical finishes that are instantly recognizable and that always denote a specific garnish. Below is a list of some of these standard garnishes.

Garnishes for cuts or joints of meat

- À la berrichonne – balls of braised cabbage cooked with slices of streaky bacon, small onions and chestnuts cooked in meat stock, and accompanied by braising liquor thickened with arrowroot.
- À la bizontine – small ovals of duchesse potato mix, glazed and filled with puréed cauliflower, with stuffed, braised lettuce halves accompanied by veal gravy.
- À la bouquetière – glazed ovals of carrot and turnip, château potatoes, buttered peas and diced French beans with bouquets of cauliflower, arranged in a symmetrical manner around the dish. The cauliflower may be mashed with sauce Hollandaise. Accompanied by a clear gravy.
- À la clamart – tartlet cases filled with peas à la française, and placed on small macaire potatoes.
- Dubarry – sprigs of cauliflower coated with mornay sauce and gratinated, and served with fondant potatoes.
- À la duchesse – oval-shaped duchesse potato mixture, glazed and coloured, and accompanied by Madeira sauce.
- À la Milanaise – spaghetti Milanaise accompanied by tomato sauce.
- Montmorency – cooked artichoke bottoms filled with buttered macedoine and small bundles of asparagus tips, and accompanied by Madeira sauce.
- À la Napolitaine – spaghetti Napolitaine accompanied by juices from the joint.
  Parmentier – parmentier potatoes sprinkled with chopped parsley accompanied by a clear gravy.

Garnishes for fish

- À la Dieppoisse – prawns and mussels in a white wine sauce.
- Doria – olive-shaped cucumber pieces softened in butter and slices of peeled lemon (used for fish prepared à la menniere style).
- À la Florentine – poached fish placed on a bed of cooked spinach, coated with mornay sauce, and glazed for service.
- Matelôte – sautéed button mushrooms and onions and heartshaped croutons of fried bread.
- À la Nicoise – sautéed tomato flesh, garlic and tarragon with anchovy fillets, black olives, capers and anchovy butter. Slices of peeled lemon are for finishing.
- À la Tourvillaise – prawns and mussels poached in white wine and button mushrooms with a shrimp sauce.

Table 2.13 Commonly used garnishes

Herbs	
Parsley	
Rosemary	
Thyme	
Mint	
Starch based	
Puff paste flowers	
Short paste tartlets	
Croutons	
Fried or toasted croutons	
Gnocchi	
Vegetable based	
Tomato – halved, stuffed, sliced	
Peppers – sliced, shredded	
Cucumber – sliced, twists	
Gherkins – sliced, formed	
Capers	
Olives – stuffed, black	
Mushrooms – sliced, diced	
Onions – silverskin, sliced, shredded	
Carrot – julienne	
Celery – fingers, flowers	
Radish – slices, flowers	
Lettuce – hearts, leaves	
Watercress – picked	
Cress – picked	
Truffle	
Chicory	
Potatoes – deep fries, sticks, wafers	
Fruit based	
Lemon – slices, segments, quarters	
Orange – peel, shredded slice, segment	
Grapes – halved	
Melon – balls, diced	
Apple – sliced, diced, sprinkled with lemon juice	
Kiwi fruit – sliced	
Nuts	
Almonds – toasted, shredded	
Walnuts – halved, chopped	
Peanuts – halved	
Water chestnuts	

CREATING THE RIGHT BALANCE

Good presentation contributes to the sale and the appreciation of the dish or product. A clean and tidy approach is essential. Good, natural light is essential as colours look different under artificial light.

When presented, each dish should give an appearance of overall balance and completeness. There are two types of balance: symmetrical and asymmetrical. In the former, all items correspond in position, size and shape on either side of a dish. An example of this is the classic preparation of dressed crab: here, if an imaginary horizontal line were drawn down the centre of the crab, one half would mirror the finish of the other half. This method of presentation is used in many different forms, some further examples include the presentation of trays of canapés, finger sandwiches, individual fancy iced cakes, sweet biscuits, a fruit tranche or flan. The pattern created by the symmetrical design appeals to the tidy, logical mind and makes the dish look appealing.

In asymmetrical dishes, the food is presented in a manner that is not as logical or organized as that in symmetrical dishes. Examples of the asymmetrical method of presentation include a bowl of fruit salad, a mixed grill, a green or mixed salad and a royal iced cake with a corner spray of flowers. This method causes the eye to rove over the entire dish with curiosity, as it is more interesting, and has more 'flexibility' and 'movement'.

Whichever type of presentation is to be used, the basic dish design principles need to be considered. These principles are that the dish should be:

- suitable for the purpose and of the appropriate colour,
- pleasing to the eye and appetizing ('it should look good enough to eat'),
- interesting by the use of suitable combinations of shape, texture and height,
- balanced and harmonious with no one item completely dominating the others.

Colour is a vital component of food presentation. To appreciate colour fully you need to be able to visualise the three primary colours of red, blue and yellow, the three secondary colours of green, purple and orange, and the intermediate colours of red/orange, light green, dark green and mauve.

Any combination of the primary colours is too harsh to be successful - this is the reason why a glacé cherry does not look 'at home' on the centre of a melon slice or half a grapefruit. However, when one matches primary colours to secondary colours (yellow to purple, red to green and orange to blue) pleasing combinations will result, which is why tomato or radish with lettuce or other green leaves is successful.

Shape is another component of food presentation. Garnishes may compliment the base shape and introduce harmony to the dish or, less welcome, a feeling of confusion is caused by too great a contrast. With a hors d'oeuvres ravier filled with mushrooms à la Portugaise, the circular shape of the mushrooms and the colour created by the sauce Portugaise create a feeling of harmony. There is no need to add any garnish at all.

Chefs practising nouvelle cuisine techniques use these presentation skills when plating the dishes ready for the customer, some of these look almost too beautiful to be eaten (see photo below).

When considering food presentation for any types of catering establishment, it is essential to pay attention fo the points listed in Table 2.14.

Presentation is at a premium with Nouvelle Cuisine.

MENUS

Every establishment that caters for the general public will have menus planned and displayed for potential customers to see. The menu is the marketing vehicle by which a catering establishment advertises and sells its dishes and services to the general public; menus should, therefore, be planned to create an attractive, appetizing image of the establishment. In welfare catering, the menu may be planned for customers who will not have the ultimate option of choosing to go elsewhere for their food; here the menus have to be carefully planned to provide not only an appetizing meal, but a good variety over the successive days or weeks.

General menu planning considerations

Costs

In every establishment, costs have to be considered and budgets kept to. Meat and fish are the most expensive forms of protein; if the budget is low, other forms of protein should be incorporated into the menu.

Staffing

The menu should reflect the staff capabilities. The content of the menu should ensure that staff satisfaction is maintained by using and extending their creative skills. If the menu is too complex for the staff, disaster will result in one form or another.

Time

There must be sufficient time available for adequate preparation and cooking of foods on the menu. The time taken to serve the foods should also be

Table 2.14 Points to bear in mind when presenting food

	Should be	Should not be
Taste	Clearly identifiable	Too strong or indistinguishable
Temperature	Either very hot or chilled	Warm or tepid
Colour	Appropriate to the dish	Overpowering or artificial
Texture	Variety of texture	All crisp or all soft and soggy
Shape	Dish complementary to the food	Not be far too large or far too small
Portion Size	Appropriate to dish size and price	Not to large or small

considered, along with the time that foods can safely and successfully be kept hot. Food spoilage is related closely to time and should never be forgotten.

Equipment

The menu should be planned so that all equipment is used to its full potential. A wide variety of cooking methods should be incorporated within the menu to ensure good equipment utilization.

Nutrition

Welfare and hospital catering units have to consider the nutritional balance of the foods offered, as in these establishments the customers may be totally reliant upon the meals provided for their nourishment. There should be a balance between protein foods, foods that contain proactive elements and those with a high carbohydrate content. Fibre is an essential part of any meal and customers are now more aware of the fibre content of their meals. Animal fat is considered by some people to be harmful if eaten in excessive quantities, and some consideration as to the fat content of the meal will be necessary. When using fresh foods, care should be taken that the nutrients are not destroyed during preparation and cooking times.

Balance

A variety of textures, shapes and sizes of foods will aid customer satisfaction. There should be a well balanced, good range of colour to create an attractive appearance on the plate.

Season

Seasonal foods should be used to create interest and to add variety of taste into the menu. These foods may also be used in small quantities to create interesting presentation effects.

Customer

The type of customer should be considered; religious and ethnic tastes should be catered for.
Adapting recipes

ADAPTING RECIPIES

When deciding upon a recipe to prepare and cook, decisions are made concerning the method of cooking involved and the choice of ingredients to be used. Think about preparing a meat pie. The following questions will arise:

- What type of meat is to be used?
- Will the meat be new or precooked when put in the pie?
- What type of sauce is to be used?
- What other ingredients will be used for flavour or bulk or both?
- Which type of pastry is to be prepared?
- Is the pie to have a base and top pastry crust or just a top crust?
- How is the pie to be finished?
- Is the pie an individual or a multiportion size?

Obviously, the answers to these questions will depend upon a number of factors, such as who will eat the pie, the ingredients that are available and compatible, and the nutritional and flavour balance to be achieved. Provided the basic prinicples of preparing and baking a meat pie are followed, then any number of variations can be prepared successfully.

Adapting recipes will be relatively simple provided that the proportions of ingredients remain the same. Many basic recipes are adapted to create quite different flavours – a good example is the basic sauce and its derivatives – béchamel, velouté and semiglace sauces. Each of these have many derivative sauces created from the basic sauce. Another example is the basic rubbed in cake mix, which can be changed by adding fruit, coconut, nibbled almonds, cocoa, orange rind, or reducing the flour content and substituting rolled oats or muesli. So it is not always necessary to have ten different recipes for small cakes. The basic plain recipe can be adapted to create ten different types of flavoured small cakes. This principle may be applied to many different types of dishes.

ACTIVITIES

- Table 2.13 lists items that are frequently used as garnishes. For each garnish, name one dish for which it is suited.
- Study the photo on page 133.
  1. Describe how (in each package) the foods have been garnished.
  2. Do you consider the garnishes appropriate in each case?
  3. Suggest alternative garnishes for the dishes, describing the effect you are endeavouring to create
- Write a list of the basic principles to be followed when preparing a meat pie.

Prepackaged snacks.

- Compile some recipe adaptations for quiche Lorraine. Indicate which of these recipes follow the healthy eating philosophy of low fat, salt and wholegrain cereals.
- Study a recipe for a meat and potato pasty (work in groups of two).
  1. Recommend changes to the ingredients to increase the nutritional value of the pasty without making a significant increase in its cost.
  2. Bake a batch of the pasties using the original recipe and on batch using your recommended changes.
  3. Analyse your results using the following methods:
     (a) your own personal preference, and
     (b) ask six individuals to taste the pasties.
  4. Compile a chart to illustrate your findings giving marks for the taste of the pastry, the finished colour, and texture of the filling - poor (1), average (2), and excellent (3).

CHAPTER 23 Financial control and accountability

PORTION CONTROL

The regulation of portion weight is part of the financial control cycle. If a four portion pie is cut into three portions, then the profit will obviously be reduced by 25 per cent. It is imperative to profitability that the portion sizes are constantly monitored and controlled. Portion control has to part in the purchasing department and then followed right through the preparation and sales cycle. Figure 2.21 illustrates this control cycle.

Portion control is exercised in every type of catering establishment. In a takeaway fish and chip shop, the chip portion size is regulated by the size of the bag for the chips, and by the size of scoop used. In a formal restaurant, the portion sizes are regulated by the service equipment used in the kitchen as the food is being prepared for presentation to the customer (ladles, scoops, and spoons are used). In a self-service area, the standard recipe will specify the quantities of the ingredients used to produce the dish, and the number of portions to be served. Thus all the stages of the preparation cycle are concerned with portion control.

While preparing foods, portion size is regulated by weight. Think of a 3 kg mix of bread dough; it is not possible to make every bread roll exactly the same size by using guess work alone. It is necessary to weigh the portions of dough to ensure that the correct number of bread rolls is produced from the 3 kg mix, and that each bread roll is of the same size. A pie cutter or mould is used when producing a large number of pies or tarts. This device will stamp out a ball of pastry into the designated size and thickness of a pie or tart. This will ensure that each pie is of the same size and that the pastry is of uniform thickness. In small establishments, this control is exercised by weighing the pastry before it is rolled out, allowing a predetermined weight of raw pastry per pie or tart mould. Dough dividers are used when preparing items such as sausage rolls, which will regulate the size of the finished product by equal lengths being cut. The same cutters may be used when preparing and cutting pasta paste to ensure strips of equal width.

Good portion control leads to profit control, which follows from accurate cost control and portion pricing. Other benefits include customer satisfaction, quality control through regulated, standard recipes and portion preparation, low or minimum wastage levels, and accurate purchasing forecasts and therefore minimum stocks held in reserve. The consequences for a catering establishment that has little or no portion control procedures will include; little chance of profit control, little or no control over food costs, poor customer satisfaction levels, low standards of staff moral and quality of work produced, a high, unpredictable wastage level, erratic purchasing quantities and a high level of food held in stock.

Below is a guide of standard weights for portions.It may be used as a reference source but remember that

Purchasing →	*Preparation* →	*Service* →	*Cashier*
Weight or size of foods and items ordered and purchased	Standard recipes standard sizes dishes, trays and tins	Ladle, scoop bowl, and plate size and capacity	Knowledge of price structure and portion sizes

2.21 The financial control cycle.

the size of portion will vary according to the type of customer that is to be served. For example, a cheese sandwich for a building worker will be a substantial item, whereas a cheese sandwich for a patient in a hospital may be light and thin.

- Soup: 4–6 portions per litre.
- Sauce: 8–10 portions per litre, or 12–16 per litre as an accompaniment.
- Fish: 100g per portion of the bone, 125–150g on the bone
- Meat: 100g per portion off the bone, 125g on the bone.
- Poultry: 200g per portion on the carcass.
- Offal: 50–75g per portion (oxtail 200g).
- Cooked meats, e.g. ham: 50g per portion.
- Potatoes: 100g per portion; new potatoes 75g.
- Vegetables: total 125g per portion (this may be mixed vegetables).
- Pastry: 25g per base or top, per portion; 25g flour per person.
- Milk pudding: 6 per litre.
- Rice (savoury): 25g per portion.
- Pasta: 25g per portion.

There are many advantages to be gained when using standard recipes within a catering establishment, such as:

- simplified staff training programmes,
- standard stock purchasing,
- minimum stock ordered or stored,
- quality control – the product is always the same (see figure 2.22).

FINANCIAL ACCOUNTABILITY

Every catering establishment, whether it be a mobile bar, a school canteen or a first class luxury hotel, has to have financial accountability. In other words, it is necessary to be able to maintain financial records that will reveal details of the daily purchases, stock held, daily sales and expenses or overheads – from these it will be possible to calculate the trading position in terms of gross and net profit.

It is not possible to run a successful business unless a sound trading position is maintained: no catering establishment is permitted to run at an operating loss. A few welfare and industrial catering units do, in fact, have a subsidy from their governing agents so that even though their selling prices are low in real terms, their accounts are balanced by means of an agreed subsidy. This method is used by some industrial units to ensure that the employees have a good and inexpensive meal available during their working day. However, the subsidy will be an agreed percentage of the annual turnover; so even in these establishments, financial accounting is essential.

The object of financial planning at any level is to provide answers to these four simple questions:

- What do we have or own?
- What do we owe to others?
- What in financial terms are we worth?
- How is the daily trading situation doing?

2.22 Standard recipe form

NAME OF DISH	No. OF PORTIONS
INGREDIENTS	SIZE/DISH/PRODUCT
	COOKING TEMP
	COOKING TIME
PREPARATION METHOD	
FINISH METHOD	
HOL/REHEAT METHOD	

Businesses exist to buy and sell goods or services to make a profit. Traders will need to identify the value of their purchases; the value of their sales; their expenses; the amount of cash in the premises and at the bank; the amount of money owed to them by customers; the amount of money owed to others by the business; the value of property and articles owned by the business; the profit made by the business over a given period of trading; and the financial position of the business on any particular day. The ultimate purpose of any business is to make a financial profit while staying in a strong, sound trading position, and the two aspects that must be separated are gross profit and net profit.

Gross profit and net profit

Take the selling price of an item, e.g. £1.00.
Deduct the initial cost of that item, e.g. 40p.
This will give a gross profit of 60p.

Should the trader wish to record this gross profit as a percentage of the selling price, it will be worked out using the following formula:

$$\frac{\text{Gross profit}}{\text{Selling price}} \times \frac{100}{1} = \text{e.g. } \frac{60}{100} \times \frac{100}{1} = 60\%$$

To be a financially stable trader it is necessary to understand and take account of the two costs involved in producing the items for sale.

For example, take a cake that cost 87p to produce and that is to be sold at £1.80. The trader will not, in fact, take a final profit of 93p because he or she has had to pay the staff costs and the premises costs to produce the cake. Staff costs include wages, accommodation, National Insurance, meals, transport, and company insurance. Overheads include mortgage or leasing repayments, insurance, rents, Community Business rate, fuel, lighting, and depreciation of equipment.

All these costs have to be evenly distributed over the costs of all items and services to be sold before the trading profit is added, if the trader is to be financially sound. Take the following example:
Selling price £1.80

Total overheads	25p
Labour costs	35p
Material costs	87p
Total costs	£1.47

Net profit = selling price = total costs
33p = £1.80 − £1.47

To find the net profit as a percentage of the selling price, all that is necessary is to use the formula.

$$\frac{\text{Net profit}}{\text{Selling price}} \times \frac{100}{1}$$

$$\text{e.g. } \frac{33}{100} \times \frac{100}{1} = 18.3\%$$

This is the percentage net profit made when selling the cake.

Pricing

It is essential to ensure that the price charged for a unit or service will be acceptable to the customer and is sufficient to guarantee an adequate net profit. If prices are unrealistic and out of step with surrounding traders offering similar services, then the potential customers will go elsewhere. They are more concerned with getting value for their money rather than the trader achieving a good net profit.

The selling price of any item is normally calculated by using the following formula:

Cost + profit will equal the selling price.

For example:
A portion of applie pie costs 36p and the profit to be added is 60%.
Remember that the total or selling price in percentage terms is always 100%. Therefore the cost of the apple pie is 40%. Therefore the selling price is:

$$\frac{\text{(Cost)}}{\text{(\% Cost)}} \quad \frac{36\text{p}}{40} \times \frac{100}{1} = 90\text{p}$$

Thus the applie pie which cost 36p per portion to make, with a 60% profit added will be sold at 90p per portion.

Costing and control terms

The following list gives a simple definition to the terms used in costing and control documentation.

- Material costs are the amounts paid for basic commodities or raw materials: food items for the kitchen and beer, spirits, mineral water and wines for the bar.
- Labour costs include wages for permanent staff and for casual labour, staff meals, staff accommodation, employer's contribution to National Insurance, and staff transport.
- Overheads are the costs of running the business after materials and labour costs have been accounted for: rents and rates; heating; lighting; and fuels; equipment depreciation; maintenance

and repair; stationery and printing; postage and telephones; cleaning and laundry costs; decorations and flowers; bank charges; interest payable; accountants fees; legal fees; insurance; advertising and special promotions; and transport costs.

- Fixed costs are costs that do not increase or decrease with fluctuations in sales: rents, rates, insurance, and depreciation. These may remain static for a period of one financial year.
- Variable costs increase in direct proportion to the sales figures. Obviously, the greater the number of meals that are prepared, the larger the volume of food that needs to be purchased.
- Unit of cost may refer to a unit cost of either a specific dish or drink; a definite event or function; or the costs incurred over a period of a day or week.
- Total cost is the sum of material, labour and overhead costs. It may relate to any unit of cost.
- Equipment – the total value of each major item of equipment and furniture is written down along with its estimated 'life'. The cost price is divided by the 'life' and the result is 'written off' in the depreciation account annually.

  For example, a solid top range cost £950
  Anticipate 'life' = 7 years
  Amount written off annually = £135.70
  (This may be expressed as 14.3 per cent per annum.)

  Therefore after 7 years, there will be sufficient funds set aside to replace the solid range.

  Naturally, a cost of living percentage is added annually to cover inflation costs. Maintenance and an estimated sum is set aside annually which comprises of the total maintenance contract costs and a sum to cover the renewal of small items not included in the equipment and furniture depreciation account. An amount is also included for building maintenance. The annual inflation costs must also be taken into account.
- Fuel costs are made up of the total cost paid for all fuel consumed on the premises, these include gas, electricity, oil, solid fuel and water. It is essential to monitor these costs, the easiest method is to use fuel consumption control sheets showing readings per section of the business so that unprecedented increases can be identified and monitored.
- Unitary cost is used to give a value to a unit so that each unit may be sold at the same price. It is used extensively in liquor sales.

  For example:

  32 tots per 26 $^2/_3$ fl.oz. bottle
  Cost per bottle = £9.78
  Gross profit required 60%
  Net profit required 15%
  Selling price must be $\frac{978}{40} \times \frac{100}{1}$ = £24.45

  $\frac{24.45}{32}$ = 76p per tot

When all these remedies have been studied, it may then be evident that the percentage net profit (selling price) on some items needs to be increased, probably

Table 2.15 Likely causes and remedies for losses

Likely causes	Remedies
Manufacturing wastage	Check purchasing/storage methods
Food preparation methods	Standard recipe sheets and production methods
High fuel costs	Staff awareness of fuel consumption costs
Increase in fuel, heating, lighting charges	Comparison charts
High raw commodity costs	Investigate supplier costs
Increase in postal and telephone charges	Substitute for high cost commodities
High staff wages	Use work study methods to analyse staff daily duties.
Sudden failure of equipment	
Increase in laundry charges	
Pilfering	

due to either increased commodity prices or an increase in staff cost or overhead costs (see page 136), or a combination of more than one of these factors.

It should be understood that it is possible to raise the prices too high and so incur customer resistance. In this case, the trader will lose customers at the expense of his nearest competitors.

Budgets

These are used to plan the financial and business operations for a given period of time.

- Capital budgets include budgets for equipment and premises.
- Operational budgets include budgets for sales, purchases and labour cost.

The aim of any budget is to increase operational efficiency, to set monetary targets within specific time periods, and to give operational objectives to be achieved within specific time periods.

Sales budget

This is used to forecast the future level of sales during a period. It is based upon actual sales over the equivalent trading period in the previous calendar year. Consideration will also be given to current sales trends and future national or local sales promotions, within that trading period.

Labour cost budget

This will estimate the labour necessary to achieve the targets set in the sales budget. Labour efficiency should be considered when setting this budget, as increased labour does not necessarily result in increased sales.

Overheads budget

This is the easiest budget to compile as many overheads are fixed annually and others are easily estimated using the equivalent trading period in the previous calendar year.

STORES CONTROL

Weekly stock sheets

This method involves the use of printed sheets on a weekly basis. All the items in stock are recorded on the sheets in alphabetical order in groups of commodities such as meat, fish, dairy products, fruit and vegetables, cereals, etc.

The purchase unit size, cost and quantity in store are all recorded. All receipts and issued stock are recorded, and at the end of the week the total stock can be calculated – stock recorded, stock issued and stock held can all be valued. This method is time consuming and only suitable for small establishments or where there is a limited range of items.

Table 2.16 Sample stock sheet

Stores	Unit	Unit Cost	Stock	Receipts				ISSUES				Total	Cost	Total	Value
Butter	kg	198p	28	50	5	10	8	11	6	8	9	57	107.73	21	41.58
Cream	litre	162p	1	9	–	1	2	1	2	3	–	9	14.58	1	1.62
Eggs	dozen	59p	12	18	2	4	3	5	2	3	2	21	12.39	309	5.31
Lard	kg	60p	40	–	3	1	2	2	–	4	1	13	7.80	27	16.20
Margarine	kg	66p	56	100	20	18	14	8	12	17	14	103	67.00	53	34.98
Milk	pint	29p	2	140	18	17	18	16	19	19	17	124	35.96	18	[illegible]
Oil	litre	72p	18	–	–	–	1	–	3	–	1	5	3.60	13	9[illegible]
													382.73		114.[illegible]

Stores bin card system

In this system the items in the store are recorded on individual record cards - one card for each item. These are hung in horizontal files so that the name of each item is permanently visible, and it is easy to 'flip up' the required card.

This system is suitable for large establishments where a stock sheet would be impractical due to the size of the operation. However, this system has the disadvantage of being time consuming to operate, and goods that are issued frequently often need their cards replaced.

Computerized stock control

These systems have become more commonly used in recent years and are, in fact, very similar to the stock sheet. Here the transactions are input into the computer and the totals will be displayed on the visual display unit (VDU). It is a fast and efficient way to keep stock records.

There are many computerized stock control systems used in modern catering establishments. They all follow the same basic paths, giving instant information to the user at any one of a number of visual display units (VDUs) around the food and beverage department. Figure 2.23 illustrates the types of areas that could be included in a busy hotel.

Figure 2.23 Areas in a hotel which might have a VDU

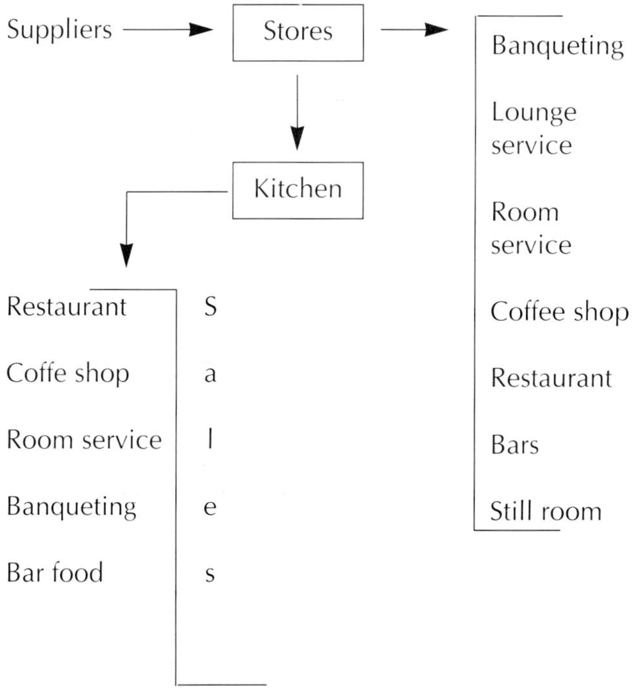

Imagine that all these movements of stock, food, meals and sales had to be recorded on paper and manually accounted for. It would be a nearly impossible task a huge number of clerks to keep an up to date financial and stock record. With a computerized control system, the main unit costs are converted and calculated automatically and instantly, these are shown on the VDU and may be printed out if required. Thus a chef may request a long list of ingredients from the storekeeper, who records the list on the computer by typing in the requested items (in some establishments the chef does this). The total cost of this list of goods is instantly recorded and deducted from the stock list.

- Request for food items from catering stores:
  1. the chef types a list of requests into the computer,
  2. the storekeeper receives a print out of the list,
  3. the computer records the list and
     (a) the cost of each item,
     (b) costs the total list,
     (c) indicates if the reordering level is reached.

Another example of instant computerized figures is the updating of stock when purchases are delivered.

- Purchases are delivered to the store:
  1. The storekeeper types in the purchases from the delivery notes,
  2. and the computer
     (a) adjusts the stock level accordingly, and
     (b) adjusts the price if this has altered.

From this it can be seen how advantageous a computerized system is to the storekeeper. It is possible to have a permanent, up-to-the-minute cost record of all stock on the premises. The stores control system will be a small part of an entire business control system used by all departments of the establishment.

ACTIVITIES

- Visit and describe one public service food establishment, and find out how the food portion sizes are controlled. Compare your findings with the portion control methods used in one of the following types of fast food establishment: a pizza house, a hamburger bar, fish and chip shop and a Chinese take away shop.

- Find the gross profit as a percentage of the selling price for the following:

Cake cost 87p	selling price £1.80
Sandwich cost 34p	selling price 98p
Four doughnuts cost 22p	selling price 59p
Filled rolls cost 29p	selling price 66p

- Study Table 2.17. Draw a graph to illustrate the price variations and describe the possible reasons for these. Take account of the anticipated customer requirements, the surroundings, the service offered, the quality of the product and the speed of service.
- Work in groups and discuss the difference between the pricing structure in a fast-food hamburger house and an average sized city hotel restaurant.
  1. Consider the following aspects associated with trading:
     (a) the type of customer,
     (b) the type of food items on sale,
     (c) the type of service, and
     (d) the time the customer spends on the premises.
  2. Indicate the probable differences in:
     (a) the type of premises,
     (b) the type of cooking equipment,
     (c) the level and numbers of qualified staff.
     (d) stock levels,
     (e) the types of foods stocked or used,
     (f) staff training needs.

Each group should produce a written summary of their discussions.

- Draw up a weekly stock sheet, enter the following data (see Table 2.18), and then calculate the total cost of all goods issued and the total value of stock held. You will have to research the current prices before you start.

Table 2.18 Receipts

Stores	Unit	Stock C/f	Recieved
Coffee – instant	kg	3	12
Tea	kg	4	14
Milk powder	kg	8	14
Caster sugar	kg	15	28
Demerara sugar	kg	10	28
Spaghetti	kg	7	10
Rice – long	kg	6	12
Flour – plain	kg	30	20

Stores issued	M	T	W	T	F	S	S
Coffee – instant	1	–	–	–	1	–	–
Tea	–	2	–	2	–	1	1
Milk powder	2	–	2	–	2	–	2
Caster sugar	3	2	4	2	1	2	1
Demerara sugar	2	1	2	–	1	2	2
Spaghetti	–	–	4	–	–	4	–
Rice – long	–	3	–	3	–	–	–
Flour – plain	6	2	3	2	4	1	1

Table 2.17 List of food prices

	School canteen	Fast foods	Train	Hotel
Beefburger in a bun	40p	98p	£1.45	£1.90
Deep fried plaice and chips	50p	£1.45	–	£3.30
Apple pie	30p	76p	85p	£1.50
Ham sandwich	35p	£1.05	–	£1.20

PART 3

Nutrition

This section will give you an understanding of the:

- cultural and social factors that influence the choice of foods,
- processes of human digestion and absorbtion of food within the body,
- effects of different nutrients and their main sources of supply to the human body,
- factors that effect the UK food industry and the methods used for the safe preservation of foods,
- effects of the diet upon the human body and the consequences of alcoholic abuse.

CHAPTER 24 Factors affecting the choice and intake of food

In the past, the choice of food was governed by the geological make-up and climate of the country where it was grown. Today, this is no longer the case, and 'what's one man's meat is another man's poison' certainly holds true in the field of eating. Within any establishment you will find that people have preferences for one food over another. Some people will refuse to eat certain products such as pork, fish or eggs for a variety of different reasons. In fact, a large number of different factors influence what we eat, why we eat and when we eat.

With increased food technology, increased personal wealth and travel opportunities, a great variety of foods are now eaten by people within their homes and while they are 'eating out'.

PHYSIOLOGICAL NEED FOR FOODS

Habit

The three main times when food is eaten are soon after rising in the morning, at or around midday, and in the evening.

Breakfast is the first meal of the day. It needs to consist of energy-giving or body-building foods in an easily digested form so that the body can utilize the nutrients during the activity that usually follows. The body will function much better if it is fuelled by freshly-digested nutrients. This is one reason why breakfast cereals are so successful - they provide the body with energy-giving carbohydrates, while the accompanying milk provides the body with body-building vitamins, protein, calcium and minerals. If sugar is added to the bowl of cereals, then the energy producing factor will rise.

Luncheon is the one meal that traditionally provides the body with its main intake of food during the 24-hour period. It was normal to eat a two or three course meal at midday while at work or school, often provided by employers at a subsidized rate. This habit started during the Second World War when it was essential to ensure that workers received good, nutritious food to enable them to contribute towards the war effort by achieving vital production targets. Nowadays, this habit of eating a large midday meal has been replaced in some areas by the preference for equally nutritious, but light meals or snacks, either provided by the employer, taken to the work place by the employee, or purchased by the employee during the lunch break from a nearby establishment. Another factor that has influenced these changes is the common use of central heating in the work place, meaning that the body does not require as much energy to keep warm.

The evening meal or dinner is the meal that symbolizes the coming together of the family after the day's activities. It is often eaten quite early in the evening for two reasons, firstly, to allow the body sufficient time to begin the lengthy process of digestion before the sleeping process starts, and, secondly, to allow a maximum length of time in the evening for other leisure activities. The evening meal is, in many cases, the main meal of the day, with the largest amount of food being eaten.

In a hotel, when guests frequent the restaurant for dinner, the meal takes the role of a social occasion, an evening's entertainment or pastime. It is important to realize that this is the opportunity to rectify deficiencies in the daily nutritional intake. It is therefore necessary to plan this meal carefully taking into account the eating habits of the rest of the day.

Hunger

The feeling of being hungry is experienced when the body has reduced its store of energy and when the stomach is empty. As food is passed from the stomach into the duodenum, the stomach becomes ready to receive further food. This process is the digestion of food and can take up to four hours, depending upon the type of food. Items that have a high fat content will take longer to pass through the stomach, whereas those high in carbohydrates will be the quickest to pass through.

Appetite is a sensation that is related to the smell and taste of foods. The salivary glands in the mouth will produce saliva when attractive food aromas are smelt, or when food comes into sight. This saliva has the effect of stimulating the digestive juices so that feelings of hunger are experienced. These feelings are accompanied by the desire to taste the food product that has caused the reaction. Feelings of hunger are experienced when passing outlets such as a bakery, where the smell of newly-baked bread is delicious, or a fish and chip shop where the smell of the chips is very likely to provoke this reaction.

The habit of eating meals at regular times during the day is beneficial to the body mechanism. Hunger and habit are closely related in many people's lives, since the body becomes accustomed to eating foods at regular intervals of time and therefore provokes feelings of hunger when the next intake of food is normally taken. It should be realized that although the feeling of being hungry is felt, the body may not always need food. It may merely be a reaction to the smell or sight of food.

Social eating

Eating is frequently part of a social event when the actual nutritional content is rarely considered by the participants. The main reason for the gathering is divided equally between the prospect of good company is congenial surroundings, and the consumption of appetizing, well-prepared and served food or drink (see figure 3.1).

3.1 The eating experience.

On these social occasions the choice of hotel or restaurant frequented is often determined by the following points:

- the geographical situation and ease of travel,
- the menu choice and cost,
- the atmosphere and other clientele,
- additional services offered such as entertainment or live music.

It is important to realize that at these social events the caterer has a moral obligation to provide foods that are as nutritionally beneficial as possible, even though the clientele will not necessarily appreciate this fact.

In a less grand manner, food is frequently part of business entertainment when a meeting is planned to include breakfast or a midday or evening meal. Here the choice of venue will be determined by:

- the geographical situation and car parking,
- quiet, discreet surroundings and good, but light, food and drink,
- and excellent but unobtrusive service.

Another form of social eating is that frequently indulged in by the younger members of society who use the numerous fast food outlets that are now in all large towns and cities in the UK. These sell such items as hamburgers, french fries, pancakes, pizzas, jacket potatoes and hot and cold beverages. Such establishments cater for customers who wish to eat their food on the premises, as well as those who wish to purchase a hot snack type meal and take it elsewhere to eat. These retail catering establishments are normally open all day and evening. Some in large cities open all night and will serve customers within a few minutes, hence the term 'fast food'.

The nutritional value of such snacks centres mainly upon carbohydrates and saturated fats, with some protein, sugar and salt as well. However, this obviously varies with the customer's choice of food. Fast-food outlets provide a socially acceptable environment for their customers to enjoy and, by this method of sales, they are able to increase their annual turnover.

A further retail area where social eating occurs is in the many coffee shops that are now trading in towns and city centres, in large department stores, in city hotels and attached to on-site bakery shops. These coffee shops cater for light or snack-type meals and are normally open for business throughout the day and in the early evening. In the case of hotel coffee shops, a complete meal service is provided for guests and passing customers in a more relaxed atmosphere than the hotel restaurant. Coffee shops that are attached to bakery retail outlets often concentrate their sales upon their own bakery products, tempting customers to purchase rolls, cakes and pastries to eat on the premises.

Impulse

Eating food on impulse, rather than need, is a common fault in the West. It is based upon a desire to taste and eat a particular food product, even though it is not necessary or beneficial to eat the product at that time. Impulse eating is governed by the:

- conscious desire to taste the product,
- known liking of the taste of the product,
- anticipated enjoyment of the taste created by the product.

Material considerations also play a role:

- the availability of the product,
- the money needed to purchase the product,
- the time required to purchase and eat the product.

The phrase 'mind over matter' is often used when trying to prevent impulse eating – it is possible to resist the temptation if one's mind is strong enough to overcome the desire to taste the product.

Manufacturers of snack items spend time and money on researching the most effective methods of advertising their products, so that they will appear irresistible to the consumer. Manufacturers will also aim advertisements at a particular sector of society in the hope that they will become future customers.

Food products such as biscuits, crisps, chocolates, sweets, salted nuts, ice creams, sweet carbonated drinks, chewing gum, are all easily available. However, the nutritional content of these products is low, and they can be harmful to health if taken in excess, especially in place of other, more nutritious, foods.

AESTHETIC INFLUENCES ON FOOD INTAKE AND CHOICE

When an customer's choice of food is analysed, it becomes apparent that this choice has been influenced by the five senses: sight, smell, touch, sound and taste. In one way or another, these will be used by the customer in selecting which food will be consumed. It does not necessarily follow that the larger range of food available, the more difficult will be the choice.

Sight

The visual effects created by foods can product a positive or a negative attitude. Poorly presented foods that lack visual appeal will not be chosen by customers. Conversely, foods that make a positive, visual impact will be more likely to be chosen. Food, whether a plain and simple dish or a complex creation, should be readily identifiable, attractively presented, appealing and colourful. Garnishes are used extensively to create attractive-looking foods. This is especially important if the consumer is ill, as in hospital.

Smell

The power of smell is a two-way process where food is concerned. Some foods can be identified by their smell, and foods that are stale can be identified by a change in their natural smell. The smell of foods being cooked can stimulate the urge to taste – in- store bakeries, back bar fast-food restaurants, fish and chip shops, beef-burger shops and pizza houses all do this to good effect to promote sales. Equally there are unpleasant smells related to the cooking of foods which may put people off food and these should be avoided.

Touch

When purchasing raw foods, touch does not often play an important role as foods are frequently bought in prepacked form. However, it does apply when buying fruit and vegetables where touch is of vital importance to determine the freshness of the product. Touch is also helpful when choosing many snack items, for example, it is easy to determine the freshness of a sandwich by touching the bread. In all dishes, there should be a variety of textures: some items should be soft, some should have a 'bite' to them, such as vegetables, and others should be crisp, such as pastry. This combination of texture makes a more interesting meal; it may be possible to provide this variety in one dish by using a combination of foods to create a course of a meal.

Sound

Where foods are cooked either separately or partically within the customer's earshot, the associated sounds can increase the customer's anticipation of the dish. For example, foods cooked on a back bar griddle can create a sizzling sound as they cook, as can foods being finished at a gueridon trolley in a restaurant.

Taste

…sense that is individualistic. It is …ell if a taste that one person likes is …recisely the same way by another … is experienced when food touches the …ipes have been created to produce an … taste and, on the whole, it is thought that …adly experienced in the same bands of … is known that extremes of flavour can be …ied, such as pepper, salt and sugar, and that it is the gradual increase or decrease of flavour that is likely to vary from person to person. It is also known that a young person's powers of taste are limited. However, as the eating experience widens so does the power of taste. It is therefore possible to season a fine sauce to perfection without spoiling the base flavour. Flavours in a dish or course of a meal should be selected and used to enhance each other; one flavour should never be allowed to dominate all the others.

If the dish does not fulfil a customer's expectations, then he or she will probably reject the dish as 'no good', 'awful' or 'poor'. It is therefore necessary to know what the customer is likely to appreciate when finishing and seasoning foods.

CULTURAL AND SOCIAL CONSIDERATIONS

It should be remembered that in a multicultural society there are many religious beliefs and restrictions of diet that the good caterer has to be aware of and consider. Orthodox members of a religion will adhere strictly to the teachings and customs of that religion, whereas unorthodox members may practise a slightly more relaxed way of life. Some immigrants have been in the UK for two or more generations and may have adopted European tastes and practices. If possible, it is always wise to check with customers in advance then any dietary restrictions can be properly catered for. The likely restrictions to be encountered are given in Table 3.1.

THE INFLUENCE OF ADVERTISING

Where did you buy your last bar of chocolate? What affected your choice? Why did you decide to buy it?

Table 3.1 Food restrictions for various groups

Hindus
Forbidden: Orthodox Hindus are strict vegans.
Unorthodox Hindus may eat dairy produce but do not generally eat beef or other red meat

Muslims
Forbidden: alcohol, pork, lard, dripping and shellfish
Animals must be halal slaughtered (a ritual killing by bleeding)
Kosher meat is acceptable in some communities
Ramadan: a time of fasting from dawn to dusk for 29–30 days in May (during the lunar month)

Sikhs
Forbidden: alcohol, beef, pork, lard and dripping
Animals must be slaughtered by a single blow to the head

Jews
Forbidden: pork, lard, dripping and shellfish
To produce kosher meat, animals must be bled to death in the presence of a Rabbi and then soaked and salted
Different utensils must be used to prepare and cook meat and milk items

Vegans (strict vegetarians)
Not eaten: animal flesh of any kind, eggs, milk, cheese or animal products
A vegan's diet is very limited and care has to be taken when balancing the protein content.
Vitamin A is found in margarine, vegetables, dried apricots and prunes;
Vitamin B is supplied through yeast extract, oatmeal, wholegrain, nuts, pulses and vegetables;
Vitamin C is found in fresh fruits and raw vegetables;
Vitamin D is supplied through margarine and minerals such as calcium, phosphorous and iron, are found in green vegetables such as watercress, cabbage, spinach, kale and nuts.

Lacto-vegetarians
Not eaten: animal flesh
Cheese, milk and often eggs are eaten
Some lacto-vegetarians will also eat fish
The diet of lacto-vegetarians is relatively easy to plan and balance, as protein is supplied by the cheese, milk, eggs (and sometimes fish).

The answers to these questions are important to the advertiser. For example, if your answer to the first question was 'the local shop', then it indicates that distribution of the brand has reached the smallest range of outlets. If your answer to the second question was because 'I had seen it advertised and wanted to try it', then the advertisement must have been successful; on the other hand, you may have answered because 'I have tasted one before and I like it', in which case the product quality and brand name is continuing to sell the product. For the third question, you may have answered because 'it looked nice', indicating that the packaging is selling the product; alternatively you may have answered because 'it will keep me going until lunch time'. In this case, the advertisement may have given you this idea by indicating that the chocolate bar was sustaining.

Customers respond to a number of factors when deciding to buy a product or service. These factors are known as the 'marketing mix'. They are basically a combination of the following:

- the product and its quality, reliability, design composition and packaging,
- the price of the product: full price, discounts, credit facilities and returnable containers,
- the promotion used: publicity, selling methods and advertisements,
- the place of sale: the distribution networks.

The advertiser

The advertiser's aim is to make the public desire the product that is to be sold. The advertisement will always be targeted at a suitable market group. Which market group this is is determined by such factors as:

- whether it is to be a business or pleasure purchase,
- the purchasing power of the prospective buyer,
- the age group if it is a pleasure purchase, and
- the business need if it is a business purchase.

Once the target market group has been identified then the price has to be settled. This will be decided by answers to the following questions:

- What are the production and distribution costs?
- What is the price of similar (quality and weight) products?
- Is there any obvious competition? If so, should it be priced low to catch the trade - a loss leader?
- What is the state of the market? Is the product a seasonal item?
- What is the shelf life of the product - for how long will the product stay fresh?

The advertiser relies on the following chain of customer action in response to an advertisement:

- the customer is unaware of the product,
- the customer is aware of the product,
- the customer expresses an interest in the product,
- the customer wants the product,
- the customer is convinced that the product is good value,
- the customer purchases the product,
- the customer repeatedly purchases the product.

Advertisement campaigns use many methods to attract sales. These include:

- the use of free samples that may be placed inside trade journals or given away at wholesale and retail outlets,
- free gifts that are given to customers in return for proofs of purchase such as product seals, caps or labels,
- coupon offers that may be included in the advertisements placed in newspapers, magazines and trade journals offering a discount off the price of the product being promoted in return for the coupon,
- competitions to promote products either by using product information in the competition solution or by giving away samples of the product to the best or first entrants,
- the promotion of products at trade exhibitions and demonstrations where they may often be tried out by potential customers.

However, the method most often used to attract sales is the well-known one of price cutting to create and advantage over competitors. A product with a greatly reduced price is sometimes referred to as a loss leader, meaning that the product price is reduced to a false, low figure where no net profit will be realized during the period of the price cut. However, when the customers buy and try out the product, it is hoped that they will be so convinced of its quality that when the price is raised to its proper level, they will continue to purchase it. A good example of this practice is instant coffee promotion campaigns.

Product loyalty is the manufacturer's most valued goal. The brand name and manufacturer's trade mark or logo are the consumer's sign that the product is good, safe, 'just as I like it' and good value for money. Some customers will always buy food items by the brand name, knowing that they are buying food of a certain quality; the price in this case is their second consideration.

Brand names are used to distinguish one manufacturer's products from another's products. They are used as a support to promotions and advertisements, and are often instrumental in maintaining and even increasing the product's market share. The product itself will become less price sensitive because of product loyalty by the consumer. Brand names often indicate the nature of the product or the name of the manufacturer.

Packaging will have some influence upon customer choice. The design should aim to be compatible with the product and the brand sign, attractive and sufficiently eye catching to ensure sales. For example, a packet of chocolate biscuits in a chocolate-coloured wrapping with light brown printing will not create any customer curiosity. However, the same packet size and shape in cream or beige with dark chocolate-coloured printing should create a sophisticated looking packet that will stimulate the customer's curiosity and make him or her purchase it.

FOOD TRENDS

Food trends constantly change, and today we are returning to healthier eating habits and increasing our use of traditional foods and methods of cooking, and tending to avoid fast-food type meals. However, the traditional British fast food is fish and chips, maybe with a few processed peas, and this meal is still a popular take-away food.

Food trends and the consumer

Caterers provide foods that they think will attract business and thereby increase their profits. Owing to increased travel facilities and international goodwill, many people now travel abroad and discover the delights of foods from other cultures and lands. This trend has caused the meals in the UK to become far more cosmopolitan, and many menus are now a combination of traditional dishes and adapted dishes from all parts of the world.

People's eating habits have changed because their life style has changed. When parents have young children they tend to sit together at meal times. As the children grow up, they take part in numerous activities, go out in the evenings and may have a part-time job as well. This often spells the end of the traditional family meal time, as the family is hardly ever in the home at the same time. It is a natural progression towards teenage independence, but causes a great change in eating habits.

The changing role of women in modern Britain has an effect upon the eating habits, too. A greater proportion of women are now able to continue their careers as well as bring up a family, so they have far less time than previous generations of women had, to cook and provide meals for the family. On the other hand, modern advances in technology have made cooking easier and faster - such appliances as microwave ovens, food processors and preset oven timers are all assets to the working mother. Frozen food is also an invaluable help fo those who own a deep-freeze compartment.

Many people such as teenagers, the middle-aged or elderly, live on their own, and they face problems with regard to their eating habits:

- poor facilities for storing and cooking food if they are living in bedsit accommodation,
- the lack of low cost, small packets of foods in supermarkets,
- the fact that eating is generally associated with social occasions can make mealtimes a depressing time,
- the lack of incentive to spend time preparing food for oneself - it is often easier to eat a snack or go without a meal at all.

NEW PRODUCTS

The catering industry is continually being tempted or persuaded by food and product manufacturers to try out newly created items. Many of these products are what are commonly called convenience foods and are frequently prepared and processed by dehydration, canning, freezing and chilling.

Manufacturers employ various methods to bring these newly-created products to the caterer's attention. They may distribute product information to the trade press, give out money-off vouchers, advertise in trade journals, give potential customers free samples to try, usually followed up by incentive price reductions agains the initial purchase of the product in question. The next stage in the product launch is to receive repeat orders. However, these will only be received if the customer actually liked the new product.

It can thus be seen that the process is a continuous circle of sales. The key to success is customer acceptance. If the customer does not like the new product, it will have a very short life. If, however, the customer likes the new product, sales will increase and another successful convenience food will be on the market.

ACTIVITIES

- Consider the reasons that may account for the success of breakfast cereals in the UK diet.
- Consider what influence the advertiser has had upon your choice of chocolate bars (see page 146). Consider the questions below and discuss your answers with the rest of your group:
  1. How is your favourite chocolate bar advertised?
  2. Does the advertisement feature people of your age group?
  3. When looking at the advertisement, do you wish that you could taste the chocolate bar?
- Working in groups of two's or three's:
  1. Identify a food product that is being actively promoted at the present time.
  2. List the methods being used to promote customer awareness of the product.
  3. Which target market group is the advertiser aiming to attract?
  4. In your opinion is the advertisement successful?
- Have you ever been in the situation of looking after yourself?
  1. Did you eat regular balanced meals?
  2. Did you eat your favourite foods regardless of balanced eating?
  3. Did you eat take-away fast foods?
  4. Did you eat snacks and sandwiches?
  5. What factors influenced your eating?
- Choose one of the nutritional aspects below and consider the impact that it is having either on your own community or nationally.
  1. Nutritional labelling on food.
  2. An indication of the nutritional content of foods in fast-food outlet.
  3. School meals - healthy eating campaigns.
  4. The increased availability of wholefoods.
  5. Farm shops and organic foods in retail shops.
  6. Healthy food articles in magazines.
  7. Nutritional leaflets in large stores

CHAPTER 25 The balanced diet

The following are important for a balanced diet:

Energy
Carbohydrate
Fats
Proteins when oxidized

Growth and repair
Protein

Regulation of body processes
Vitamins
Minerals
Water

THE RECOMMENDED DAILY INTAKE (RDI)

The Department of Health and Social Security compiled a report on recommended daily intake of food and nutrients for groups of people in the UK in 1991. The recommendations apply to healthy people.

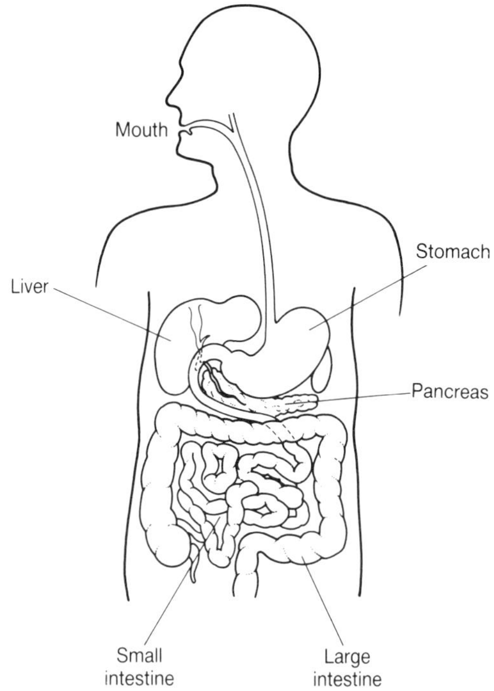

3.2 The alimentary canal.

However, no two people's requirements are exactly the same as we all expend varying amounts of energy. If a person is eating more energy producing food than is required by his or her body, the excess intake will be stored by the body, thus increasing the risk of becoming overweight and, worse still, obese.

Different individuals have different food energy needs, it is for this reason that some people eat far more food than others. The caterer needs to be aware of these facts and should plan meals so that those people who eat the smallest portions of food, still receive their RDI. In general terms, if the energy needs have been satisfied after eating a mixed diet, then the other nutritional requirements are likely to have been achieved too. The secret is therefore to supply food that is satisfying and sufficiently varied to ensure that the RDI is achieved. The key to success here is the word varied, as many different foods as possible should be offered to ensure a full range of nutrients is present in that meal.

THE DIGESTION OF FOOD

Digestion is the name given to the breakdown of large, insoluble molecules found in food into smaller, soluble molecules that are capable of absorption into the bloodstream. It involves both the physical breakdown of compounds (for example, by the teeth) and their chemical breakdown (for example by enzymes), and occurs in different regions of the digestive system (alimentary canal).

The alimentary canal

The alimentary canal can be split into five separate areas, each playing its own important role in the digestion and absorption of food (see figure 3.2).

Mouth

In the mouth, food is chewed by the teeth and the food fibre broken up. Chemical breakdown commences through the action of saliva on the food. Produced in the mouth, the saliva moistens food, begins the conversion of cooked starch to maltose, and stimulates the digestive juices in the stomach. The tongue forms the food into small parcels which are then swallowed and passed down the foodpipe (oesophagus) to the stomach.

Stomach

In the stomach, food is churned around by the muscles of the stomach wall. It becomes more liquid as it is mixed with gastric juices: pepsin converts protein to peptides and rennin starts the digestion of milk protein.

Duodenum

In the duodenum, digestion is completed and absorption commences: bile neutralizes mild irritants and emulsifies fats while the pancreatic juices get to work: trypsin converts peptides to amino acids (the component molecules of proteins); lipase converts fats to fatty acids and glycerol; and maltase converts maltose to glucose. Iron is absorbed and transported to the liver and spleen.

Ileum (small intestine)

In the ileum, food is absorbed into the blood stream through villi (finger-like projections) that make up the large surface area of the small intestine.

Colon (large intestine)

In the colon, water is absorbed from the foods that are not required by the body and is passed to the kidneys, where urine is collected and expelled from the body. Fibre from the foods helps to stimulate the colon to expel unwanted solid matter from the body.

THE STORAGE AND USAGE OF NUTRIENTS

After food has been digested it is the nutrients (basic constituents) of the food that are absorbed in the intestines. These can be divided into five basic groups, each playing an important role in keeping the body healthy.

Carbohydrates

Most carbohydrates are broken down into simple sugars, for example, glucose, fructose and galactose. They are then stored as glycogen in the liver and muscles, or converted into fat and stored. Carbohydrates are required by the body to provide energy and for tissue formation and maintenance.

Fats

Fatty acids and glycerol are stored in the liver where some may be transferred into energy if required. Excess fat is stored around the body as a layer under the skin and round the organs.

When required by the body to produce energy in the form of heat, fats are transported back to the liver, converted and then taken to the body tissues where they are converted into carbon dioxide and water. Heat is then produced.

Proteins

Amino acids are stored in the liver. Here, any part of the amino acid that contains nitrogen (which is not required by the body) is removed, conveyed to the kidneys and excreted in the urine. The remaining amino acids are used for the repair and production of body tissues.

Vitamins

Vitamins are substances required by the body in small quantities. They are essential to the body as a deficiency leads to specific diseases or disorders. Unfortunately, the body is not able to make vitamins (except for vitamin D from sunlight), so they must be taken into the body by eating appropriate foods.

Vitamins may be divided into two classes: fat soluble vitamins (vitamins A, D, E and K) and water soluble vitamins (vitamins C and B complex) (see Table 3.2).

Minerals

Minerals are inorganic elements that are absorbed by the body, some of which have important physiological or biochemical roles to play.

Calcium

This is required for the production and repair of teeth and bones. In order to be used, the body has to have a quantity of vitamin D and phosphorus to enable the calcium to be absorbed (via the villi in the ileum) into the blood stream. The use of calcium is controlled by the parathyroid glands, which are found near the base of the neck.

Sodium (sodium chloride = common salt)

This is an essential element of most body fluids. Without it the salt balance of the body is upset, leading to muscular cramp.

Table 3.2 Some important types of vitamins

Fat soluble vitamins

Vitamin A (retinol)

Source	Found in the liver in significant quantities, retinol can be made from the yellow pigment carotone found in red and yellow vegetables such as carrots and tomatoes. Another useful source is dairy produce (milk, butter, eggs and cheese) and also in dark green vegetables. Vitamin A is added to margarine.
Function	Retinol is essential for night vision and the maintenance of healthy skin and surface tissues.
Effects of deficiency	Deficiency leads to difficulty in vision in dim light (night blindness) and to a general deterioration in eye health. Severe deficiency could cause actual blindness. Children need vitamin A to ensure healthy, well-formed bones.

Vitamin D

Sources	Found mainly in the liver of fish and in fish oils. Also the body has the ability to make vitamin D through the action of sunlight upon the skin. Vitamin D is added to margarine and is also present in eggs and butter.
Function	Essential since it affects the body's ability to absorb and utilize calcium and phosphorus, and has a direct effect on the growth of bones.
Effects of deficiency	Deficiency results in poor bone growth, and severe cases lead to poor bone structure in children known as ricketts.

Vitamin K

Sources	Main supply comes from dark green vegetables, the best source being spinach.
Function	Necessary for the normal formation of clotting factors in the blood.
Effects of deficiency	It is not known whether there are any long-term effects in adults. However, young children have difficulty with natural blood clotting factors, so wounds take longer to heal. Intestinal bacteria can synthesize vitamin K, so a deficit in adults is unlikely.

Water soluble vitamins

Vitamin C (ascorbic acid)

Sources	Most is supplied to the body from the consumption of fresh fruit and vegetables, including potatoes; small amounts are also present in milk. Blackcurrants, citrus fruits and strawberries contain signficant amounts of vitamin C. Vitamin C is not a very stable nutrient. It is easily destroyed once vegetables have been peeled or chopped, and it is water soluble. The washing or soaking of vegetables in water will lead to a reduction in the vitamin C content of the food. It is also destroyed by the application of heat, so cooked vegetables contain less vitamin C than raw vegetables. Owing to the relatively large volume of potatoes eaten in the UK, they are considered to be one of the major sources of vitamin C in the British diet.
Function	It is necessary for the maintenance of healthy connective tissue.
Effects of deficiency	Deficiency causes wounds to heal more slowly and for gum disorders to occur. Severe deficiency will result in scurvy.

Vitamin B_1 (thiamin, aneurine)

Sources	Found in animal and vegetable foods, milk, offal, pork, eggs, vegetables, fruit and whole grains. Some breakfast cereals have vitamin B_1 added. Like vitamin C, vitamin B_1 is unstable, being destroyed by water and heat so it is lost when foods are cooked. Since vitamin B_1 is added to flour (other than wholemeal flour) bread is another important source. It is also present in yeast extract.
Function	Used by the body to release energy from carbohydrate.
Effects of deficiency	Results in disease called beri-beri.

table continued overleaf

Table 3.2 continued

Water soluble vitamins

Vitamin B_2 (riboflavin)	
Sources	Main source is milk, although others include offal, yeast extract, eggs, cheese and fortified breakfast cereals. Since vitamin B_1 is destroyed by ultra violet light, milk should be kept out of sunlight and in cool surroundings.
Function	Important for the growth of skin and helps maintain healthy skin.
Effects of	Rarely seen in the UK, but results in sores on the skin.
Vitamin B Nicotinic acid	
Sources	Main sources are meat, meat products, breakfast cereals, bread and flour that have been fortified, vegetables and milk.
Function	Helps the body utilise the energy derived from other foods.
Effects of deficiency	Deficiency will result in a skin complaint called pellagra, which causes the skin to become scaly or dark.
Vitamin B_{12}	
Sources	Found in animal products, especially liver. Significant amounts are present in eggs, cheese, milk, meat, fish or yeast extract.
Function	Acts with folic acid to divide cells in the bone where blood is formed.
Effects of deficiency	Deficiency leds to anaemia and nerve cell deterioration. Vegans need to be careful with their intake of this vitamin as it is found almost exclusively in animal products.
Folic acid	
Sources	Present in offal and raw green leafy vegetables. It is also found in pulses, bread, oranges and bananas. Folic acid is destroyed when heat or water are applied, so few foods can actually provide the body with folic acid if cooked.
Function	Acts with vitamin B_{12} to divide cells.
Effects of deficiency	Causes anaemia (deficiency of the required level of red blood corpuscles in the blood).

Note: Vitamin E is found in most foods, especially vegetable oils, cereal products, eggs, lettuce and green vegetables. This means that deficiency is very unlikely in man, although the effects of such a lack have yet to be fully determined. In some mammals, deficiency leads to sterility and/or muscle degeneration.

Potassium

This is essential for growth, especially of muscles and red cells. The use of both this and sodium is controlled by the suprarenal glands found next to the kidneys. These secrete hormones that ensure that sodium and potassium are retained in the body.

Iron

This is absorbed into the blood stream in the presence of ascorbic acid (vitamin C) and then taken into the liver. It is essential for the formation of blood haemoglobin (red blood corpuscles), which carry oxygen around the body.

Iodine

This is essential for the formation of the hormone thyroxin which is made by the thyroid gland, and controls the metabolic rate of the body. Iodine is stored in the thyroid gland.

PROTEINS

Proteins consist of a combination of substances known as amino acids. They are an essential part of the human diet as they are necessary for the growth and repair of tissue in all parts of the body. If the body has an excess of protein, it can be converted into glucose and used by the body to provide energy. there are two types of amino acids: those called 'essential amino acids' cannot be made in the human body and must therefore be present in foods; and those called 'nonessential amino acids', these are also required by the body but, if necessary, they may be made from other amino-acids in the body.

Sources of protein

Most animal products – meats, fish, milk, eggs and cheese – are a good source of 'high biological value' proteins. The ideal intake of protein is a mixture of high and low biological value proteins (see figure below).

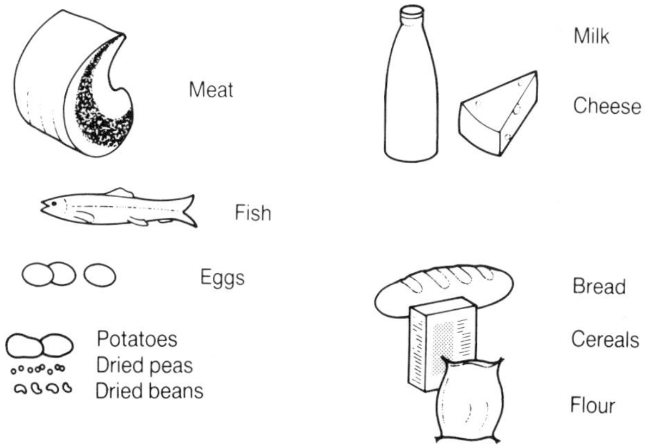

3.3 Sources of protein.

Textured vegetable protein

Because of the expensive nature of animal products and the growing appreciation of the vegetarian in our society, there are now products available that resemble meats that have, in fact, been carefully manufactured from plants such as the soya bean. These are known as textured vegetable protein (TVP), and are usually fortified with the essential minerals and vitamins – iron, thiamin and riboflavin – normally found in meats.

TVP can be used entirely on its own as a meat substitute, or as a supplement to animal protein when a proportion of fresh meat is replaced by the TVP. This has the advantage of reducing the cost of the finished dish and of maintaining the nutritional value.

FATS

Unsaturated fats

Unsaturated fats fall into two groups: monosaturated fats and polyunsaturated fats.

Monosaturated fats are rich in monosaturated fatty acids, the best-known and most frequently used fat in this group being olive oil. These fats are not bad for the heart and have a neutral effect on the arteries and blood.

Polyunsaturated fats are rich in polyunsaturated fatty acids, and it is thought that they help lower the level of cholesterol in the blood (especially those taken from oily first origins) so that arteries will not become blocked. They may also reduce the risk of blood clots. Polyunsaturated fats are found in herring, corn and sunflower oils, and in other oils made from vegetables, seeds and nuts, and in margarines that are labelled high in polyunsaturated fats (see figure below).

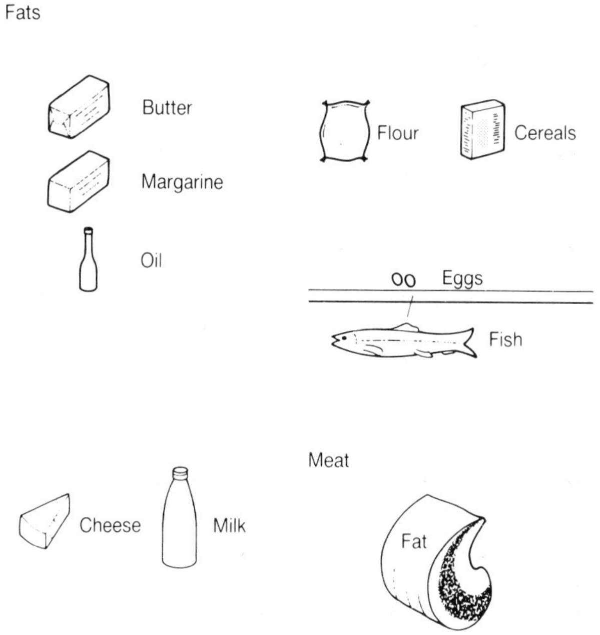

3.4 Sources of fats.

Saturated fats

These are mainly animal fats from meats, dripping, lard and fats from milk, cream, cheese and butter. They are also found in palm and coconut oil, which are commonly used in the manufacture of biscuits and cakes. Margarines that are manufactured using hydrogenated (hardening oils) are also classed as saturated fats – the harder the margarine the more saturated fat it contains. Saturated fats increase the level of fat in the blood, and are linked with the risk of coronary heart disease. It is necessary to reduce the amount of saturated fat used for cooking and preparing foods. At present, 60 per cent of the adult population of the UK has blood cholesterol above ideal limits.

SUGARS

The only natural sugars in foods are sucrose, which is found in sugar beet and cane; lactose, which is found in milk; and fructose, which is present in fruits and honey. However, sucrose and glucose have been added to many food products. Glucose syrup is not classed as sugar by food manufacturers, and so it is not always identified as such on food labels.

Sugar sensitivity

It is well known that some people have what is called a 'sweet tooth'. For such people, cutting down on sugar intake is very difficult without the aid of artificial sweeteners.

It is often thought that chocolate and sweets are suitable rewards for good behaviour or progress by children. However, sweets should not be given in this way since it will establish the habit of eating sweets. Furthermore, people frequently crave for chocolate or sweets at certain times of the day, for example, at morning coffee break, after lunch or during the journey home at the end of the working day. Since chocolate bars are easily bought, this craving is easily satisfied, unless strong willpower is used to break the habit.

It is possible to become accustomed to less sweet foods, and to consciously choose savoury items in preference to sweet foods, for example, many people prefer a yeast or beef extract spread instead of jam or honey.

However, it should be remembered that sugar is a preservative used in many food processes such as canning, jams and freezing. Sugar also speeds up the working of baker's yeast and helps to ensure a well-risen, even dough when bread making.

There are artificial sweetening agents available in tiny tablet or fine granular form. These products are used by many in the place of sugar to help reduce the calorie content of meals. These artificial sweetening agents are frequently used by manufacturers of fruit-flavoured squashes and fizzy drinks.

SALT

The Nutritional Advisory Committee on Nutrition Education (NACNE) report recommends that the daily salt intake should be reduced by 3g per day. Salt is needed in small amounts ($^1/_2$g per day) for the effective working of the heart. A reduction in the salt intake into the body will reduce the risk of developing high blood pressure. Persons whose blood pressure is high are three times more likely to suffer from heart attacks and four times more likely to have heart failure than persons whose blood pressure is normal.

Since it is the sodium in salt that can cause problems, it is necessary to monitor the sodium content in foods. One quarter of a teaspoonful of salt (6g) contains 2.4g of sodium.

Salts in foods

The amount of salt present in a dish at the time of eating will be much higher than the total salt content of the constituent parts for two reasons: first, salt is added to the foods during the preparation by the chef, and second, salt is added to foods during manufacture by the food manufacturer. Some foods are naturally high in salt, for example, shellfish.

The following foods are known to be high in sodium:

- Tinned foods: baked beans, peas, new potatoes, sweetcorn, and instant potatoes.
- Dairy foods: dried milk, butter (salted), Cheddar cheese, Edam, margarine, Stilton, Danish blue, Camembert and processed cheese.
- Nuts: salted nuts and peanut butter.
- Beverages: cocoa powder, Horlicks and rosehip syrup.
- Fish: smoked fish, prawns, shrimps, cockles, winkles, crab and lobster.
- Tinned fish: salmon, crab, tuna and pilchards.
- Meat: bacon, gammon, tongue, ham, luncheon meat, chopped pork, black pudding, salami, sausages and salted silverside.
- Tinned meat: corned beef, frankfurter sausages and meat balls.
- Sauces and pickles: piccalilli, sweet pickle, tomato ketchup, bovril, marmite, Oxo cubes, tinned and dehydrated soups, tomato purée, French dressing, mayonnaise and salad creams.
- Cereals: Allbran, cornflakes, Rice Krispies, oat-cake, spaghetti tinned in sauce, breads and porridge.

The sodium/potassium balance affects blood pressure. It is relatively simple to increase the daily intake of potassium and the following foods are good sources of potassium:

- Fruit: apples, bananas, dates, grapefruits, oranges, prunes and raisins.
- Vegetables: beans, Brussels sprouts, cabbage, corn on the cob, peas, peppers, potatoes and radishes.

If vegetables are boiled in salted water the potassium content will be reduced.

FIBRE-RICH FOODS

Dietary fibre is essential for good, healthy living because it is used by the body in the large intestine to absorb water, and it aids the expulsion of waste products from the body. Constipation is the name given to the infrequent or difficulty evacuation of the bowels. This is often caused by lack of sufficient fibre in the diet and by insufficient water intake.

Sources of fibre

The following foods are all sources of dietary fibre:

- whole grain cereals, such as wholemeal bread or brown rice, are rich in fibre,
- peas and beans are a good source of fibre, particularly their skins,
- root vegetables and potatoes are all high in fibre. The skins are a very useful source of fibre, and are often eaten on new potatoes or baked jacket potatoes,
- green leaf vegetables and fruits such as cabbage, spinach, oranges and apples, all provide the body with dietary fibre, but in fairly small quantities.

A diet rich in fibre need not be dull and uninteresting; it can include all the usual dietary components but, in addition, will include high-fibre foods (see Table 3.3).

Table 3.3 High and low fibre foods

High fibre foods	Low fibre foods
Wholemeal flour	White refined flour
Wholemeal bread	White refined bread
Brown rice	White refined rice
Baked jacket potatoes	Boiled peeled potatoes
Whole grain breakfast cereals	Refined grain breakfast cereals
Dried stewed fruits	Canned fruits
Wholemeal pasta	Refined 'white' pasta
Peanuts and raisins	Potato crisps or sticks

To keep healthy, it is recommended that 30g of fibre be taken into the body daily. This can be provided by many different foods, for example, two Weetabix or a small packet of dried apricots.

ACTIVITIES

- Visit a local health food supplier and take note of the varieties of TVP products that are available and their price. Compare these with the cost of equivalent fresh meat products.
- Plan a series of main dishes for a vegan. Use high protein foods that a vegan can eat and use TVP products. With each dish, include a short description of its main ingredients and method of preparation, cooking and garnish.
- Discuss the difference between saturated and unsaturated, particularly polyunsaturated, fats.
- List ways in which caterers can use polyunsaturated fats in their daily foods production methods.
- Suggest ways in which the salt intake can be reduced without reducing the palatability of dishes and meals.
- Next time you reach for the salt pot at the table stop and instead of sprinkling the salt directly over the food, sprinkle it onto an empty plate. Then see exactly how much you would have added. Remember one eighth of a teaspoon is all that an adult needs per day.
- Look at the nutritional chart on boxes of breakfast cereals and decide which ones provide the highest amount of dietary fibre.

CHAPTER 26 Nutritional value of foods

It is necessary to provide a varied diet in order to ensure that the body remains healthy. Different foods contain different nutrients, and it is necessary to be aware of which types of foods can supply what types of nutrients to plan a balanced diet.

MEATS

Most people have a natural appetite for meat and poultry which contains about 20 per cent of high biological value protein, 5 per cent fat and 1 per cent minerals, the rest being water. Some cuts of meat may contain a much higher percentage of fat, such as minced beef or lamb, belly of pork and streaky bacon. Cultural or religious preferences may dictate that particular types of meat are not eaten, and in these cases foods can be substituted to ensure a good balance of nutrients ar eaten.

Meat-based products are regulated by the Meat Products and Spreadable Fish Products Regulations 1984. These regulations list minimum edible meat in prepared meat products. For example, a beefburger must contain a minimum of 65–80 per cent meat overall. Chopped meats must contain 65–90 per cent meat overall. Cooked meat pies must contain a minimum of 25 per cent meat. Sausage meat must contain a minimum of 65 per cent meat, of which 50 per cent must be lean meat. Many manufacturers make meat products that contain more than the minimum requirement. In the case of all-meat products, it is advisable to read the product label prior to selection and purchase. It is also possible to purchase low-fat sausages; these have a lower saturated fat content and often a higher starch content than the normal sausage.

The white meat of poultry has a very low fat content as low as 5 per cent, and this makes it a valuable and nutritionally acceptable protein food. Remember that the method of cooking meat will have a significant effect upon its nutritional value.

In recent years, there have been 'nutritional scares' concerning poultry affected by *Salmonella* bacteria and, more recently, beef affected by *Bovine spongiform encephalopathy* (BSE). Provided that meat is purchased from reputable suppliers who are inspected by Environmental Health Officers (EHO) and who deal with cattle and poultry that have been passed as disease free by veterinary inspection, then there should be no cause for alarm. To date there is no proof that BSE in cattle can be passed onto humans.

FISH AND SHELLFISH

White fish

The fibres of white fish are held together by very little connective tissue that causes the fibres to stay in 'bundles' or 'flakes' when it is cooked. Hence the expression 'flake the fish' often used in cookery instructions. Each fibre is short and of a structure that enables it to cook far more quickly than the muscle fibres found in meat.

Some flat, white fish, such as plaice and sole, are made up of very fine, short muscle fibres that take only a few minutes to cook. Such fish are called delicate fish as they easily break up and overcook if they are not protected from fierce heat. The flesh of white fish contains less than 5 per cent fat. It is therefore a useful source of protein that may be easily digested.

Oily fish

These contain anything between 5 and 20 per cent unsaturated fat. Oily fish such as herring, salmon, mullet, trout and sardine contain approximately 5 per cent fat, while those that contain up to 15 per cent fish oil include fish that are associated with the 'oily' description, such as salmon, pilchard, herring, sardine, whitebait and tuna. Oily fish contain the fat-soluble vitamins A and D. Small oily fish, such as sardine and whitebait, are often eaten whole or semi-whole so the bone are a source of calcium. This also applies to tinned oily fish such as salmon, pilchard or tuna.

The flesh of oily fish comprises muscle fibres held together by connective tissue; this muscle contains protein and iodine. The high fat and/or oil content of these oily fish make them less palatable than white fish. Since oil fish have a full flavour and may be purchased either fresh or tinned, they are frequently added to create flavour in both hot and cold dishes. Oily fish should never be fried because of their high fat content - they are usually either poached or grilled.

Shellfish

These are a rich source of protein but are also high in cholesterol, and have very compact, firm flesh which causes it to become hard due to the protein coagulating. Undercooking, on the other hand, is equally undesirable as shellfish are likely to contain harmful bacteria that must be destroyed by the application of heat.

If shellfish have come into contact with contaminated water they become contaminated themselves and unfit for human consumption. For this reason, shellfish should only be bought from a well known supplier.

EGGS

Hens' eggs are a useful commodity because of their availability and nutritional value:

- the shell is porous and made from calcium carbonate,
- the yolk is held in suspension by a white, fibrous protein membrane,
- the yolk contains protein, fat (high cholesterol), iron and vitamins A, D, B (Riboflavin) and calcium
- the white albumen is thick around the yolk and is composed of water and protein.

DAIRY PRODUCTS

Milk

Milk is almost a complete food. Cows' milk contains proteins, fats, carbohydrates, minerals and vitamins, but has no fibre content. It is an emulsion - 87 per cent water with minute droplets of fat suspended throughout. There are 22.2 g (3.8 per cent) of butter fat in each pint of milk. It is this butter fat that gives milk its creamy taste and that is used in the manufacture of butter and cream.

The fat content of milk can be reduced to 0.1 per cent producing skimmed milk. Such milk will contain all the other useful nutrients, but not the saturated fats. Semi-skimmed milk is also produced, which contain 2 per cent fat.

Lactose is the sugar found in milk. This sugar produces lactic acid when it comes into contact with lactic acid bacteria, causing milk to go sour.

The minerals found in milk are calcium (in an easily absorbed state), phosphorus, potassium, sodium and chloride. Traces of iron, magnesium and sulphate are also present. Milk provides one of the main ources

of calcium in the diet, the other source being cheese.

Milk is an important source of vitamin B_2 and also contains vitamins A and B and smaller quantities of vitamins C and D. Vitamins A and D are found in the fat content of milk, and so they are only trace nutrients in skimmed milk. Small quantities of vitamins B and C are found in the water content of the milk and so will remain in skimmed milk.

Table 3.4 Composition of cows' milk per 100 g

Nutrient	Pasteurized	Semi-Skimmed	skimmed
Protein	3.3	3.4	3.4
Fat	3.8	2.0	0.1
Carbohydrate (Lactose)	4.7	5.0	5.0
Minerals	0.6	0.5	0.5
Water	87.6	89.0	90.0

Milk products

Condensed milk

The milk for condensed milk is homogenized and then heated to 80C for 15 minutes. Sugar is usually added and the mixture is boiled to evaporate the water content until it is two and a half times more concentrated then whole milk. It is then sealed in tins.

Evaporated milk

Is unsweetened milk, and is twice as concentrated as whole milk. It is sterilized in tins at 115.5°C for 20 minutes.

Yoghurt

Yoghurt when prepared commercially contains all the nutrients of the milk from which it was made. Usually, it is manufactured from low-fat milk with added skimmed milk powder so that the finished yoghurt is lower in fat and higher in protein than whole fresh milk. Yoghurt can be obtained in a large variety of flavours: some fruit and sweetened and some especially prepared as low fat or low sugar varieties. It is therefore essential to read the nutritional analysis of each yoghurt purchased to identify its value.

Ice cream

Ice cream contains protein, calcium and B vitamins, sugar, fat and water. It is produced from milk and cream products with sugar flavourings, colourants and permitted stabilizers and emulsifiers added. Dairy ice cream is produced from milk and milk fat. Other ice cream may be made from skimmed milk and vegetable fats.

Cream

Cream is produced by either skimming off the fatty layer which comes to the top of the milk, or by using a mechanical separator. It is treated in similar ways to normal milk using pasteurization, homogenization, sterilization and ultra-heat treatment.

Nutritional value of cheese

The nutritional value of cheese varies according to the type of milk used. However, cheese is highly concentrated food that provides protein, fat, minerals and vitamins.

Most cheeses are high in saturated fats. It therefore follows that if less cheese is used in meals, the amount of saturated fat is reduced. The stronger cheeses, such as mature Cheddar, can be used in small quantities to produce the required flavour without adding too much saturated fat to the meal.

Table 3.5 Nutritional value of cheese (per cent)

	Cheddar	Cottage cheese	Edam
Protein	26.0%	13.6.%	24.0%
Fat	33.5%	4.0%	23.0%
Carbohydrate	–	1.4%	–
Minerals	3.4%	1.4%	3.4%
Water	37.0%	78.8%	44.0%

Table 3.6 Fat content of cheese

Type of cheese	Fat (%)
Cream cheese	47
Stilton	40
Cheddar	33
Brie	27
Camembert	27
Edam	25
Processed	25
Low-fat Cheddar	17
Cottage cheese	4

VEGETABLES

Numerous varieties of vegetables are grown in the UK and many more are imported, so there should be no cause for repetitive use of vegetables on menus. The same vegetable can be served in many different ways.

If vegetables are to be served cooked there will be inevitably be a loss of nutrients. Some nutrients are destroyed by heat, for example, vitamin C, and some are water soluble, for example, vitamins A and D. Tuber vegetables such as potatoes contain starch, fibre and vitamin C. Green leaf vegetables are rich in vitamin C and minerals – mainly iron and calcium. Root vegetables provide starch and fibre and carrots contain vitamin A. Legum vegetables provide protein, fibre and vitamins B and C. Owing to the fibrous and starch content of many vegetables, they are frequently used as part of a meal to provide 'bulk' and to provide a sustaining element to a meal. Potatoes and other vegetables high in starch or fibre are used in this way.

Flavour

The flavour properties of vegetables are exploited by chefs during the preparation of the majority of savoury dishes. Each vegetable has its own unique flavour, for example, it is easy to distinguish between the flavour of onion and garlic and leek even though all these are categorized as 'bulb' vegetables. The flavours often mingle together to produce a medley of flavours. This technique is used in such dishes as ratatouille. When storing vegetables, care should be taken to prevent strong flavoured vegetables from contaminating delicately flavoured vegetables, for example, onions should not be stored next to asparagus.

Texture

If vegetables are served raw in salads, they will provide a crisp texture to a meal. Some vegetables however still provide this crispness even if they have been subjected to heat in the form of quick cooking using the deep fry, stir fry or saute´ methods of cookery. Even boiled vegetables should have a slight crispness (al denté) to them when served if they are cooked for the minimum number of minutes.

Colour

One of the most useful properties of vegetables is their enormous variety of colour. Even the commonly known cabbage may vary from deep purple through to a very pale green. The selection of a particular variety is frequently determined by its colour rather than its flavour or nutritional value. The colour of vegetables is exploited when preparing such dishes as vegetable terrine, flans and salads where contrasting colours are essential to create an attractive dish.

Garnishes

Many types of vegetables may be cut or moulded to produce attractive garnishes for dishes. These may be cooked or raw and include such items as straw potatoes for a roast chicken garnish; deep fried onion rings used as a steak garnish; turned barrel shaped root vegetables for a ragout garnish; and julienne root vegetables as a consommé garnish. Raw vegetables can be cut to produce rosettes, fans, strips and other shapes that are used to garnish food.

Salads and hors d'oeuvre

With the growth in popularity of salads in general, vegetables are increasingly being incorporated onto salads of many types. They are usually used raw, but a few are blanched first, for example, peppers, where blanching takes away their bitter flavour, without reducing the crispness. Another example is the blanching of tomatoes to enable the skins to be removed more easily. Raw vegetables lose some of their nutrients and crispness if prepared too far in advance of their service time.

Safety

Soil is likely to contain the harmful bacteria *Clostridium perfringens*, which means that vegetables should be prepared carefully in a separate area of the kitchen away from other foods to reduce the risk of cross-contamination. Mushrooms especially cultivated to be eaten are obviously safe; however, any uncertainty about the edibility of fungi should result in the fungi being discarded immediately. If potatoes are exposed to sunlight during growth, they will produce a poisonous substance in their flesh, which can be identified by a green colour and a bitter taste. Any such potatoes should be discarded.

FRUIT

Fruits should be used when they are just ripe. Under-ripe fruit may prove difficult to digest, especially if eaten raw. Over-ripe fruit will quickly begin to deteriorate due to bacteria or moulds which will alter the fruits' flavour and upset the digestion.

There is a large amount of water in fruit – up to 85 per cent in some cases, such as melon, and this gives fruit its refreshing quality. When eaten raw, fruit will provide the body with vitamin C and roughage in the form of cellulose from the fruit structure; when cooked this vitamin C is largely destroyed by the heat and lost in the liquid, which may be either water or syrup. Fruits contain a good supply of fruit acids, which increase as the fruit ripens. These acids, particularly pectin, help in the setting of jams.

Numerous tropical fruits are now becoming increasingly popular, many of which are eaten raw – examples of such imported fruits are kiwi fruit, mango, pawpaw, pomegranate and star fruit. Imported fruits result in a large variety of fruits available all the year round.

NUTS

Nuts are the edible seeds produced from a tree or plant. Generally, nuts are not eaten in very large quantities, although with the increase of vegetarian and wholefood natural diets, their consumption is on the increase. This is due mainly to their nutritional content and versatility: nuts contain fat in the form of oils and also protein which makes them invaluable to the vegetarian.

FLOUR AND CEREALS

Wheat is produced in the UK and imported from Canada. The Canadian wheat is of a harder type and has a higher proportion of protein (gluten), which is used to produce the strong flour used for bread making.

Flour production

The milling process may be summarized as follows:

1. The grains of wheat are cleaned by sieving to remove small seeds and dust particles.
2. The clean, dry grains are passed between rollers which crack and remove the outer layer.
3. The removed bran and aleuron layer is returned to the rollers to remove the remaining endosperm particles.
4. The endosperm is crushed to produce semolina.
5. The semolina is rolled between smooth rollers to reduce it to a fine, white powder thus producing flour.

Table 3.7 Types and characteristics of flour

Flour type	Characteristics
Flour type	High in gluten, made from Canadian wheat
Soft white	73% extraction rate, made from ground endosperm
Self-raising	Soft, white flour with raising agent incorporated
Wholemeal	100% extraction rate, high in vitamin B, iron, calcium, fat and cellulose
Wheatmeal	80% extraction rate Brown flour
Stoneground	Brown flour produced by stone grinding to preserve its flavour, high in vitamin B content
Wheatgerm	Mix of strong white flour and cooked wheatgerm, malt flavoured
Starch-reduced	Starch level reduced by 'washing', useful for low-starch diets
Gluten-free	Starch is 'washed' out and the starchy liquid is then dehydrated to produce a gluten-free flour, soya flour added to increase protein content

The milling process can be modified to produce more or less flour from the wheat. The percentage of flour obtained is called the extraction rate of flow. For example, wholemeal flour has an extraction rate of 100 per cent, which means that the whole grain has been converted into flour. As the extraction rate is reduced, the proportion of the bran in the flour decreases, so producing a light coloured flour. White flour has an extraction rate of 70 per cent and is produced from the crushed endosperm.

In the UK, the nutritional value of flour is increased by the addition of thiamin, nicotine acid, iron and calcium. However, 100 per cent wholemeal flour does not need any additional nutrients.

PULSES

Pulses contain protein, B vitamins, iron, potassium and dietary fibre. Soya beans contain protein made up of all the essential amino acids, it is therefore a complete protein. Most other pulses contain protein that though incomplete in itself, can usefully combine with proteins formed in other grains, seeds, nuts or milk. they are therefore a very useful source of protein to groups of people who do not eat animal protein.

Dried beans need to be soaked for up 12 hours before gentle simmering, stewing or braising takes place. In order to add variety as well as nutrients to the diet a good variety of pulses should be used. They are readily available, easily stored and cooked and offer varying contrasts of colour and shape.

Popular pulses

Aduki beans

Small, dark-red or brown beans with a nutty flavour. They may be used in soups, savoury or sweet dishes, and need a soaking time of 4 hours followed by a cooking time of 1 hour. They have a high protein content.

Table 3.8 Nutritional content of different types of bread

Bread type	Protein	Fat	Carbohydrate	Calcium	Iron	Vit. B	Vit. B	Niacin
	(g)	(g)	(g)	(mg)	(mg)	(mg)	(mg)	(mg)
Wholemeal	9.0	2.2	45.0	100.0	2.5	0.25	0.05	3.0
White	7.8	1.7	50.0	100.0	1.7	0.20	0.03	1.4

Blackeyed beans

Small, cream-coloured beans with a black and yellow 'eye'; they have a smooth creamy flavour and may be used in soups, casseroles and burgers. They need to be soaked for 12 hours and then boiled for 15 minutes followed by simmering for 1 hour. They have a high protein folic acid content and vitamin B content.

Black kidney beans

Large, black-skinned kidney-shaped beans. They are tender and slightly sweet to the taste. They may be used in soups, salads, casseroles and bakes. These beans require 12 hours soaking followed by 15 minutes boiling, and then $1^1/_2$ hours simmering. They have a good protein and folic acid content. To ensure that these beans are safe to eat, they must be boiled rapidly for the first 15 minutes of the cooking time to destroy a toxin found in the outer skin.

Butter beans

These are large, creamy-white beans. They have a floury texture and a mild but distinctive flavour. They are useful in salads, savoury dishes and in soups. Soaking needs to be for 12 hours and simmering for $1^1/_2$ hours. They have a high sodium and very high potassium content. Carribbiani beans and Lima beans are smaller varieties of the popular butter beans.

Chickpeas

These are hazelnut-shaped, golden brown beans that have a distinctive nutty flavour. They may be used in salads and casseroles after soaking for 12 hours and simmering for 3 hours. They have a high vitamin A, vitamin C, iron, calcium and magnesium content.

Haricot beans

There are two varieties: the small, white type and the green flageolet beans. The small, white Boston beans are the type used to produce the well-known baked beans. The larger white haricot beans are used in casseroles and bakes. They require 12 hours soaking followed by $1^1/_2$ hours simmering. Haricot beans are high in dietary fibre and are a useful source of protein, calcium, magnesium and potassium.

Lentils

These may be obtained as whole or split, and either the red variety or the larger green variety, or the Indian brown lentil. Lentils are used to prepare soups,bakes, burgers and stews. They require no soaking and only 30 minutes simmering to produce a soft purée. Lentils provide a good source of protein, iron and zinc to the diet.

Marrowfat peas

These are large peas with a blueish tinge. These are the peas that are used to create the well known 'mushy peas' dish, and may also be used to make soup. They require soaking for 12 hours followed by 1 hour's simmering time. They have a very high vitamin A content, as well as a good protein and dietary fibre content.

Mung beans

These are small, dull green-coloured beans that may be used whole or split. When sprouted, they are commonly called beansprouts. They need 8 hours soaking and 45 minutes simmering time while cooking. They are used in soups and casseroles, and are a good source of protein, and dietary fibre, and have a high iron content.

Pinto beans

These beans are speckled brown beans that turn pink when cooked. They need 12 hours soaking and $1^1/_2$ hours simmering. They are used in salads, soups and other savoury dishes, and have a very high level of carbohydrate, vitamin B and phosphorous content.

Red kidney beans

These are shiny red beans, and they require vigorous boiling to destroy a toxin found on their skins. The beans are used in the Mexican dish chilli con carne. The beans require soaking for 12 hours followed by 15 minutes vigorous boiling and then $1^1/_4$hours simmering. They have a good protein content.

Split peas

These may be whole or split and may be green or yellow in colour. They lose their shape in cooking, like lentils, and so are used for soups, casseroles and burgers. Yellow split peas are used to make the traditional pease pudding that is served hot with boiled beans. They require no soaking and may be cooked by simmering for 30 minutes. They have a good carbohydrate content and a very high level of zinc.

Soya bean

These small beans are grown extensively in North America and Canada because of their high nutritional value. They contain a higher proportion of complete protein, fat, B vitamins, iron, calcium, magnesium and phosphorous than any of the previously mentioned pulses, and very signficant quantities of carbohydrate, folic acid and potassium. They do not contain any dietary fibre, vitamin B_{12}, vitamin C, vitamin D, vitamin E or zinc. To cook soya beans, they require soaking for 12 hours followed by one hour vigorously boiling and a further 3 hours simmering. They are versatile and may be used successfully in stews, braises or salads. Soya beans are made into soya flour, soya milk and soya oil.

Bean sprouts

These are grown from fresh beans (mung beans) and are used extensively for stir fry dishes. They should be thoroughly rinsed before being cooked.

Soya flour

This is twice as rich in protein as wheat flour, and is used in cakes and to bind sauces.

Tofu

This is soya cheese and may be used in a variety of ways.

Miso

This is made from a fermented mix of rice and soya, and used on broths, soups or as a garnish for fish with vegetables.

Soy sauce

This sauce is manufactured from soya beans, wheat flour and salt, and has a similar use to meat extract and will mature with age. It may be used as part of a marinade, as a seasoning agent or as a table condiment.

Textured vegetable protein (TVP)

This is manufactured from defatted soya flour and results in a dried product with an 8 per cent content with a chewy texture. It may be processed to resemble meats and comes in chunks, flakes and granules. TVP may also be produced from the soya bean protein which produce a product that is more like real meat in texture. Because TVP is in dried form it is easily stored. TVP is a meat extender when it is used to extend real meat dishes or as a meat substitute for vegans.

ACTIVITIES

- Contact your local hospital, school or old people's home and find out which milk products are normally used for the preparation of (1) food dishes, and (2) hot beverages. Discuss the reasons why these products are being used. Are there any milk products currently not in use that you think could be usefully introduced in the chosen welfare establishment?
- Make a collection of six recipes that use pulses in place of animal proteins and compile a chart to show the main nutrients provided by each recipe. Which of your recipes will provide the most well balanced dish for a Vegan?

CHAPTER 27 The food industry

THE AGRICULTURAL INDUSTRY IN THE UK

Since the end of the Second World War, there have been many advances in the farming industry and in the farmer's ability to produce good crops and breed animals. Farmers now achieve higher crop yields and better livestock products with fewer diseases from pests. The benefits to the consumer are a plentiful supply of good quality foods at a price that most can afford.

The role of the MAFF

In 1955, the Ministry of Agricultural Fisheries and Food (MAFF) was created. This Ministry is responsible for carrying out government policies on agriculture, horticulture and fisheries in England, and for policy on a wide range of trade and food issues for the whole of the UK.

The agricultural industry has a growing need for technical advice and assistance. This is provided by the Agricultural Development and Advisory Service (ADAS).

Today's agricultural industry consists of approximately 250,000 businesses that occupy nearly 80 per cent of the land in the UK. MAFF is closely concerned with the European Economic Community (EEC) negotiations to create a single European market in 1992, and with the revision of the food laws. MAFF administers the livestock subsidy schemes and grants given for capital investment, schemes to encourage alternative uses of land, and countryside conservation schemes. Another aspect of MAFF's work is that of sea fisheries management, marine conservation, and checking for signs of illegal contamination and waste disposal around the coasts.

MAFF has to ensure that there are adequate supplies of foods, and its food divisions make sure that the foods are wholesome and free from contamination. It is always looking to identify problems in the manufacture and processing of foods. This aspect of its work is co-ordinated by the Steering Group on Food Surveillance. Additives, labelling and nutritional contents are all controlled by MAFF. Furthermore, research into the diseases and pests that attack plants and animals, is all controlled by MAFF, who also run the State Veterinary Service.

Fertilizers

Fertilizers are used by farmers to supplement naturally occurring minerals found in the soil. The most important plant food minerals are nitrogen, phosphorous and potassium. Phosphorous and potassium are present in the ground, but there is frequently a deficiency of nitrogen. Fertilizers are manufactured to rectify this balance of plant food. They may be manufactured to suit specific plant requirements, thus increasing the crop yield.

Agro chemicals

Chemicals are used by farmers to protect their crops from insects, diseases and weeds. These chemicals are known as pesticides, and fall into the following categories:

- herbicides control the growth of weeds so that the crops can grow and flourish without being smothered by weeds,
- fungicides control fungal growths on the crops, both during and after the crop has been harvested,
- insecticides control insect pests that will either eat or spread disease to the crop, and they are used to combat the threat of beetles and weevils on wheat and barley grain during storage,
- rodenticides control rodents that may damage growing and stored food crops.

Pesticides are strictly controlled and are not used unless they are considered to be safe to the farmer, the consumer and the environment. Their use has undoubtedly increased food production to its present healthy levels but, even so, there is still a loss of crops due to insects, rodents and weeds.

Food surpluses and the EEC

The EEC Common Agricultural Policy (CAP) has created a strong supply of food throughout the

European Community. The excess of food produced above the amount of food required and exported results in a surplus of food (commonly referred to as food mountains). Butter, wheat, beef and milk powder are some of the foods of which the EEC has a surplus. To reduce surpluses, the EEC has set limits on the production of foods - by setting quotas, and by reducing the price paid to farmers for these types of foods.

Organic farming

Organic farming is the production of foods without the use of chemical fertilizers and pesticides, and without the use of livestock growth regulators and food additives.

Organic farming advocates the use of traditional methods of crop rotation, mixed farming to increase natural manuring, organic materials that have not necessarily been produced on the farm such as mushroom compost and seaweed liquids, short fallows to control weeds, and mineral fertilizers such as rock phosphate and potash, calcified seaweed, fishmeal and bonemeal to enrich the soil. Crop rotation, mechanical cultivation and the use of plant-derived insecticides are used to control weeds and pests.

Organic farming has increased recently in the UK because of the farmer's respect for the environment, the demand for healthy, wholesome foods without additives and chemicals, and because of over production caused by some nonorganic methods, causing farmers to think again about their industry.

Organic produce is often thought to be healthier than conventionally produced foods. However, it is very difficult to prove these theories scientifically. It is known that some pesticides will remain in the soil, and these are strictly controlled so that their effects are not detrimental to the foods and the environment.

The difference between organic and conventional foods is very difficult to detect. However, organic foods are gaining in popularity in spite of the fact that they are generally considerably more expensive than conventional foods, and are not so widely available.

Organic sheep farming

Some farmers have developed a range of dairy sheep products that are becoming increasingly popular, since they are produced using the organic farming philosophy. Sheep are able to live off grass and hay, and they can manure their own pastures so there is no need for artificial fertilizers.

The dairy ewe provides meat, milk and wool. The flocks comprise spring and autumn lambing groups of ewes so that milk is available throughout the year. Once the lambs have been reared, the ewes can be milked for a further 2–3 months.

Sheep's milk is comparable to cow's and goat's milk, and is an alternative for persons who are allergic to cow's milk. It has a high fat content (5.79 per cent compared with 3.5 per cent in cows and 3.9 per cent goats milk), and has a slightly sweet, rich taste. It is sold in frozen 1 pint packs that have a 4 month deep freeze life.

Not only is sheep's milk available in its natural form, but it is also used to produce a variety of yoghurts and cheeses, for example:

- Duddleswell is a full-fat, hard cheese which has been matured for 2 months. It is a smooth, creamy-coloured cheese with a shelf life of 4 months.
- Sussex Slipcote is a full-fat, soft cheese with a distinctive flavour. Available plain, with herbs and garlic or cracked peppercorns it has a shelf life of 21 days at 5°C.
- Halloumi is a full-fat, medium hard cheese with a touch of mint. Rindless and folded in the Mediterranean tradition ,it has a shelf life of 3 months at 5°C.
- Feta is a creamy-soft, white cheese with a tangy taste. Its shelf life is 3 months at 5°C.

THE PRESERVATION OF FOODS

Many fresh foods are treated to prolong their 'safe life', that is, to keep the food edible and safe, while inhibiting the growth of the harmful bacteria that cause food spoilage. However no preservation method will create a food that will have an endless, safe life. Most preservation methods have some effect upon the food itself, either a change in its taste, such as the caramel-like taste of sterilized milk, or a change in its texture, such as in a dried apricot. Changes in colour and nutritional content may be less easily detected. Some foods only become palatable through the preservation process, such as crab apples used in crab apple jelly. The presentation of food by heat, refrigeration, dehydration, preservatives and vacuum packing is covered elsewhere in this book.

Preserving by acids

Picked vegetables are an example of foods that are preserved by the use of vinegars (acetic acid). Some sauces such as salad cream and salad dressings also act in the same way. Fruit drinks contain citric acid, which keeps them free from spoilage bacteria and moulds. Carbonated drinks that are fizzy contain carbonic acid, which also keeps harmful organisms at bay.

FOOD ADDITIVES

A large majority of manufactured foods contain additives of some kind. There are over 3,000 permitted food additives, which are controlled by regulations drawn up by the Ministry of Agriculture Fisheries and Food (MAFF) and legalized by the Food Act 1984. When an additive has been approved by the European Community, it has an 'E' in front of its number. Additives are grouped together and numbered according to their function.

Preservatives

These are used in foods to prolong their keeping quality or shelf life. Preservatives are used to prevent the growth of bacteria, yeasts and moulds which cause fresh foods to decay and spoil. Examples include:

E200 (sorbic acid) in soft drinks, fruit yoghurt and processed cheese.
E203 (calcium sorbate) in frozen pizza and flour confectionery.
E219 (methyl 4-hydrozybenzoate) in sodium salt, beer, jam, salad cream, soft drinks, fruit pulp, fruit-based pie fillings, marinated herring and mackerel.
E227 (calcium hydrogen sulphate) in dried fruit, dehydrated vegetables, fruit juice, fruit syrups, sausages, fruit-based dairy desserts, cider, beer, wine, prepared, vacuum-sealed raw potatoes and biscuits.
E232 (sodium biphenyl-2-yl oxide) in surface treatment of citrus fruit.
E234 (nisin) in cheese and clotted cream.
E239 (hexamine) in marinated herring and marinated mackerel.
E252 (potassium nitrate) in bacon, ham, cured meats, corned beef and cheese.
E283 (potassium propionate) in bread, flour confectionery and Christmas pudding.

Antioxidants

Antioxidants are used to slow down the normal process of food decay that occurs when food is exposed to oxygen in the air. A well-known example is the rancid flavour of fat that has been allowed to become stale. Antioxidants stop oils and fats and fat soluble vitamins from combining with oxygen and going rancid. Examples include:

E300 (L-ascorbic acid) in fruit drinks, to improve flour and bread dough.
E304 (6-0-palmitoyl-L-ascorbic acid) in Scotch eggs.
E307 (synthetic alpha-tocopherol) in cereal-based baby foods.
E310 (propyl gallate) in vegetable oils and chewing gum.
E320 (butylated hydroxyanisole (BHA)) in beef stock cubes and cheese spread.
E321 (butylated hydroxyanisole (BHT)) in chewing gum.
E322 (lecithins) in low-fat spreads and emulsifier in chocolate.

Emulsifiers and stabilizers

Emulsifiers mix together ingredients such as water and oil that would normally separate. Stabilizers prevent them from separating out again. They also add to the smooth texture of foods and help to keep bakery products fresh for longer periods of time. Examples include:

E400 (alganic acid) in ice cream and soft cheese.
E401 (sodium alginate) in cake mixes.
E405 (propane-1-2-diol alginate) in salad dressings and cottage cheese.
E405 (agar) in ice cream.
E407 (carrageenan) in quick setting jelly mixes and milk shakes.
E310 (locust bean gum) in salad cream.
E312 (guar gum) in packet soups and meringue mixes.
E413 (tragacanth) in salad dressings and processed cheese.
E414 (gum arabic) in confectionery.
E415 (xanthan gum) in sweet pickle and coleslaw.
E460 (microcrystalline cellulose) in high fibre bread and grated cheese.
E461 (methylcellulose) in low-fat spreads.
E470 (sodium, potassium and calcium salts of fatty acids) in cake mixes.
E474 (sucroglycerides) in edible ices.
E477 (propane-1-2-diol esters of fatty acids) in instant desserts.
E481 (sodium stearoyl-2-lactylate) in bread, cakes and biscuits
E482 (calcium stearoyl-2-lactylate) in gravy granules.

Colours

Colours are added to foods to enhance their natural colour and to compensate for colour that has been lost during the processing process. Examples include:

E100 (curcumin) in flour, confectionery and margarine.
E101 (riboflavin) in sauces.
E102 (tartrazine) in soft drinks.
E110 (sunset yellow FCF) in biscuits.
E120 (cochineal) in alcoholic drinks.
E122 (carmoisine) in jams and preserves.
E127 (erythrosine) in glace cherries.
E142 (green S) in pastilles.
E150 (caramel) in beer, soft drinks, sauces and gravy browning.
E153 (carbon black) in liquorice.
E161(B) (annatto, bixin, norbixin) in crisps.
E162 (beetroot red) in ice cream and liquorice.
E163 (anthocuanins) in yoghurt.
E171 (titanium dioxide) in sweets.
E175 (gold) in cake decorations.
E214 (ponceau 4R) in dessert mixes.

Sweeteners

There are two kinds of sweeteners, those that have intense sweetening properties, and are therefore used in very low quantities; and those known as bulk sweeteners that have the same sweetening properties as sugar.

Intense sweeteners

Acelsulfame potassium in canned foods, soft drinks and table top sweeteners.
Aspartame in soft drinks, dessert mixes, drink mixes and sweetening tablets.
Calcium saccharin in soft drinks, cider, sweetening tablets and table top sweeteners.
Sodium saccharin.
Saccharin.
Thaumatin in table top sweeteners and yoghurt.

Bulk sweeteners

Hydrogenated glucoce syrup.
Isomalt.
E421 (mannitol) in sugar-free confectionery.
E420 (sorbitol) in sorbitol syrup, sugar-free confectionery and jams for diabetics.
Xylitol in sugar-free chewing gum.

Unclassified additives

There are many more additives than the ones listed above and these are added to foods for a variety reasons, and include the following:

E260 (acetic acid and 262 sodium acetate) Acidity regulators used in pickles, salad cream and bread. They contribute to the flavour and protect against mould growth.
E263 (calcium acetate) A firming agent also providing calcium, which is useful in quick set jelly mix.
E270 (lactic acid) Protects against mould growth and used in salad dressings and soft margarine
E290 (carbon dioxide) A carbonating agent for fizzy drinks.
E325 (sodium lactate) in jams, preserves, sweets, flour and confectionery.
E333 (monocalcium citrate, acid flavouring, buffers, sequestrants, emulsifying salts) in soft drinks, jams, preserves, sweets, UHT cream, processed cheese, canned fruit, dessert mixes and ice cream.
E337 (potasium sodium L-(+) tartarate, acid flavourings, buffers emulsifying salts, sequestrants) in soft drinks, biscuit creams, sweets, jams, dessert mixes and processed cheese.
E341 (calcium tetrehyosphate diorthophosphate, calcium hydrogen orthophosphate, tricalcium diorthophosphate) as a firming agent, anti-cake agent, raising agent, cake mixes, baking powder and dessert mixes.
E621 (sodium hydrogen L-glutamate (Modosodium glutamate, NSG)) is naturally found in Japanese seaweed called seatango. It is commercially prepared from sugar beet pulp and wheat gluten. It is a flavoured enhancer for protein foods and used in packet soups, chilli sauce, pork pies and pork sausages.

Problems

Certain additives have been identified as the cause of behavioural problems, or the cause of aggravated medical symptoms, such as hyperactivity in children and medical dangers for asthmatics or aspirin-sensitive people. The hyperactive children support group formed in 1977 lists the following additives as those that are recommended to avoid for hyperactive children.

E102 (tartrazine)
E104 (quinoline yellow)
E107 (yellow 2G)
E123 (amaranth)
E124 (ponceau)
E127 (erythrosine)
E128 (red 2G)
E132 (indigo carmine)
E133 (brilliant blue FCF)
E150 (caramel)
E151 (black PN)
E154 (brown FK)
E110 (sunset yellow)
E120 (cochineal)
E122 (carmoisine)
E155 (brown HT)
E210 (benzoic acid)
E211 (sodium benzoate)
E220 (sulphur dioxide)
E250 (sodium nitrate)
E251 (sodium nitrate)
E320 (butylated hydroxyanisole)
E321 (butylated hydroxytoleune)

There are some additives that are not suitable for the very young: E212, E213, E214, E215, E216, E217, E218, E219, E310, E311, E312, E621, E622, E623, E627, E631, E635.

It is advised that those persons that are asthmatics or who are aspirin sensitive should avoid all the additives listed above.

ACTIVITIES

- Find out the symbol used by organic farm producers to signify that their produce is organically grown.
- Conduct a survey to find out the difference in price, quality, and availability of organic produce in your area.
- Compile a list of snack foods that contain the additives listed as dangerous for hyperactive children. Now try and find acceptable alternatives that these children could be offered in their place. (You will have to look in shops and stores to find out the foods with additives in them, the nutritional labels will show the additives).
- Can you list some additives that are found in foods that have not been in the previous pages? Why are these added to foods?
- Think of the lists and types of additives previously described. Can you divide these into desirable additives and list the reasons why these are necessary, and then list the additives that we could manage safely without. For example, we do not have to eat brightly coloured sweets, do we?

CHAPTER 28 Healthy eating

The Nutritional Advisory Committee on Nutrition Education (NACNE) produced a report in September 1983, and the Diet and Cardiovascular Disease Report by the Committee on Medical Aspects of Food (COMA) was produced in 1984. These two reports came to similar and quite striking conclusions on the consumption of fats, salt, sugar and fibre rich foods.

FATS

The consumption of saturated fats should be decreased by as much as 30 per cent, with the exception of infants and children under the age of 15 years and those persons who already have a very low saturated fat intake. This reduction in saturated fats will have a significant effect upon the intake of cholesterol. It is necessary to reduce the amount of cholesterol in the blood stream because it will cause the arteries to become blocked with fatty deposits, and damaged arteries can lead to heart attacks and strokes. Cholesterol is essential for the functioning of normal cells, but only a very small amount is required. The caterer can take the following steps to reduce the amount of saturated fats in foods:

- use polyunsaturated fats and oils,
- serve chicken, turkey and fish more frequently than red meats,
- use leaner cuts of meat and trim off visual fat from meat,
- use moist methods of cooking that do not require extra fat,
- use skimmed milk (fresh or dried)
- make low-fat sauces not by the roux method,
- serve low-fat cheeses (if a hard cheese is necessary, use a strong flavoured variety so that a smaller quantity is required),
- use less cream, prepare pastries and cakes with a low fat content.

SALT

The consumption of salt should be reduced by 3 g per day. High salt intakes are likely to produce high blood pressure, which can lead to heart attacks and strokes. To reduce salt in food, the caterer should:

- use a salt substitute or simply remove salt cellers completely,
- use less salt in the kitchen,
- increase the use of other natural flavourings, such as herbs and spices,
- check the salt content of commercially prepared products such as table sauces, meat and vegetable extracts, breakfast cereals, smoked fish and meats, sausages, processed cheeses and tinned meats.

SUGAR

The consumption of sugar should be reduced in order to reduce the risk of obesity and dental care. There is a risk of coronary heart disease in the overweight person. If young men are just 10 per cent overweight, they have a 30 per cent higher risk of heart disease than their slimmer colleagues. Overweight persons also run the risk of hypertension, increased plasma cholesterol and diabetes mellitus. To reduce the amount of sugar used, the caterer should:

- sugar substitutes, commonly known as sweeteners,
- use canned fruits in natural juice instead of syrup,
- use jams that have a sugar reduced content,
- serve breakfast cereals that have no sugar added,
- fresh fruit as an alternative to gateaux, margarines, etc,
- serve sugar-free baked beans and tomato ketchup,
- use natural yoghurt and add fresh chopped fruit,
- offer artificial sweeteners in place of cubed sugar for beverages.

FIBRE RICH FOODS

In the UK, the average adult consumes approximately 20g of fibre per day. It is recommended that this should be increased by 50 per cent to 30g per day. Dietary fibre has a positive effect upon the levels of blood cholesterol and glucose, it therefore results in a reduction of the risks of heart disease. The caterers can increase the amount of fibre in foods by:

- using some wholemeal flour in cooking,
- offering generous portions of vegetables, potatoes, rice, etc,
- making sure the customer has the choice of wholemeal bread if white bread offered.

ILLNESSES AND NUTRITION

Diabetus mellitus

The title is derived from Greek words meaning syphon and sweet, and explains an important feature of the disease - the passing of large amounts of sweet urine. The condition has been known for thousands of years, but only in the last hundred years have doctors known that damage to cells in the pancreas causes diabetes. These cells - called the islets of Langerhans - produce insulin which is the hormone that controls sugar in the body and allows it to be converted to energy. If sugar in the blood cannot be converted to energy, the amount builds up in the blood (high blood sugar) and is excreted into the urine. The lack of insulin means sugar cannot be converted to energy, and so the body breaks down fat and protein instead to produce energy. People with undiagnosed diabetes lose weight due to this breakdown of fat and protein. Using fats for a sugar source produces ketones, which can be detected in blood and urine (the smell of acetone or pear drops on the breath is often noticeable). If a large amount of ketone is present, the patient may go into a coma.

Diabetes is usually divided into insulin dependent or type I, and noninsulin dependent, or type II. Insulin dependent diabetics may be treated by diet alone, or by diet and tablets, which are not insulin but which have the chemical effect of lowering blood sugar. Diabetics are advised to have a healthy diet similar to everyone else, but they should take care to control their intake of refined sugars which lead to a rise in blood sugar. In practice, the means avoiding sugar and sweets, but eating bread, potatoes, fruit and cereals. They should also try to control their weight. Diabetics should not be overweight and so must cut back on fatty foods, for example butter and chips. Diabetics are also advised to increase fibre-containing foods, for example, wholemeal bread, wholegrain cereals, fruit and vegetables. Diabetics should aim to live a normal life, but they do need to be aware of what they eat and maintain a regular routine.

Hypertension

Raised blood pressure or hypertension is difficult to define. In healthy young adults systolic blood pressure is about 120 mm mercury (M3). Systolic refers to when the heart is contracting and pushing blood out into the blood vessels. The diastolic pressure is about 80 mm Hg. The diastolic phase is when the heart muscle is relaxing. So normal blood pressure is usually expressed as 120/80. If the diastolic pressure goes up, it suggests hypertension: 100mm M3 is moderate, whereas 120mm is severe. Blood pressure tends to rise with age, so 160/90 for an old man is not a problem, but would be if the person is young. To reduce hypertension, it is important to reduce obesity (being overweight) and to reduce the amount of salt in the diet.

Stroke or cerebrovascular accident (CVA)

'Stroke' is the popular name given to the interruption to the flow of blood to the brain. Patients may be left with speech or swallowing difficulties, and passing weakness in a limb to profound paralysis. Following a stroke, a patient may have to learn to feed him or herself, and food that is suitable for such people includes thick soups and custards, mashed potatoes, vegetables, ice cream and jellies.

Heart attack and coronary thrombosis

Narrowing of the blood vessels supplying the heart leads to angina, a severe pain following exercise, which may pass when the body is rested. A blockage of the blood vessels supplying the heart leads to a heart attack or coronary.

The narrowing of the arteries is caused by deposits of cholesterol in the vessels, like the furring up of pipes in hard water areas. It is recommended that people with high blood cholesterol try to reduce their dietary intake by eating less saturated fat in their diet, and by increasing the proportion of polyunsaturated fats. It is also important to achieve the correct body weight - obesity is another contributory factor in heart disease. People at risk are also advised to have plenty of natural fibre in their diet - oat and wheat products are beneficial - and to avoid highly salted foods. Smoking is another serious risk factor in heart disease and should be stopped.

Coronary heart disease (CHD)

Six people die every minute from CHD. It is the major cause of premature death (under 64 years) in men, and the second major cause of early death in women.

Death rates from CHD are higher in some parts of Britain than others. The rates are highest in N. Ireland and Scotland. Many different factors affect the likelihood of an individual developing heart disease. Family history, age and sex are risk factors which we cannot alter. However, there are risk factors which can be reduced by the individual. The main ones are not smoking, taking regular exercise (for 20 minutes, three times a week), and eating a low-fat, high-fibre diet.

There are many ways of modifying cooking methods and recipes to bring them into line with current recommendations on healthy eating, and some of these have already been mentioned.

Smoking is another major risk factor for heart disease. The caterer should ensure that a no smoking seating area is available for customers who choose not to smoke. Table signs are often available from the health promotion department of your local health authority or from Action on Smoking and Health (ASH).

The caterer can play a major part in educating the customer on healthier food choices - 20 per cent of all meals are consumed outside the home. The crucial factor is that the customer has a choice. The appearance on a menu of chips or roast potatoes is not a choice since both these items are high in fat.

A good example of giving the customer a healthy choice of foods can be found in the Heartbeat Award Scheme. This scheme is run nationally by the Health Education Authority and locally by Environmental Health Departments. Contact your local authority and check if they are operating the scheme in your area (see figure 3.5 opposite).

WEIGHT REDUCTION

Numerous people try in vain to lose some weight by going on a slimming diet. Many magazines and daily papers publish diet information all designed to encourage people to change their eating habits drastically in a bid to lose unwanted pounds. It is not only vanity that motivates a person to try to lose weight, though it does play an important part; it is a well-known fact that being overweight can cause permanent damage to the heart and be detrimental to one's general good health by causing high blood pressure. Most slimming clubs are designed to allow a variety of foods to be eaten while at the same time reducing the calorific value of the food eaten. Thus only low calorie foods are recommended. Any diet should include sufficient fibre to keep the body functioning regularly. At the same time, this fibre will help to reduce the feelings of hunger that make dieting so difficult to maintain.

The number of calories that the body requires to keep the body processes going varies from person to person. The base metabolic rate (BMR) is the number of calories the body need to keep functioning. If too many calories are eaten, then the surplus will be transferred as fatty deposits around the body and will result in an increase in weight. These fatty deposits must be used up if weight is to be lost. However, losing weight is not easy from a physiological point of view. The body stores fat for emergency supplies, and it takes a long time to transfer it back into usable body energy. Glycogen is also stored away by the body. It is a high energy fuel that can be released easily should the body be short of energy-giving food. This is the first source of food to be used up by the body when the calorific value of foods eaten is reduced. Since glycogen is stored with water in the body, weight may be lost by using the glycogen and excreting the water. This does not, however, reduce the layers of fatty deposits.

A reduction in foods eaten leads to a reduced BMR, and this reduces the amount of energy that the body needs. So, again, the fatty deposits will not be used up. What then is the answer?

Exercise over a period of several weeks combined with a controlled diet will have the desired effect of encouraging the body to use the stored fats, which will cause a weight reduction. However, this is a slow process and there is no quick or easy way to lose weight.

Heartbeat Award
CHECKLIST

Name of Catering Establishment: ______________________________

____

Type of Establishment: ______________________________

Name of EHO: ____________________ Date of Inspection: ____________________

Three criteria must be fulfilled to achieve a Heartbeat Award:

1. **GOOD STANDARDS OF HYGIENE**

- Premises must comply with the law ☐
- Good hygienic practices must be adopted ☐
- A minimum of 30% of food handling staff should be trained & qualified ☐

2. **PROVISION OF NON-SMOKING AREAS**

- A minimum of a third of the eating area must be clearly designated as non-smoking ☐

3. **PROVISION OF HEALTHY FOOD CHOICES**

- Where categories listed in Column A are offered then the healthier alternatives in Column B must be available and featured on the menu

Approximately one third of the total number of dishes in each section of the menu should fulfil the criteria in Column B.

COLUMN A	COLUMN B	
Meat, Poultry or Fish	A meat, poultry or fish dish prepared with a minimum of fat, served without a high fat sauce. All ingredients should have the visible fat removed during preparation.	☐
Vegetarian dishes	Any vegetarian dish prepared with a minimum of fat. This would not include dishes containing pastry, which were cooked by frying, or contained large amounts of a high fat cheese, e.g. cheddar, stilton.	☐
Vegetables/Salads	Some vegetables and salad to be available served without butter, creamy sauces or any oil based dressings.	☐
Rice/Pasta/Potatoes	Jacket potatoes, boiled potatoes, boiled rice, pasta, chappati or pitta bread served without added fat.	☐
Bread and Butter	Wholemeal alternatives must be available for all bread items. Low fat spread or polyunsaturated margarine as an alternative to butter.	☐
Desserts	Fresh fruit, fresh fruit salad, sorbet, low-fat yoghurt, fruit cobbler (fruit topped with scone) or other low fat, high fibre dish. (Pastry items would not be acceptable).	☐
Set meal/meal of the day	Low fat alternatives to high fat items should be offered, or an alternative 'healthy' set meal.	☐
Soft drinks/alcoholic drinks	Unsweetened fruit juices, mineral water, low alcohol or alcohol free options.	☐
Cream/whole milk for drinks	Semi-skimmed or skimmed milk.	☐
Cakes/Pastries	Low fat/sugar cakes, e.g. apple or carrot cake, currant bun (no icing), low fat sponges	

NOTE: If one or more of the food categories does not apply to a particular caterer, it may be disregarded.

If your establishment has failed to comply with any of the above criteria, a Heartbeat Award will not be given. However, if you feel you can fulfil all the criteria in the future, and would like to be reconsidered for the award, please contact your environmental health department.

This checklist is the national minimum standard required for a Heartbeat Award and higher standards may be applied locally.

LOOK AFTER YOUR HEART

3.5 Heartbeat Award checklist.

EFFECTS OF ALCOHOL UPON THE BODY

The consumption of alcoholic beverages is an acceptable custom in the Western world, and many religions accept its consumption in moderation. However, it is known that abuse of alcohol by over consumption can cause many dangerous changes in the body: alcohol passes from the mouth into the stomach where some is absorbed into the bloodstream. The rest (around 75 per cent) passes to the small intestine where it is then absorbed into the bloodstream. It is therefore true to say that, if the stomach contains food, this will slow down the effect of alcohol.

Blood carries alcohol around the body to the liver where it is gradually broken down at the rate of one unit per hour (one unit is equivalent to a half-pint of beer or 16 gill spirit. The liver changes the alcohol to acetaldehyde, a poisonous substance that quickly turns to acetic acid (vinegar). This then changes further to water and carbon dioxide, giving off heat in the form of calories during the process. Since the liver cannot store the calories derived from alcohol, it has to burn them up immediately. Other food is therefore stored as glycogen and then as fat; hence the expression 'alcohol makes you fat'.

It should be remembered that alcohol contains no vitamins, proteins or other nutrients. Thus a person who has consumed an excessive amount of alcohol over a long period of time may become fat and yet be under-nourished. Since alcohol tends to dull the appetite, such people become accustomed to alcohol instead of food and therefore may neglect to eat foods with vital nutrients. Even if food is eaten, it is known that alcohol reduces the body's ability to absorb vitamins C and D.

Alcohol is often referred to as an anti-depressant, used as a quick pick-up, or taken to give 'Dutch courage'. In fact, heavy intakes of alcohol make a person more anxious and tense even if they experience a calming effect while drinking.

Alcohol is a drug that is a depressant, not a stimulant. The reason why alcohol can cause people to become loud, noisy and uninhibited is because it depresses the higher centres of the brain first. The inhibition cells in the brain, which contain fat molecules, become more porous when affected by alcohol. This results in a disruption or slowing down of messages to and from each cell. Thus the powers of co-ordination, memory, thought and feeling are all slowed down after alcohol is consumed. Slurred speech and unco-ordinated movements can easily and quickly result.

The frontal lobes of the brain are the next areas to be affected by alcohol. These lobes are responsible for self-awareness, co-ordination, advanced planning and impulse control. It is these abilities that are quickly affected by alcohol.

The cerebral cortex is affected next, where complex reasoning, powers of calculation, perception, advanced skills and memory are affected. The regulatory powers of basic alertness, breathing and heart rate are also affected if a stage of alcoholic poisoning is reached, and the dangers that this will bring are obvious.

The consumption of alcohol

When a neat spirit is introduced into the stomach, the latter secretes mucous to protect its lining, which slows down the rate of alcohol absorption. At the same time, the pyloric sphincter (the valve at the base of the stomach used to regulate the flow of blood out of the stomach) closes, keeping the alcohol in the stomach. Only 20–25 per cent of the alcohol can be absorbed into the bloodstream, so the immediate effects of consuming 'neat' alcohol can be reduced.

However, the reverse is the case if a spirit is drunk with carbonated mineral waters or 'mixers' – the alcohol will be absorbed very easily and quickly into the bloodstream. Also, the pyloric sphincter will not close since the alcohol is diluted, so the latter will pass into the small intestine quickly, allowing the remaining 75–80 per cent to be absorbed. It therefore follows that the idea that drinking spirits with 'mixers' will reduce the effect of the alcohol is not valid in reality.

Generally, women are more quickly affected by alcohol than men, this is partly because, on average, they are lighter in weight than men and because of their different physical make up. Water accounts for approximately 60 per cent of a man's body weight, whereas in women the equivalent figure is only 50 per cent. One unit of alcohol is therefore more concentrated in a woman's body.

Women also tend to absorb alcohol more quickly than men and are liable to reach peak blood alcohol levels more quickly. This means that, in general, women get drunk faster, but at the same time they sober up more quickly. Women are also more liable to physical damage from alcohol abuse, particularly liver damage and some cancers, and during pregnancy the fetus can be harmed. The combination of alcohol and drugs - those prescribed by the doctor for medical reasons - can be very dangerous. Attempts should be made to ascertain the likely effects of alcohol if under medical care of any kind.

The hangover

The 'morning-after' feeling, commonly referred to as a hangover is symptomized by a raging thirst, splitting headache, a feeling of nausea, sensitivity to sudden noise and bright lights, and a general feeling of lethargy.

These symptoms are caused by the alcohol still in the body and by the effects on the body that the alcohol has caused. Dehydration is one of the major causes of these symptoms - it occurs because the alcohol has a diuretic effect, that is, it stimulates the body to expel urine. The result is a deficiency of water in the body when the hangover starts. A large quantity of water drunk after too much alcohol is a help, but it will not always be in time to prevent dehydration, because it takes time for the body to absorb the water into the blood stream.

Another contributory factor to a hangover is loss of real sleep. Excess alcohol tends to dull the brain thereby not allowing the body to rest naturally. A third cause of these unpleasant feelings is the presence of congeners (all the other items in the alcoholic drink except for the alcohol and water, for example, colour pigmentation, traces of metal from the brewing process, or traces from the oak maturing barrels). The EEC permits a total of 27 additives to be included in wines without their being declared on the labels. In general, dark, sweet beverages contain more congeners than lighter ones. Thus a vodka drinker may have less of a hangover than a whisky or brandy drinker. However, this is only a generalization - there will, unfortunately, be exceptions. For example, isobutanol is present in both alcoholic and low alcohol drinks, and for some people the presence of isobutanol will in itself produce hangover symptoms.

Another factor in a hangover is the reduced blood sugar level. This is caused by the alcohol's stimulation of insulin production in the pancreas, which in turn results in feelings of tiredness, depression and irritability. The old saying of 'feeding a hungover person with a large cooked breakfast' results from this low blood sugar factor. Other remedies that are often offered also help to raise the level of sugar in the blood.

Consequences of alcohol abuse

Liver damage

Liver cirrhosis is caused by excessive quantities of alcohol being drunk and having to be converted into acetic acid and then water but the poisonous substance acetaldehyde may linger in the liver if there is too much alcohol to be dealt with at any one time. This can cause damage to the tissues in the liver itself, and the more tissues that are damaged, the less effective the liver will be.

Alcoholic hepatitis is an even more dangerous condition, as many of those that suffer from severe liver damage of this level will never recover. It is often a fatal condition. People who have been dependent upon alcohol for ten years have a 1 in 10 chance of dying from cirrhosis.

Brain damage

It is not known how well the brain can cope with a regular intake of alcohol. However, it is thought that an intake of between 21 and 50 units of alcohol per week for men, and between 14 and 35 units for women, can cause a reduction in the brain's powers of concentrating and accuracy of thought. The skills connected with reasoning and problem solving can become less effective if the body is subjected to a large, regular dose of alcohol.

The digestive system

With a large alcoholic intake, the mucous lining of the stomach can become damaged and inflamed, and may even cause the lining to bleed. This may cause nausea, but the damage to the stomach can be repaired naturally by the body if the intake of alcohol is stopped for a period of time. A prolonged time of alcohol abuse can cause a duodenal ulcer.

The heart

It is known that heavy drinkers (those who drink more than 70 units per week), run a serious risk of heart disease (cardiomyopathy). Many more people drink less than this, but regularly drink above the safe limits and so run the risk of suffering from high blood pressure and the likelihood of a stroke.

Maximum recommended consumption

In normal pub measures, the drinks mentioned each contain the same amount of pure alcohol: one unit. The approximate number of units in various drinks are shown in Table 3.9.

Table 3.9 Approximate number of units in various drinks

Drink	Quantity	Units
Beer and lager	$\frac{1}{2}$ pint	1
	1 pint	2
Export beer	1 pint	$2\frac{1}{2}$
Strong beer and lager	$\frac{1}{2}$ pint	2
	1 pint	4
Cider	$\frac{1}{2}$ pint	$1\frac{1}{2}$
Strong cider	$\frac{1}{2}$ pint	2
Spirits	$\frac{1}{6}$ gill	1
Sherry and port	1 small glass	1
Vermouth	1 standard measure	1
Wine	1 standard glass	1

The maximum *recommended* consumption per week is 14 units for women and 21 for men (the young and elderly should keep to a lower limit than these). It is advisable to keep below these figures. At this level, there should be no risk of long-term health damage as long as alcohol consumption is not drunk in one session.

Men who drink 22 to 35 units and women who drink 15 to 21 units a week are above the recommended safe levels and damage to the body may occur.

Men who drink 36 to 49 units and women who drink 22 to 35 units are very likely to damage their bodies. The damage will occur slowly and not become obvious until permanent injury has developed. By this time, there will be the added danger of alcohol drinking becoming habitual. This is a dangerous situation for any adult to be in, and there should be a reduction in drinking of alcohol immediately.

ACTIVITIES

- Discuss why you think people drink alcohol. Consider the following points:
  1. social drinking,
  2. drinking to relieve bordeom,
  3. drinking to 'keep up with the others',
  4. drinking as a 'pep-up'.
- Compile a series of questions that could be used in a survey to ascertain people's drinking habits and, in particular, their reasons for drinking alcohol.
- Use your survey on a specific group of people and then analyse your findings.

PART 4

Food service

This section will give you an understanding of the:

- diverse nature of the food service sector and the factors that influence consumer choice,
- the factors that affect the efficiency and customer acceptability of food service outlets,
- social skills and personal attributes of successful food service personnel,
- methods employed to ensure efficient food and beverage control,
- service of alcoholic and nonalcoholic beverages in formal and informal situations, including vending.

CHAPTER 29 Food services outlets

The service of food takes place in a large variety of establishments and in many different manners or styles. For example, there is food served in a formal manner in a luxury five-star hotel and food served in a corner fish and chip shop. In both establishments, the same basic principles need to be understood and carried out even though the actual method of service is quite different, as is the customer expectation and price charged. The maintenance of a scrupulously clean environment has already been covered.

PRINCIPLES OF FOOD SERVICE

The service of food at the correct temperature

The temperature of the food is an important factor. If a dish is designed to be served cold, then it should be served straight from the chilled shelf or cabinet - 3°C is the recommended temperature as it ensures that pathogens cannot multiply and that customer acceptance is assured because the food tastes chilled when eaten. Similarly, hot foods should be served at a safe temperature of above the minimum of 70°C. During the time it takes to serve hot food, it will cool slightly, therefore when the food leaves the kitchen it should be well above 70°C (nearer to 90°C). This way the food still tastes hot to the customer and, more importantly, pathogens cannot multiply and cause any harm.

The creation of a welcoming atmosphere

Customers will appreciate a welcoming atmosphere in any type of food service establishment. Some restaurants are so formal and daunting that potential customers are discouraged from entering. Customers are the 'bread and butter' of every food service establishment, and without them there would be no need for the restaurants. It therefore follows that all customers should be welcomed and positively encouraged.

The promotion of the establishment's services

Very often the food service unit is part of a larger complex, for example, a restaurant or coffee shop situated within a hotel or department store. In both these outlets, it is easy to imagine ways that the customers using the restaurants may be made aware of other facilities that they could be tempted to use. Food service staff have a duty to promote the facilities, goods and services within their establishment as part of the team spirit. This ability to be aware of the establishment as a whole, and to help to promote it in every conceivable way, is an essential attribute for successful employees.

The creation of a selling environment

The word encourage is the single word that all food service personnel should remember. To increase sales, it is frequently necessary to positively encourage customers to choose a further course, another beverage or an untried dish. This encouragement should never be so forceful as to result in frightening customers away, but should be subtle and persuasive without giving the least hint of being domineering or aggressive. The prime selling aid is, of course, the menu. This should be clearly and attractively written, illustrated, and displayed so that the potential customers are encouraged to enter. Food service staff are often asked to describe a dish from the menu for the customer; this description is either going to sell the dish or discourage a sale. The use of such words as a rich sauce, a hint of garlic, a tangy lemon soufflé all make the dish sound attractive.

SUMMARY OF FOOD SERVICE UNITS

Commercial sector

Restaurants

Most hotels have restaurants that are open to nonresidents. There will be a menu displayed outside many hotels so that passers-by can stop and read it and, hopefully, feel tempted to step inside. Hotel restaurants are used for the service of traditional English breakfasts, sometimes for Continental breakfasts too, followed by luncheon service and finally the service of dinner in the evening. Thus, the restaurant in a hotel is a department that never stops and that is busy every day of the year.

Restaurants may also be situated inside department stores where shoppers are encouraged to sit and relax while out shopping. These restaurants often serve morning coffees, afternoon teas, light snacks, and full meals during shop opening hours. Many restaurants are independently situated in self- contained premises in commercial, retail and tourist areas of cities, and in the countryside.

Some restaurants specialize in a particular type of food, these include the many popular steak houses, vegetarian restaurants, Chinese and Cantonese restaurants, Indian restaurants, 'pizza houses', and Italian, Greek and Spanish restaurants. It is possible to find many different types of restaurant in many major cities and towns. Restaurants are also found at airports, in roadside service complexes, and on ocean liners.

Public houses are increasingly opening restaurants adjacent to their beverage sales areas, to attract both their regular clientele and passing trade. Most restaurants are licensed to sell alcohol to their customers, as wines, beers and spirits are an acceptable compliment to a meal.

Take-aways and fast-food restaurants

The phenomenal growth of take-away and fast-food restaurants in the last twenty-five years is partly attributed to the fast pace of inner city life, the flexibility of the working week, the increased affluence of the younger generation of citizens, and the influence of USA customs in the UK. Many of the types of restaurants already discussed offer a take-away facility for customers.

Well-known examples of take-away restaurants include Chinese restaurants, Indian restaurants, fish and chip bars, pizza houses, and hamburger or fried chicken restaurants. In all these cases, the food is ordered from a service counter by the customer who then pays the bill, and waits only a few minutes for the food, which is in appropriate containers and bagged ready for carrying away. A few of these types of establishments offer a telephone order and/or delivery service for their customers. An important consideration is the temperature of the take-away food when the customer receives it in the restaurant, and the temperature at which the customer eats it when at home. Some aluminium foil or cardboard containers do not retain the heat very satisfactorily. These fast-food establishments may be found in retail and leisure areas and in roadside service complexes, airport customer areas, railway stations and at some outdoor events, where mobile hamburger and hot dog stalls are traditionally seen.

Coffee shops

These have grown in number and popularity over the past ten years. First, they are found in hotels where they serve the dual purpose of providing hotel guests with an informal atmosphere in which to enjoy a snack, a hot beverage, or a light meal, instead of the formal restaurant atmosphere and service (see figure 4.1). Hotel coffee shops are usually situated in sight of passers- by, so that nonresidents can be aware of their existence and, hopefully, be attracted inside. Normal trading hours usually fall from 7 am through to 10 pm.

The second area where coffee shops are found is in modern shopping precincts and areas around retail shops. Here they are often adjacent to, or part of, a retail baker's shop, and trading follows the retail shopping times within the surrounding area. They attract shoppers and other persons who work nearby and wish to have a light or snack type luncheon, or a drink and a pastry. Hot and cold beverages are served, as are salads, filled rolls and sandwiches, light, cold sweets, pastries and cakes. Coffee shop service may be self-service or a simple form of table service with the accent on clean, wholesome, light meals and snacks served in an informal manner and offering value for money to the customer. Many have seating and tables outside their premises or in a specially screened area in shopping precincts, in order to attract customers.

WINES

White — Bottle / ½bottle / Glass

	Bottle	½bottle	Glass
Arc de Triomphe – Medium dry	5.50	2.90	1.00
Muscadet Cuvée Madeleine – Fresh, crisp and dry	6.50	3.45	1.25
Liebfraumilch Wedding Veil – Fruity and famous	6.50	3.45	1.25
Soave Classico – Dry crisp wine of Italy	6.50	3.45	
Frascati superiore A light dry wine	6.50	3.45	—
Mouton Cadet Blanc – the house of Rothchilds, dry	9.25	4.95	—

Red

	Bottle	½bottle	Glass
Arc de Triomphe – dry, red	5.50	2.90	1.00
Côtes du Rhone; dry, full flavour	6.50	3.45	1.25
Claret, Cafe Royal; from the cellars	6.50	3.45	
Chianti Classico, all the flavours of Italy	6.50	3.45	—
Mouton Cadet Rouge, a modern classic	9.25	4.95	—

Organic Wines

Grown without chemical pesticides or fertilizers, these wines are for a new healthy way of living.

	Bottle
Poggio Alle Roche, Il Palagio Italian medium dry white wine	7.50
Albaric – Vin de Table – A well rounded robust red	6.95

Champagne

	Bottle	½bottle
Ayala Château d'Ay Brut NV	15.00	8.60

BEERS

Dortmunder Siegal Pils	£1.20
Dortmunder Export	£1.40
Double Diamond	£1.00
Carlsberg Export	£1.10
Heineken	£1.40
Grolch	£1.60
Budweiser	£1.20
Germania (Low alcohol lager)	£1.00
John Courage Pale Ale	£1.00
Smithwicks Premium Export	£1.20
Guinness Extra	£1.00
Cider	£1.00

COCKTAILS!!!

Dry Martini
Stirred, not shaken, gin and dry vermouth, over mountains of ice and served with a twist of lemon and an olive — 2.95

Bloody Mary
Vodka, tomato juice, spices and lemon juice served with a celery stick. — 2.95

Harvey Wallbanger
A long cool blend of vodka and fresh orange juice with a dash of galliano — 3.50

Pimms No 1 Cup
That essential mix of Pimms No 1 and lemonade built on ice and garnished with lots of dressing! — 3.50

Planters Punch
Dark rum, orange juice and pineapple juice; Long, strong but very fruity — 3.50

St James Park
Fresh banana, passion fruit juice, peach juice and coconut cream; a special alcohol free cocktail "Sober and out", captain — 2.25

SOFT DRINKS

Coca Cola	0.60
Perrier (200ml)	0.90
Ashbourne, sparkling	0.60
Lemonade	0.60
Apple Juice	0.90
Milk	0.50
Squash	0.50

A full range of Spirits and mixers are also available.
Spirits are served in ⅙ gill and multiples thereof.

TO START

Freshly Squeezed Orange Juice — 1.45

Prawn Cocktail
With brown bread and butter — 2.75

Melon In Season
With ground ginger — 1.50

Egg Mayonnaise
On a salad base with Marie Rose and mayonnaise — 1.95

Crispy Mushrooms
Dusted with parmesan cheese, offered with lemon and tartare sauce — 1.95

Tomato Soup
With croutons — 1.00

Minestrone Soup
With parmesan cheese — 1.00

PASTA

These dishes include a side salad.

Maccheroncini Bolognese
Freshly poached pasta strips with a beef sauce and freshly grated parmesan — 3.75

Agnellotti
Lamb filled parcels of pasta in a rich tomato and basil sauce — 3.75

Vegetable Lasagne
Layers of pasta, aubergine and courgettes — 3.75

ANYTIMERS

Open Pastrami Sandwich
With spears of dill pickle — 2.75

Open Roast Rib Of Beef Sandwich
With fresh creamed horseradish — 2.75

Open Prawn And Egg Sandwich
With Marie Rose sauce — 2.75

Open Smoked Scottish Salmon
With lemon and capers — 3.75

Toasted B.L.T.
Smoked back bacon, lettuce and tomato with mayonnaise, coleslaw and crisps — 2.75

Steak Sandwich
Quarter pound (uncooked weight) grilled on french bread with salad, coleslaw and crisps — 4.25

Toast, Butter And Preserve — 1.10

Omelettes
Your choice of plain, cheese, ham, mushrooms or prawns. Served with french fries or jacket potato and grilled tomato — 3.75

All Day Farmhouse Breakfast
Two fried eggs, three rashers of bacon, sausage, grilled tomatoes, mushrooms and fried bread — 4.50

Beefburger
6oz (uncooked weight) served in a sesame seed bun with dill pickle, coleslaw and french fries — 3.75

Cheeseburger
6oz (uncooked weight) topped with cheese served with dill pickle, coleslaw and french fries — 3.75

SIDE ORDERS

Garlic Bread	0.95
Mushrooms	0.95
French Fries	0.75
Garden Peas	0.75
Side Salad	1.10
Jacket Potato	0.75

MAIN DISHES

Served with french fries or jacket potato and a grilled tomato; side salad with your choice of dressing.

Chicken Kiev
Filled with garlic butter, rolled in breadcrumbs and deep fried — 6.25

Sirloin Steak
Half pound (uncooked weight) chargrilled — 6.95

Valentine Lamb Steak
Grilled and served with mint sauce — 5.75

Breaded Veal Escalope
Shallow fried in butter — 6.45

English Calf's Liver
With onions and grilled bacon — 6.45

Grilled Whole Plaice
With lemon and tartare sauce — 5.75

Deep Fried Scampi
With lemon and tartare sauce — 6.25

SALADS

Served on an abundance of seasonal salad leaves

Avocado Pear And White Crab Meat
With 1000 island dressing — 4.95

Tuna Fish And Prawn
In garlic mayonnaise — 4.95

Cottage Cheese And Fresh Pineapple
With grapefruit and walnuts — 4.25

Chefs Chicken Salad
Served with your selection of dressing — 4.95

SWEETS

Soft serve dairy ice-cream
with your choice of delicious toppings Strawberry, Raspberry, Chocolate or Caramel — 1.20

Raspberry Rippler
Crushed raspberries with soft ice-cream — 1.85

Banana Midori
Twists of ice-cream with banana and chocolate sauce flavoured with melon liqueur — 1.85

Caramel Nut Sundae
Lashings of ice-cream flavoured with Strega topped with toasted hazelnuts and rich caramel sauce — 1.85

Fruit Topper
Fresh fruit rippled with ice-cream — 1.85

Fresh Fruit Salad
Naturally good — 1.55

Blackcurrant And Apple Pie
Traditional English fruit pie — 1.55

Profiteroles And Chocolate Sauce
A light choux pastry filled with whipped cream and chocolate sauce — 1.85

Rich Chocolate Gateaux — 1.55

Sorbets
A mouth watering mixture of water ices, cassis, mango and papaya — 1.85

Cheese Board
Mature English Cheddar, Lymeswold and Double Gloucester, served with an apple, Bath Oliver biscuits and butter — 1.50

CREAMIEST TEA IN TOWN

(3.00pm–5.30pm)

More than a tradition, an absolute delight!

Afternoon Tea
3.25
Choose from:
Hot scones
Choice of preserves
Clotted cream
Your choice of freshly made cream gateaux

A choice of:
Darjeeling, Ceylon, Earl Grey, Assam, Grosvenor Blend

We would be delighted to arrange a selection of our cakes to take away.

MORNING COFFEE PASTRIES

Danish Pastry	0.65
Scones with butter	0.95
Old fashioned biscuits	0.75

BEVERAGES

Pot of Coffee
(Per person)
Decaffeinated coffee, if preferred — 0.90

Pot Of Tea
(Per person)
Choice of Darjeeling, Ceylon, Earl Grey, Assam or Grosvenor Blend — 0.75

Mug of Chocolate
Hot chocolate topped with whipped cream and a chocolate flake — 0.85

Glass Of Cold Milk — 0.50

Giant Frosted Milk Shakes
Vanilla, strawberry or chocolate — 1.35

Iced Tea — 0.90

Iced Coffee
Served with whipped cream and topped with nuts — 0.90

FOR CHILDREN

All children under 14 may choose from the main menu at half price.

Our menu for younger children is also available.

Open daily from 11.00am–12.30am

Breakfast available from 6.30am–10.00am

All prices include VAT. Gratuities are at your discretion.

RSPH 10/87

4.1 An example of coffee shop menu.

Catering for the travelling public

There are three main areas in this sector of catering: firstly, roadside service complexes situated at arterial road junctions and or motorways; secondly, railway catering, at railway stations and on trains; and, thirdly, sea ferries and airport and airline catering. These outlets experience common problems: the need to feed customers quickly; irregular customer flow patterns; irritability among customers caused by the stress of travelling; large groups of customers all wishing instant service; customers ranging from young babies to the elderly; and customers who do not understand the English language and who are unfamiliar with the currency and food.

Most of these outlets are designed to be self-service units, the traveller is able to see what is available and how much each item costs. In addition, self-service allows staff to stock display and service areas during slack trading periods, and then to be available to serve when the flow of customers increases.

The type of food and beverages available will vary slightly according to the time of day, but the traveller will find a good variety of snacks, salads, and full, hot meals generally available. On trains, the food is either snack or sandwich type service that the traveller can purchase from the buffet car or from a roving trolley, or a more comfortable table service served by stewards. On aeroplanes, the cabin staff serve complete meals on trays to the passengers, and silver serve the first-class passengers. At roadside complexes, some food is available from vending machines at night time, when travellers can purchase filled rolls and snacks or beverages as required without requiring the services of food service personnel (see opposite) On sea ferries, there are formal restaurants and self-service style areas for an informal type of service.

A refrigerated vending machine.

The main criteria for food service in all these transport sectors includes the following points:

- to provide a clean uncomplicated form of service,
- to provide the traveller with a calm, welcoming atmosphere,
- to sell snack items and food that is well packaged and clearly labelled and priced,
- to provide an adequate number of well-appointed public conveniences (not forgetting the very young and those in wheelchairs),
- to provide comfortable seating and tabling for travellers with space for light luggage, pushchairs and refuse bins, not forgetting good visibility of departure screens and clocks.

Industrial catering

Whether subsidized or a profit-making concern, many of these catering units are now modelled on their retail counterparts, and, indeed, do compete favourably with them. Gone are many of the old, dismal, self-service counters in large, plain dining areas. In their place are light, modern, self-service units and seating areas, tastefully decorated and partitioned by movable half screens and plants to create a restful atmosphere for the employees. Speed and value for money are the main criteria in these units, and healthy eating is increasingly encouraged, and so an increasing number of units serve salads, wholemeal products, pasta and pulse-based dishes, and low fat products.

The service will normally be self-service or, sometimes, a simple form of table service. Executives and directors may have the use of a more formal type of service when entertaining visitors and clients, this will be a traditionally styled silver service dining room, and may only be used when there are visitors to be entertained. In this type of food service, the customers quickly become 'regulars' and the food service staff will soon get to know their names and food preferences. This should be encouraged as it promotes the caring attitude that is so vital in successful food service. Menu 'fatigue' should be avoided by providing a varied selection of dishes, the use of theme menus, and special events.

School meals service

Like industrial catering units, the school meals service has been modernized in recent years. The majority of units are self-service areas where children may select their own choice of food; this will encourage them to become more discerning customers in the future. Here again the customers soon become known to the food service staff, and this helps to create a friendly atmosphere where the children may relax after their morning lessons. Time is restricted here also, so speed and a simple form of self-service is essential. In these units, the food service staff are often the same staff that supervise the children after their lunch, and who may also have helped to layout the dining room prior to service. The menus in modern school dining rooms reflect the national awareness of the healthy eating programme, and regional preferences will also be reflected in the choice of dishes served to the children.

Factors influencing the customer's choice of establishment

The factors that influence the customer's choice of establishment fall into three main areas: aesthetic, monetary, and palatability.

Aesthetic factors

These include such things as the 'atmosphere' of the restaurant; is it: Calm, but friendly? Efficient, but with time for all its customers? Peaceful, but busy? The atmosphere is a vital part of the restaurant's ultimate success. Different types of customer will be looking for different types of atmosphere. The decor, the method of service and the use of background music all help to create the restaurant's atmosphere.

Monetary factors

The cost of the food offered will be decisive in influencing the customer's choice of establishment. This is the reason why it is common practice to display menus outside restaurants: apart from tempting potential customers with the description of dishes on the menu, the customer will want to see that the food is value for money.

Palatability expectations

Even though people have their own individual preferences, it is generally accepted that classic dishes on a menu will be presented in a standard manner and have a distinctive, recognizable flavour. For example, if cod a l'orly is on a menu, the customers will expect the cod fillet to be an acceptable size, to have been coated in a batter and deep fried. They will also expect a well-flavoured tomato sauce to be offered as an accompaniment. Customers think they will like a particular dish on the menu, so they decide to enter the restaurant.

Presentation and the guest

Whether or not a customer returns to an establishment is determined, to a great extent, by the degree of satisfaction that he or she felt on leaving the establishment after his or her first visit. This is why presentation and service are so important in the hotel and catering industry. If food is badly served, by poorly trained and clumsy staff, in an inhospitable environment, then it can be guaranteed that the guest will not return.

ACTIVITIES

- Compile a map of your local town centre, to show the position of food service outlets. List the menus and prices charged at each of these outlets. Discuss which types of customers are being targetted at each outlet. Do you think they are being successful?
- Study the types of food service outlets in the UK and list the functions common to each.
  From your list try to determine the difference between a meal purchased at a fish and chip shop, and the cod a l'orly and pommes frites purchased in a hotel restaurant. (You might consider such things as: the method of presentation and service, the cost to the customer, and the decor and atmosphere of the establishment.)

CHAPTER 30 Business promotion

There are many considerations that need to be addressed before planning or launching a new catering establishment or simply updating an existing establishment. A faster form of service with less formality, fewer staff employed and tight target costs is a recent trend. Catering is a very competitive business and all aspects must be carefully considered. Below are some examples that must be thought about in the initial planning stages.

INITIAL PLANNING

The image and atmosphere of the establishment

Think of two licensed public houses. There are bound to be many similarities between the two, for example, they will both serve alcoholic and nonalcoholic beverages. But, equally, there will be differences in their structure, decor, finish, and style. To a large extent, these differences will determine the image and, therefore, the atmosphere of the establishment. There are establishments with a very correct, formal team of staff and a formal style of decor and service. Here the atmosphere will be quiet and reserved. A relaxed and informal atmosphere will be created in the establishment decorated and fitted to resemble the interior of a barn or a galleon. the traditional English pub will have a unique atmosphere, dictated by the history of the premises and its local customs and customers. Here the passing trade will be attracted to sample this traditional atmosphere.

The type of customer to be attracted

Different social groups, ages and types of customers will be attracted towards different types of establishments. In order to attract the type of customers planned for, it is necessary to analyse what it is that these customers are expecting to find in an establishment, and then to ensure that their expectations will not be disappointed. For example, young customers require and expect a certain type of popular music at a fairly high volume; elderly customers will like a quiet atmosphere with no music; while middle-aged customers might appreciate subdued, light music. Sound will also vary with the volume of light provided - generally the louder the music, the dimmer the lighting seems to be. Restaurants may use subdued, light music as background sound, but this should never be so loud that it causes the customers to raise their voices. In addition, the cost of the meals or beverages on sale will determine the type of customer that will be attracted.

Transport and accessibility

All establishments with a licence to sell alcoholic beverages need to consider the availability of transport facilities. It makes no financial sense to invest time and money on a remote premises and then open to the public and find that no customers arrive as they are reluctant to drink and drive. Hotels and restaurants in cities have a slightly different consideration. Theirs is one of car parking. In central city commercial areas, car parking is always a problem, parking meters always seem to be full when potential customers drive past, so the establishments with car parking are those that will have a good trade. When advertising the establishment, it is always a good idea to include details of car parking, even if it is only to mention a public or multistorey car park. Another point to mention is the availability of public transport, particularly in central city areas.

Competition

This needs to be investigated both within the immediate vicinity and further afield. In the case of a fish and chip shop, the eating habits and traditions of local residents will determine how often they buy fish and chips. If a second shop opens, the effect will be to split this trade between the two shops - on the face of it, not a very satisfactory situation. However, with forethought, the fish and chip shops could consider the following types of questions. Firstly, would a sit down and eat on the premises area be a selling point, or would deliveries for functions and individual households be worthwhile? Would an increased range of products for sale serve to increase customers both locally and regionally?

Local situation

A knowledge of local conditions such as market days, customs, annual feasts, holiday periods and events may have a considerable effect on the trading pattern if, for example, market day is Wednesday, which would make Wednesday the busiest trading period in the week.

Geographical situation

Weather can have a dramatic effect on the catering establishments in this country. It is advisable to study the local weather pattern to ascertain the average temperatures and rain fall or even snow fall for the trading seasons. This, gives some base for future planning. Take Scotland as an example. For some months of the year, the snow makes winter sports very popular - this will obviously increase the potential for guests in highland areas, whereas in low land areas it may have the opposite effect. Another area where this information is vital is the seaside resorts. These are busy when the weather is fine, and slack when it is cold and wet. This has obvious trading and planning implications. Many coastal resorts and individual establishments have spent a considerable amount of time and money recently to provide attractive all-weather recreation areas for their guests - these areas and establishments are the ones that will remain profitable in the future. Most all-weather facilities are sports and leisure based - large swimming pools with artificial waves, a children's area, water play areas and slides, pool side coffee shops, squash courts, all-weather tennis courts, indoor bowls centres, fun fairs and go karts, for example. All these points must be considered as the weather is very unpredicatable and unreliable.

Customer facilities

Once it has been decided which type of customer the business is to attract, the next decision to make is the facilities that are to be available for those customers. Is the establishment to be licensed? What will the trading hours and periods be? What cost brackets will the customer be expected to pay? And, crucially, are suitable permanent, casual or part-time staff available to enable you to run the business at the planned costs? At this stage an accountant is employed to produce a business plan, which will determine whether the whole project is feasible.

Unique selling points (USPs)

A USP list gives the selling points that a business actually has. These can then be compared with the USPs of the most likely business competitors. From this information, the promotional strengths of the business can then be identified. The following serves to illustrate this.

USP for Hotel A:

- function rooms for 500 persons seated and 750 persons informal;
- car parking for 150 vehicles, car valeting service available,
- accommodation for 680 persons and all rooms have en suite facilities,
- all rooms fitted with GPO lines, telex and fax facilities for residents,
- secretarial facilities,
- membership of the leisure club is free for all residents (includes squash courts, gymnasium and pool plus jacuzzi and solarium),
- two bars and one restaurant open to nonresidents.

USP for Hotel B:

- three types of restaurant all open to nonresidents,
- golf club membership for residents at a nearby facility,
- public car parks 150 yards away,
- twice weekly live entertainment.

If these two USPs are compared, it will become apparent that Hotel A can cater for business clientele, whereas Hotel B is better equipped to concentrate on the restaurant and entertainment end of the market.

The product statement

From the USP it is possible to devise a product statement. This takes the form of a short advertising description that contains details of all the hotel's facilities as listed in the USP. For example, the product statement for Hotel A might read as follows:

> The hotel is conveniently situated in the centre of the town with its own car park with space for 150 vehicles. All of the 680 rooms have full, en suite facilities with panoramic views of either the city or the Victoria Park area. Ideal facilities are available

for business conferences, functions or exhibitions, in function rooms that can take up to 750 persons. Full business communications are available and so are secretarial facilities. GPO telephones, telex and fax facilities are in all the rooms. Membership of the fully equipped leisure club for residents is an automatic bonus where the business traveller may relax at the end of the day. The restaurant offers a full à la carte choice for dinner, as well as a superb carvery at luncheon offering a full choice of traditional and continental cuisine.

ADVERTISING

Once the business is established, then the actual marketing may commence. This means that potential customers need to be made aware of the facilities and services that are available at the establishment. Advertising comes in many forms, but it is always designed to influence behaviour in some form or another. It will have an informative element which is designed to entice the customer to try the product or service, and a confirmation element which will reassure the customer about his or her decision to try the produce or service. It is very difficult to ascertain how effective advertising actually is, and so, any advertising plan needs to consider the most suitable advertising media.

Broadcast media

Local radio stations are administered by the BBC and the IBA; these have well-defined catchment areas and are, therefore, the ideal media for food and beverage advertising. Some run focus features on different establishments, which allow for a few minutes broadcasting time at a reasonable cost or even at no cost. This type of advertising needs to be well designed and scripted, so that the potential customer has all the facts at the end of the broadcast.

Television is a media that requires a large investment and specialist consultants. It is a media used by national or international operators for special promotions, such as off-peak weekend breaks. Its main advantage is its visual appeal, something not possible with radio. One way that the independent hotel or restaurant can sometimes obtain television coverage is to invite their local television company to send a reporter and camera crew to a promotional event, such as a new restaurant opening, or a sponsored charity event. This way the event could be featured on the next local news bulletin.

Print media

National newspaper advertisements are expensive and not normally suitable for individual establishments. However, local newspapers are the ideal media. There are some newspapers that are delivered free to every household - here the advertiser will know that his or her message will reach a large number of potential customers. Specialist periodicals are expensive and usually not used by small, independent hotels and catering establishments unless a special promotion coincides with the readership of a particular periodical. For example, a wine appreciation and gourmet weekend could be advertised in food topic periodicals. Some newspapers have regular features on restaurants. This is a good easy way to achieve a mention provided the restaurant receives a favourable review.

The Yellow Pages is a good media for the small, independent operator. Many potential customers use this directory to seek out information about where to eat and what type of foods are likely to be offered for sale.

Small leaflets may be another print media that is suitable for independent hotels and restaurants. The cost of reproduction is not normally prohibitive, the success, however, is dependent upon distribution. One method is to have the leaflets delivered with the free local newspapers. Another is to negotiate delivery with a local distributor, the disadvantage here is that the actual date of delivery is not normally guaranteed. Leaflets targeted at specific potential customers, such as local businesses, can be sent by post. This method of advertising has advantages over the newspaper advertisement, for example, the use of colour printing and photographs, the use of graphics and diagrams, text may be printed on both sides, and the leaflet may be kept by the recipient without the bother of having to cut it out of the newspaper and becoming irritated by newsprint ink on the hands.

Other media

Posters are an expensive form of national advertising, as are cinema advertisements, and like national newspapers and television are best left to the professionals. However, advertisements on public transport are worth consideration. In cities where there are underground systems, there is an abundance of advertising space available. Buses, too, give the opportunity to advertise an establishment, both with posters inside and on the exterior of the bus. Advertisements on the exterior will receive more notice, but the information must be limited and clear in the interests of road safety.

Specialist guides are published at regular intervals, and inclusion in these is a good marketing strategy. These guides are published to appeal to a particular type of consumer or by a specific organization. A few examples are listed below:

Egon Ronay Good Pub Guide
Egon Ronay Lucas Guide
Automobile Association Guesthouses, Farmhouses and Inns in Britain
Automobile Association Hotels and Restaurants in Britain
British Hotel Restaurants and Caterers Association Hotels and Restaurants in Britain
English Tourist Board Hotels in England

These types of guides are available in public libraries or for purchase. Local chambers of trade and regional tourist boards also publish their own guides. These are the ones in which most independent hotels and restaurants will advertise.

SALES IN THE FOOD AND BEVERAGE SECTOR

All staff are responsible for sales in the food and beverage sector. So far the role of those with primary responsibility for the marketing policy (the sales and marketing director or unit manager) has been considered. However, others have a secondary responsibility - to ensure that direct sales remain high. These people include the food and beverage manager, unit supervisor, reception personnel and food service personnel.

Motivation is a word that is important to everyone, as this is very much tied up with staff working and a feeling that they are part of a team and that their efforts are important and recognized by the management. There are three aspects which make an efficient sales team.

Activity

The need to serve quickly is a critical part of sales, particularly in fast-food and popular food establishments. One hold up on a cafeteria style service will quickly affect the entire service run. Likewise a slow cashier can cause chaos in a similar type of establishment. These 'hold-ups' will quickly lead to customer dissatisfaction and therefore decreased sales.

Style

This is concerned with the nonverbal communication used by members of staff towards customers. These may have a considerable effect upon the sales: below Table 4.1 gives methods of nonverbal communication - both positive examples that should be encouraged, and negative examples that should be discouraged.

Efficiency

An efficient member of the team will be that member of staff who produces maximum work output in relation to work expended. Other words used to describe an efficient member of the team could be capable, effective, skilled and able. Such a team member will be able to work efficiently with other members of the team, and work to company guidelines with the minimum of supervision.

Faults identified through the use of direct sales analysis should be corrected. Staff training, increased supervision, or more drastically, staff redeployment to a less critical part of the organization are all remedies. Once the team members are made aware that direct sales analysis is to be undertaken and, in particular, the style component, it can act as a form of team motivation, which can be further encouraged by incentives such as employee of the month awards, increased sales target team awards, and sales personality certificates.

POSITIVE MERCHANDISING

All merchandising must aim to stimulate customer response and therefore increase sales.

Visual merchandising

Catering is a very visual business, and so much of the merchandising is, based on visual responses. A customer is more likely to be tempted to purchase a sweet course if he or she is able to see the range of sweets on offer, whether at a cafeteria service point, or a sweet trolley in a restaurant, or even good colour photographs adjacent to his or her menu description. This last example is often used for ices as, for obvious reasons, these cannot be displayed.

Beverages can also be effectively displayed to encourage sales: pictures of cocktails, specialist coffees and beverages can increase customer demand. Another method is the one that enables the customer to see his or her food being prepared, an example of this is the sandwich bar where the customer selects his or her choice of bread or roll, filling and garnish.

Table 4.1 Examples of nonverbal communication

POSITIVE		NEGATIVE
Smiling face Interested expression Sparkling eyes	**Facial expressions**	Frowning face Sulky looking face Eyes downcast
Standing up straight Looking at the guest Hands at the side	**Posture**	Slouching against the wall Gazing into space Hands in pockets
Clean uniform Clean-smelling breath Clean hands and nails Clean, neat hair No body odour	**Personal**	Unpressed, stained uniform Breath with unpleasant odour Hands stained, bitten nails Untidy, greasy hair
Calm, clear voice Correct pronunciation Friendly, persuasive voice Speaking in an even tone	**Vocal**	Muddled speech Unclear pronunciation Aggressive, sullen tone of voice Speaking quietly or loudly
Standing near guest, but still retaining a space as a sign of respect	**Space**	Crowding or towering over guest Keeping guest at a distance
Hands and fingers kept still Breathing quietly Keeping mouth still	**Sound**	Clicking fingers Coughing and sniffing Clicking pen top Jingling money Chewing gum
Sufficient time spent at each task Never glancing at clock or watch Walking calmly around restaurant	**Time**	Hurrying the guest Keep looking at the time Dashing around the restaurant Dawdling around the restaurant

A further form of visual merchandising is a full restaurant or even a long queue at a fast-food establishment. It is a sign that the food is good if it is proving to be so popular, and more customers will be tempted in to try it for themselves. Conversely, an uninviting atmosphere and poorly presented foods may cause customers to curtail their visit prematurely.

Aromatic merchandising

The most well-known example of this form of merchandising is the fish and chip shop. Here the aroma of the fried fish and chips acts as a stimulus to the passerby, who has the desire to taste the fish and chips. The hot dog stall is a another example.

Coffee shops often use this form of merchandising, especially when they are attached to bakeries. There, the smell of the baked goods coupled with the smell of freshly roasted and ground coffee are irresistible to many passers by.

Unnecessary and sometimes unpleasant aromas of food can have an advere effect on sales. Customers in restaurants do not, on the whole, appreciate the aromas of foods being cooked. Similarly, they do not appreciate the sounds and noises sometimes associated with the cooking of foods.

Written merchandising

The prime written merchandising tool is the menu. Customers should never be made to feel inferior, uncertain, or insecure when they look at the menu. Their first sight of the menu is often outside the restaurant cafeteria or fast-food establishment. It is here that the customer will make the decision to enter the premises or to look elsewhere. Table 4.2 lists the points that are important for both the restaurateur and the customer.

Table 4.2 Types of hotel guests and their expectations

Restauranteur's	Customer's
• List of dishes for sale	• Do I like the dishes that are for sale?
• Brief description of each dish	• What does each dish contain?
• Cost per meal or per head	• How much will I have to spend?
• Type of service	• Type and estimated time to be served with the meal
• Service charge	• Is there any additional charge?
• Licensed to sell alcohol	• Is the premises licensed?
• Attractive, clear menu offering a balanced selection of dishes at competitive prices	• Is the overall impression favourable?
• Establishment's name and address	• Where am I?

In cafeteria service, establishments will sell their products by an illustrated menu and by a brief description of the dishes for sale. Many establishments that cater for people at work have a healthy catering policy. Here the menu will indicate the nutritional content or value of each dish. The price of each item for sale is clearly indicated for customers to see as they pass through the service area. Bread rolls and sandwiches are individually packaged and labelled with a clear description of the filling and the price. In catering for people at work units, the majority of customers are 'regulars' and menu fatigue needs to be avoided – changes in the style of menu presentations, posters giving details of the special promotions, and theme menus all help to reduce this.

Tent card are an effective and inexpensive form of written merchandising. The cards are used to draw the customer's attention to special dish or beverage promotions, and are placed along bar tops, on tables and, sometimes, on the restaurant tables. The tent cards may be provided by the manufacturer of a product or caterers can produce their own.

Brochures are used frequently to sell function trade and conferences. The prospective customer can take a brochure away after the initial enquiry and study the alternative room layouts, the different menus and costs involved. Brochures are detailed, explicit and concise, and include a contact name and telephone number so that queries can be promptly answered. Book matches, drink mats, place mats, cocktail sticks and toiletries in the cloakrooms, with the establishment's name and address printed on them, are designed as 'take-aways' which extend the merchandising further afield.

Spoken merchandising

Spoken merchandising refers to the form of words that are used to persuade customers to purchase dishes and beverages (see also pp. 176 and 180). At all times, a positive sales style must be used. Some training and practice is required to ensure that this positive sales style is used by all food and beverage staff at all times. It applies as much to the formal restaurant as it does to the fast-food, takeaway unit. Table 4.3 gives some examples of spoken styles:

Table 4.3 Examples of spoken styles

Positive sales style	Minimum sales style
• Here is our menu. I will leave it for your perusal.	• Do you want to see the menu?
• What have you selected?	• What are you going to order?
• Which hors d'oeuvre have you decided upon?	• Are you having an hors d'oeuvre?
• There is a good selection of vegetables on the menu. Which shall I order?	• Do you want any extra veg?
• I will bring the trolley so that you can select a sweet.	• Would you like a sweet?
• Which type of coffee would you like?	• Are you having coffee?

FOOD SERVICE PERSONNEL

Personal hygiene

All food service staff should have high standards of personal hygiene as they are in close contact with both the customer and food. The same standards of personal cleanliness as are laid down for all food preparation personnel apply to food service staff. Hair should be clean, neat and able to remain in place while working without having to touch it. Hands should be clean, and nails short and clean and never bitten. Any grazes or cuts should be covered with a waterproof dressing. Uniforms should be cleaned and well pressed. Shoes should be a plain design and safe with a good shine. Food service staff need to be alert and aware of their customers, so staff should not be over-tired.

Wherever food is served there are common factors that must be followed by all food service personnel. These factors include:

- the hygiene of the premises and food,
- the cleanliness of the equipment and personnel,
- the temperature of the food, either piping hot or chilled,
- the service must be efficient and genuine.

The restaurant 'team'

Each member of staff in the restaurant is dependent on others for the successful completion of the tasks in hand (see table 4.4). A helpful, positive approach leads to greater staff cohesion and motivation. A negative, unhelpful attitude may lead to poor service and disenchantment, and may even give the restaurant a bad name. The positive and negative attitudes shown by staff are summarized in Table 4.5.

ACTIVITIES

- Make a collection of advertisements for food service establishments in your area. Which advertisements caught your eye the most? Why was that? Discuss the different styles used in these advertisements.
- Draw up a list of positive sale styles for a fast-food takeaway establishment
- Discuss the four factors that should be followed by all food service personnel. How many more can you add to the list?
- Compile a list of personal hygiene points to be given to new food service recruits when they arrive for their induction training.
- The diagram opposite indicates the various actions that food service staff in a formal restaurant will perform before a guest is served with his or her first course. Compile a list of sentences that the food service staff could use at each of these actions.

Table 4.4 The members of staff that comprise a traditional restaurant brigade

Position	Responsibilities
Restaurant manager	Organizes and supervises the work in the restaurant
Head waiter	Supervises the service staff; greets customers
Station head waiter	Responsible for an area; receives guests; takes menu orders
Wine waiter	Takes order for alcoholic drinks and serves them
Station waiter	Responsible for the service of food to a group of tables
Junior waiter	Assists station waiter; clears and carries out some service duties

Table 4.5 Attitudes of staff

Positive	Negative
Works with other staff	Works for oneself only
Helps other staff	Displays unfriendly attitude to staff
Completes own duties	Does not show initiative
Works in a friendly manner	Displays a lack of respect for staff
Shows initiative	Shows rudeness to staff
Respects other staff	
Shows politeness to senior staff	

Guest arrives
↓
Reservation?
↓
Seat guest
↓
Offer aperitif
Call wine waiter
↓
Place butter on table
Offer bread rolls
↓
Offer water
Serve water
↓
Take and record order
↓
Ask if require wine
Call wine waiter
↓
Serve guest with selected first course

CHAPTER 31 The food service area

PLANNING A FOOD SERVICE AREA

Before any decisions are taken about such things as furnishings and equipment for a restaurant, it is essential to establish some important factors.

The customer flow

A premeal reception area will eliminate customer congestion in the doorway to the restaurant. Customers will be able to make their choice from the menu and then be shown to their table when ready. This area may also be used for after meal coffees and afternoon teas.

Reception billing

The reception/billing position should be positioned near the restaurant entrance, with a clear view of the restaurant and the service points.

Service points

These are the service station points used by a team of service staff: - areas between the kitchen and the customer where the team works. These stations should be strategically placed so that each team has a clearly defined area of the restaurant to serve.

Staff flow

The flow of staff through and around the restaurant needs to be considered, because unnecessary walking will cause staff to become over tired and the customers to become aggravated by a continuous stream of staff walking past their table. By repositioning service stations, it is possible to reduce the distance a team walks while keeping their area serviced.

Seating arrangements

A variety of seating arrangements will create more interest in a restaurant. For example, the use of fixed wall bench seating with free standing chairs opposite gives flexibility of seating choice for the customer. A raised seating area with steps to create a terraced style is also an effective way to break up a large area. Screens, plants and alcoves can also be used to create more intimate seating arrangements.

Restaurant atmosphere

This will be created by a combination of the effects of the decor, the seating arrangements, lighting, the soft furnishings and the staff. It can be a bright, cheerful, relaxed restaurant or a quiet, more formal type of establishment (see page 2).

Once these factors have been considered, it will be possible to select the furnishings and fittings for the restaurant. It should always be remembered that the height of the seats and the tables must be compatible, and that the versatility of lighting will help create a comfortable restaurant for the customer.

FURNISHINGS

All furniture and furnishings need to be chosen carefully as they can have a dramatic effect on the ambience of the restaurant. General points to consider include:

- the type of service and therefore the space required for side tables, service points and trolleys,
- the existing features and furnishings, colour textures and image of the restaurant,
- versatility of use to allow for different types of service and function catering,
- complimentary colours, materials and lighting that will produce the desired effect both during day light hours and at night.

All furniture should be chosen with consideration given to ease of cleaning and maintenance. Poorly cleaned and badly maintained furniture will be unacceptable to the customer, and can lead to a decline in general hygiene standards in the restaurant. It will be necessary to decide the following points during the planning stage:

Combination of fixed bench and free chairs.

Floor fixing table and chairs.

Tables

- Should tables be fixed to the floor or free standing?
- Do they need to be stackable, extendable or linkable?
- Are they to be a variety of shapes and sizes or all identical?
- What height should they be?
- Are they strong enough for their anticipated uses?

Seating

- Should the seating be all chairs or a mixture of bench seating and chairs?
- What height should the seats be?
- Should the chairs be with or without arms?
- Should the chairs be padded or fully upholstered?
- Should the chairs be stackable?
- Should the seating be fixed or moveable?
- Should the surfaces be fully washable or merely wipeable?

Service points

- Should the service points be traditional style sideboards or merely service-top ovens?
- What purpose will these service points serve?
- Is there to be any equipment or cutlery to be kept at these points?
- Should they be fixed or mobile?

Trolleys

- What is to be the function of the trolleys?
- Will they be suitable?

Chilled sweet trolley.

- Are they the correct height?
- Do they run smoothly?
- Are removable shelves required?
- Is there a suitable storage place for the trolleys when they are not in use?

Linen

- Is it in keeping with the restaurant style and colour scheme?

- Is the quality acceptable?
- Is the requirement for 'real linen', or disposable 'linen', or mats?
- Should it be purchased or hired under a contract arrangement?
- Are there suitable storage facilities for the linen?
- Is the linen to be laundered at the premises or under a laundry contract?
- How much will each option cost?

The materials used in the making of furniture, floorings and linens can be found in Part 1.

TABLEWARE

It is necessary to decide what type of tableware is going to be suitable by considering the following:

Flatware

- What should be the size and shape of the flatware?
- What is the style of service?
- What is the number of covers and turnround time in the restaurant?
- Are adequate storage facilities available?
- What is the size and shape of the machine racks for washing?

Cutlery

- What should be the size and style of the cutlery?
- Should it have grooves and patterns?
- Should it be silver or stainless steel and have handles of bone or steel?
- How versatile should it be?
- What is the number of covers and turnround time in the restaurant and the style of service?

Glassware

- What is the design and style of the premises?
- Should the glassware be of strong toughened glass?
- Should it be plain for easy cleaning and handling?
- What is the correct size?
- Is it easily replaced?
- What is the percentage of each type of drink sold?
- Is it versatile to use?
- What is the range of drinks sold?
- What are the washing methods and rack sizes, storage facilities and shelf heights?

Table 4.6 The sizes and uses of cutlery

Cutlery	Example of uses
Soup spoons	Soup when served in a soup plate
Fish knives and forks	Fish courses, fish as a main course, hors d'oeuvres
Large knives and forks	Entrées and main courses
Large forks	Singly or for omelettes, gnocchi, macaroni
Large forks and dessert spoons	Spaghetti dishes
Dessert spoons and forks	Sweets served on plates
Dessert spoons	Consommé served in cups, breakfast cereals, porridge
Small knives and forks	Dessert fruit course (ornate design)
Small knives	Toast, cheese and biscuits, bread rolls and butter
Dessert knives and forks	Savoury course
Tea spoons	Tea, morning coffee, ice cream in coupes, oeuf en cocotte
Coffee spoons	Demi-tasse cups of coffee
Small tea knives	Afternoon tea service
Sundae spoons	Ice cream sundaes
Table spoons	Service of food onto the plate, afters with a large fork
Pastry forks	Pastries served with midmorning coffee

4.2 Types of tableware.

Tableware contributes greatly to the creation of an attractive environment for the customers as the style must be carried throughout the restaurant.

Tableware materials

- *Earthenware*: glazed earthenware is used in the hotel and catering industry. Fireproof earthenware for oven-to-table ware is used for microwave items and bar snack trade.
- *Stoneware:* this is commonly used in the industry. Vitreous stoneware is durable and able to withstand the daily use of a busy restaurant.
- *Porcelain: v*itrified porcelain has the necessary strength for the industry. It is lighter in weight than vitrified stoneware, has a finer finish and is shock resistant. It may be used in microwave ovens, and has a hard, glaze finish that may be cleaned in a dishwasher without damage.
- *Stainless steel:* this is a metal alloy containing 13 per cent chromium. Staybrite contains nickel and chromium. It is widely used for cutlery tableware and for service ware. It should not stain unless it has been overheated or misused.

Table 4.7 The capacity and uses of tableware

Type	**Capacity shape**	**Uses**
Side plate	15 cm round	Bread, pastries, cheese
Entrée plate	20 cm round	Hors d'oeuvre, single fish dish, desserts
Dinner plate	25 cm round	Main meals
Grill plate	30 cm oval	Whole fish, grill specialities
Soup plate	23 cm round, shallow	Soup, Irish stew
Bowl	13 cm round, deep	Breakfast cereals
Dessert plate	13 cm round, deep	Desserts and hot puddings
Soup cup/saucer	Circular	2 handles, consommé
Tea cup/saucer	5 to the litre	Tea, morning coffee
Demi-tasse cup	10 to the litre	Coffee after lunch and dinner

- *Silver:* in tableware 'silver' refers to a metal that has been silver plated. It is used for ornate tableware and cutlery. It is more expensive than stainless steel tableware, and has the disadvantages of becoming tarnished easily.

Trays

Wine waiters use trays or silver salvers for serving and clearing drinks. A 25cm round salver for serving drinks and a larger round salver for clearing drinks. Salvers are covered with a napkin before use to prevent the glasses from slipping.

Rectangular trays are used for the service of breakfast and afternoon teas. In some establishments, rectangular trays are used to carry foods from the kitchen servers to the sideboard in the restaurant.

GUESTS WITH SPECIAL NEEDS

Catering for guests with special needs calls for a caring attitude from the food services and foresight in the planning stages of the establishment. The food service staff should always maintain as normal a service as possible, but with some definite considerations, particularly when the guest first comes into the restaurant.

Elderly guests

- There must be easy access to the table and cloakrooms.
- An upright chair or a chair with arms may be required.
- Elderly guests should be seated at a table with a good view of everything, and not in a dark, alcoved area.

Blind guests

- There must be easy access to their table.
- Blind guests should be seated in an area where no one sound or smell is dominant.
- They should be seated at a table where there is no danger of them being bumped or knocked by other persons passing the table.
- Facilities must be provided for their guide dog and cane, away from direct heat; the dog will appreciate a drink of water.
- If they are unaccompanied by a sighted person, the food service staff should always offer to read the menu to the guests.

Deaf guests

- Deaf guests should be placed where they can see as much as possible.
- They should be seated in a region with even sound levels.
- They should be situated away from loud speakers, music outlets or telephones.
- Staff should always listen carefully to the guest's voice. It may not be very clear.

Wheelchair guests

- Easy access into the restaurant is essential.
- Adequate turning space for the wheelchair is necessary.
- Guests in wheelchairs should be placed at a table where they can see as much of the restaurant as possible.
- If they should never be tucked away in a corner or close to the exit area.

TRADITIONAL ANCILLARY DEPARTMENTS

The traditional ancillary departments are found in larger hotels and restaurants to prevent noncooking tasks from being carried out in the kitchen, which would cause unnecessary congestion at peak times of the day, and disruption to the food production team of staff.

The still room

This still room is found in large establishments next to the restaurant and near to the lounge area. In small establishments, part of either the service room or the kitchen may be equipped to provide still room services. The purpose and function of the still room:

- Breakfast time: preparation of tea, coffee, beverages, provision of toast, breakfast cereals, butter and preserves.
- Luncheon time: preparation of coffee, provision of butter pats, brown bread and butter and melba toast.
- Dinner/evening service: provision of coffee, butter pats, melba toast and bread and butter and after dinner mints.
- Midmorning: provision of coffee and pastries.
- Afternoon: preparation of tea and sandwiches and the provision of scones and cakes.
- Evening: provision of hot drinks, tea, coffee and milk-based beverages.

Equipment likely to be found in the still room includes a boiling water point, a steam injection unit for milk, coffee making equipment, toast making equipment, a refrigerator, cash and billing equipment, sandwich making equipment, a washing up machine or area, refuse disposal facilities, a storage area for food commodities, and cutlery, crockery, trays, napkins and doilies.

The advantages of the still room:

- staff are able to keep a stock of items ready for the food service staff to collect,
- condiments, sugars, etc., are not subjected to the steam and heat of the kitchen,
- morning coffee and afternoon tea trade can be expanded without the use of the main kitchen,
- cash and stock can be controlled separately to give precise costings.

The service area

The service area is between the kitchen and the restaurant, which keeps the food production free from the washing up of tableware, collection of linen and disposal of restaurant refuse. The advantages of the service area:

- food service staff can obtain replacement linen, cutlery, crockery and glassware as and when they are needed,
- all used tableware from the restaurant can be deposited in the service room before the staff enter the kitchen to collect their next order; clean items can be collected at the same time.

DEALING WITH ACCIDENTS AND SPILLAGES

If any item is inadvertently dropped or split, it must be attended to without delay. Always endeavour to cause the least amount of disruption to the guests. A stiff brush and long handled pan can be used to sweep up dry items from the carpet (a suction cleaner would cause too much noise). With liquid spillages on the floor, any surface liquid should be mopped up with a cloth (this is all that can be done until the end of service). If liquids are spilt on the table cloth, it may be possible to mop up any lying on the surface, and then tidy the table by placing a clean slip cloth over the stained or wet area so that the guests may continue their meal.

Spillages upon a guest should be dealt with as soon as they occur. The guests will probably want to sponge off his or her own clothing, but will need a jug of hot water and a clean cloth. The accident should be recorded in the complaints book so that the restaurant manager can deal with any complaints by the guest concerned, or any claims for compensation or dry cleaning costs.

Dealing with complaints

The following steps should be followed:

- find somewhere quiet to speak to the complainant,
- seat the complainant down in a comfortable chair,
- ask for the *actual* fact or action that has caused the complaint,
- encourage the complainant to analyse exactly what has caused them to become upset,
- always deal with complaints promptly (never act in a hurry), and keep calm.

Remember the following:

- Calm: keep an evenly toned voice and listen.
- Care: show that you care about the complainant; ascertain the complete facts.
- Conclusion: is the complaint valid? What is the best way to compensate the guest?
- Consequences: are there likely to be any repercussions? Who or what is responsible? What further action is necessary?

SERVICE AREA – HYGIENE AND SAFETY

The hygiene and safety rules are as relevant to the food service personnel as they are to the food preparation staff (see pages 23–36). However, there are some specific points that should be remembered. The process of washing table and glassware is the vital point where the chain of infection is broken. Poor or unnecessary handling and storage after this cleansing process will result in unclean items being used. A check list of hygiene rules for any food service areas follows.

Storage

- Service dishes should be stacked in piles with their lids inverted.
- Flat dishes should be stacked according to their size and with the top dishes inverted.
- Glasses should be stacked up-side-down on shelves according to their type.
- Cups should be stacked on trays with a clean tray on top.

- Plates should be stacked in plate warmers or on shelves according to their size.
- Cutlery should be sorted and stored on covered trays or in drawers.
- Jugs and pots should be stored on shelves with their handles on the right-hand side.

Handling

- Use trolleys or trays to minimize handling.
- Polish up cutlery when it is to be laid on the table to eliminate recontamination.
- Only use scrupulously clean glass and food service cloths.
- Only pick up cups by the handles and never touch the rim.
- Handle glasses by their stem, handle or base only; never touch the bowler rim.
- Wash hands frequently; never lick fingers.

General

- Never permit tableware to be used in the kitchen.
- Wash flower vases separately and store them on a tray away from tableware.
- Descale tea and coffee pots regularly.
- Empty and wash cruets and sauce containers regularly.

RESTAURANT CLEANLINESS

The procedures for preparing a room have already been described (see page 62). In most establishments, the restaurant is thoroughly cleaned after the end of the breakfast service and before the lunch trade commences - this may be the duty of the food service staff or the housekeeping staff. In either case, it is the responsibility of the food service team to ensure that the premises are clean and ready for the next service. So although the main cleaning is carried out in the mornings, there will be minor clearing tasks to be completed at the end of the other meal services. One task that has to be done at least three times per day is suction cleaning the floor areas. Customers will not like to see crumbs and fluff on the carpet. Other items requiring attention are the chairs, tables, window sills and other flat surfaces. The restaurant cloakrooms and toilets are not usually cleaned by food service staff because of the risk of cross-contamination, but it is their responsibility to check their cleanliness before each service time. Micro-organisms multiply in dust and fluff, and therefore it is essential for staff to wash their hands after the cleaning process has been completed and before any tableware is handled.

ACTIVITIES

- List the materials from which the tableware at your place of work of training are made.
- Look at the storage methods used and make suggestions for improvements, bearing in mind what you have learnt about hygiene and the contamination of food.
- Compile a daily cleaning schedule for your place of training food service area.
- Indicate the cleaning agent and equipment to be used. And the approximate number of minutes each task should take.
- List the areas that need special attention when cleaning food service units. Why do these areas need this attention?

CHAPTER 32 Food service in restaurants and hotels

TRADITIONAL FORMS OF SERVICE

The most traditional forms of service found at a catering establishment are summarized below.

Table d'hôte

In table d'hôte service, the full range of cutlery is laid and then removed should the guest not require that particular course. The menu has a fixed price, offering some choice within each course. The advantage of this type of menu is that it gives the customer a choice of dishes while eliminating the wastage that could occur when an extensive menu choice is used. It may also reduce the size of the kitchen brigade (see figure 4.3).

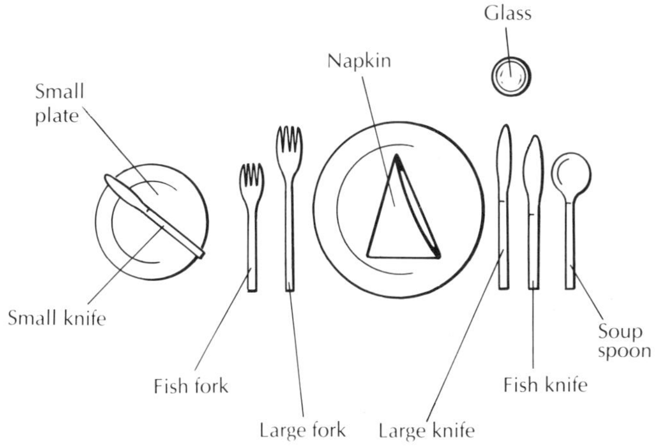

4.3 Table d'hôte setting.

With this setting, the cutlery for the meal is laid, any cutlery that is not required will be removed course by course as the meal progresses by the food service staff.

À la carte

With à la carte service, the minimum crockery and cutlery are laid on the table, additional items being set as the guest's choice dictates (see figure 4.4). The choice of food is extensive and usually priced by each item (see figure 4.5 p. 196). An à la carte meal is often more expensive than table d'hôte, and tends to be prepared to order, so the service time is lengthened. These factors mean that the use of microwave ovens, convection ovens, and high- pressure steamers in the kitchen is essential.

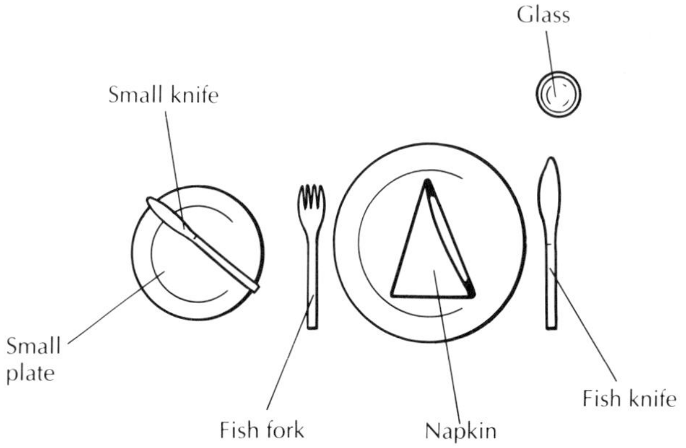

4.4 À la carte setting.

Here any cutlery that the customer will require is brought to the table on a napkin-covered plate by the food service staff, and laid in the appropriate place, course by course, as the meal progresses.
In the above settings the final touches include cruets, butter dishes and flowers placed on the table.

Gueridon service

Gueridon service is a form of personalized service often used with an à la carte or a table d'hôte menu. The food is carved or finished at the table or on a trolley placed adjacent to the guest's table. A gas or spirit lamp is used to finish cooking portions of poultry, meat or fish. Some dishes are completed in this way with a flambé technique or sauce. Vegetables may be tossed in butter or sauce, and then served to accompany the meal. Sweet dishes and fruits may be finished in a similar manner, one of the most popular being the preparation and service of crêpe suzette.

This form of service required a skilled team of staff

APPETIZERS

MOUTAIN CURED HAM WITH EXOTIC FRUITS £9.80

COLLAGE OF LOBSTER, SALMON AND ST. JACQUES £14.10

MARINATED DUBLIN BAY PRAWN SALAD WITH A CITRUS VINAIGRETTE £15.85

TARTARE OF FRESH AND SMOKED SALMON WITH CUCUMBER AND CAVIAR £12.25

BALTIMORE CRAB CAKE WITH SPRING ONIONS AND WHOLEGRAIN MUSTARD SAUCE £10.45

GLAZED ASPARAGUS, MUSHROOMS AND ARTICHOKES SCENTED WITH SORREL £8.65

POACHED CHICKEN ROULADE WITH A GREEN LENTIL SALAD AND HERBS £7.65

AUBERGINE MOUSSELINE WITH A TOMATO BUTTER SAUCE £6.90

BAY SCALLOP NAVARIN WITH GINGER AND CORIANDER £11.45

SOUPS

RICH BEEF CONSOMME FLAVOURED WITH CEPS £5.25

CREAM OF POTATO AND LEEK WITH CHIVES, CHILLED OR WARMED £4.70

SEAFOOD MINESTRONE, ROUILLE SAUCE AND CROUTONS £5.55

FISH

DOVER SOLE, GRILLED OR PAN FRIED £18.30

CASSEROLE OF RED BREAM AND SOLE WITH CUCUMBER AND DILL £14.65

BRAISED MONKFISH IN BROUILLY WINE WITH HOMEMADE NOODLES £12.15

LIGHTLY SMOKED TOURNEDO OF SALMON ON A FONDUE OF LEEKS WITH TRUFFLE £16.95

SOUFFLED FILLET OF TURBOT WITH SAFFRON AND SWEET PIMENTO £19.35

MEDALLION OF SCOTCH SALMON SIMMERED WITH VEGETABLES £15.45

MEAT

ROSETTES OF BEEF CRUSTED WITH SESAME £15.40

"TOURTIERE" OF NORFOLK DUCKLING WITH WILD CRANBERRIES £16.15

FILLETS OF SOUTHDOWN LAMB WITH BAKED COURGETTES AND AUBERGINES £14.70

CURLS OF CALVES LIVER WITH GLAZED ONIONS, MUSHROOMS AND BACON £13.60

CHATEAUBRIAND ROASTED WITH HERBS AND GARLIC, SAUCE FOYOT (TWO PERSONS) £35.50

CUSHION OF MIGNON AND SWEETBREAD OF VEAL FLAVOURED WITH SAGE £15.50

SUPREME OF CORNFED CHICKEN WITH A RED ONION "CONFIT" £13.25

GRILLED SIRLOIN STEAK, BEARNAISE SAUCE £15.85

SELECTION OF MARKET VEGETABLES
OR
SEASONAL SALADS

4.5 An à la carte menu.

who can complete the cooking of the foods as the menu and form of service dictates. The staff must also be able to present and serve the foods in an efficient and pleasing manner to the guest and have the air of quiet confidence and *salesmanship*.

Plated/family service

Plated family service is a less formal form of service used with table d'hôte menus. Here the main meal is presented to the customer on a 25 cm diameter plate. The accompanying vegetable dishes are placed on the table, thus allowing the guests to help themselves. Full plated service promotes a fast, efficient service and ensures good portion control. Plated service is used in cafeterias where customers may help themselves or be served by staff at the hot plate.

PREPARATION OF THE RESTAURANT – MISE EN PLACE

Before each service time, the food service staff have to spend time ensuring that the restaurant is clean, prepared and ready for the customers. It is normal practice for the restaurant tables to be laid up ready for the next meal at the end of each shift. For example, after luncheon they are laid up for dinner, and after dinner they are laid up for breakfast. This enables the in-coming team of staff to prepare the accompaniments and the room without having to relay all the tables. It also ensures that the restaurant always looks tidy and ready for service. Since many guests have breakfast in their rooms, the number who come down to the restaurant is often low. The service is therefore usually finished by approximately 9.30 am, which gives ample time for a thorough clean before the cycle begins again with luncheon.

Order for laying the table

- First, the table and chairs should be positioned appropriately for the anticipated number of customers.
- The table cloth is then carefully laid over the table, allowing an even drop each side.
- The cover plates, and side plates are placed on the table, one per customer.
- The appropriate cutlery is polished and placed at each side of the cover plate.
- The glasses are polished and placed to the top and right of each place setting.
- Finally, the napkins are folded and placed appropriately for each setting. The cover plate may then be removed.

The sideboard

In order to serve correctly, promptly and efficiently, it is necessary to equip the sideboards with any service equipment, cutlery and crockery that the waiting staff will require throughout the meal service.

Condiments

These should be cleared and refilled daily, and care should be taken to ensure that they sprinkle or pour as effectively as possible. Never refill cruets while they are damp as this will cause the salt or pepper to clog the cruets, making sprinkling impossible.

Butter pats

These should be kept refrigerated until required as they will soon become soft and runny. If cut butter is used, it is normally kept in a container of iced water as this will prevent the pats from sticking together. Butter pats are normally prepared in the still room.

RESTAURANT SERVICE

Taking the order

Three copies of the guest's order are written on a triplicate pad. In addition to the foods they have ordered, it is necessary for the service staff to record the date, table number and number of guest's.

- The top copy is for the kitchen so that the order can be prepared.
- The duplicate copy is for the cashier so that the bill can be prepared.
- The flimsy copy is kept by the waiter so that the orders can be checked.

It is essential for food service staff to remember that their job is to sell the items on the menu to the guests. In order to do this, they must have a good knowledge of the menu, a pleasant manner and patience to allow the guests to make up their own minds about their choice, without becoming impatient. If a guest orders only a main course dish, it is possible to increase sales by asking if a preliminary course is required or, perhaps, an additional portion of vegetables or a salad to be served with the main course. Even though the food service staff are there to sell the food, they must never become too 'pushy' or overbearing or this could have a detrimental effect on future business. The kitchen brigade and the restaurant brigade are dependent upon each other, neither will be able to work effectively without the other.

Serving the guest

Before the first course is served, there it time to make any necessary adjustments to the guest's cutlery, so that when the food is ready it may be served without delay. If any additional cutlery is necessary, each place setting is adjusted by a member of the staff removing unobtrusively any unwanted cutlery and putting the new cutlery in its place.

When the first course is ready, it is collected from the kitchen service area by one or more members of the station team and taken to the sideboard. Here the dishes are checked and then they are served to the guests.

For small parties of up to seven or eight, it is usual to serve the hostess first, followed by the other ladies, then the remaining guests and, lastly, the host. If dishes or plates are hot, a clean food service cloth may be used by the food service staff.

Food service rules

Below is a list of the basic food service rules that should be practiced at all times.

- Greet customers cheerfully and assist with their coats.
- Be ready to sell to customers by the use of positive speech.
- Record orders clearly and pass these on to the kitchen and cashier promptly,
- Never assume that you know what a regular customer will choose, always wait for them to decide.
- Remember that you are there to serve the customers, keep your eyes on all the tables in your station and maintain a pleasant facial expression.
- Adjust cutlery by using a napkin-covered plate, and avoid reaching in front of a customer.
- Serve customers with food from their left-hand side where ever possible.
- Clear all plates and glasses from the customers right-hand side, as quickly and quietly as possible.
- Serve all beverages from the customer's right-hand side.
- Crumb down the table between clearing the main course and serving the sweet course using a napkin-covered plate and a folded, food-service cloth or clean napkin.
- Change ash trays as soon as possible if they have been used.
- Never allow a thumb or finger to come into direct contact with the food.

- When serving from a trolley or side table, always position yourself so that the customer has a clear view of the proceedings.
- Try to anticipate the customer's requirements.
- Offer appropriate accompaniments for the customer's meal.
- Never appear to hurry or harass your customers.
- Always maintain a clean, neat and tidy personal appearance and an orderly work station.
- Say 'goodbye' to your customers when they are ready to leave and assist them with their coats.
- Work as a member of a team, and be as helpful as you can.

Breakfast service

A less formal type of service is frequently offered in the restaurant for breakfasts (see below). Most continental breakfasts are served to guests in their room, so the number eating in the restaurant will normally be considerably less than the total number of residents. As soon as the guest is seated, check the room number so that the bill can be made. Then offer the guest a choice of tea or coffee. Many establishments have a self-service arrangement for breakfast where the guests select their own choice of fruit juice and cereals followed by their choice from hot foods at a heated serving area or trolley. This type of service means that guests can be accommodated with the minimum of formality and time. The food service staff will be there to assist the guests, to serve beverages and to clear their tables between courses.

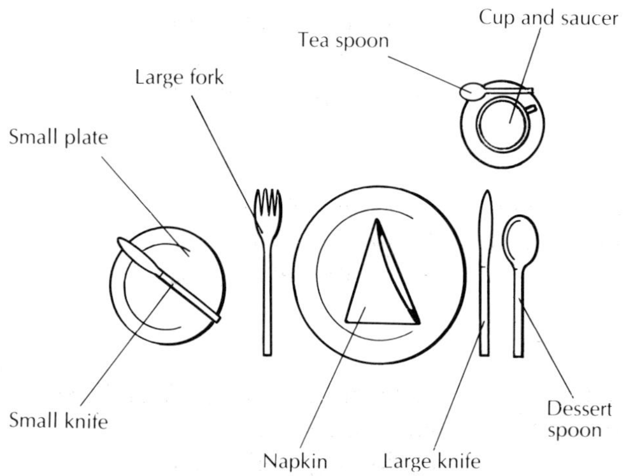

4.6 Breakfast setting.

LOUNGE SERVICE

Lounge service staff are concerned with the service of morning coffee, afternoon tea and alcoholic and nonalcoholic beverages throughout the day. The alcoholic beverages are obtained from the dispense bar, adjacent to the restaurant, while all other items for this service are prepared in the still room. Orders are taken at any time during the day and recorded on a triplicate order pad so that guests can be billed appropriately.

Lounge service is there for the guests' convenience, and so the staff should be as quiet and unobtrusive as possible. Guests will sometimes sit and talk in groups for a considerable length of time - they should never be made to feel they should get up and move. All service and clearing should be carried out so that guests are comfortable at all times.

Some hotel lounges have a pianist or guitarist playing in the afternoon and evening. If this is the case, guests should be asked whether they wish to sit near the music or some distance away.

Foods and beverages for lounge service are served by lounge food service staff using trays and, maybe, a trolley for cakes and pastries. All trays are carried to the guest, where the items are placed upon the table. It is customary for tea and coffee to be served by the pot for this type of service. The guests may then pour for themselves if they wish.

ROOM SERVICE

The purpose of a room service department in a large or luxury hotel is to provide guests with a complete service of meals, snacks and beverages in their rooms or suites at all times of the day or night. It is organized from an area adjacent to the kitchens and still room, or from a completely separate area with just the main meals coming from the main kitchen.

Trolleys and trays are used for room service and are all laid up ready so that the service is prompt. The trays are clothed, laid and then stored in racks; the trolleys are clothed ready to be laid. The side flaps are lifted when the trolley has been delivered and the items on the trolley are then adjusted so that the meal can be served properly.

At the guest's door, the room service staff will knock on the door and await instructions to 'come-in'. The trolley is wheeled into the room and placed in a suitable area, either near the table or window where the guests can sit comfortably. The trolley may be adjusted to act as the dining table itself, or the guests can sit at the table already in the room. Once the table has been laid, the meal should be shown to the guest to check that the order is correct. The room service staff should then leave the room quietly and, as with restaurant orders, the order is recorded in triplicate.

CARVERY SERVICE

Carvery service is becoming increasingly popular. They are found in hotels, out-of-town public houses and city restaurants, and are usually open for luncheon and dinner service and are normally licensed to sell alcohol (see figure 4.7, p 200).

The service is a combination of self-selection from the carvery area for the main roast meat course, and the guest being served by restaurant staff for their first and sweet courses. A typical service sequence could be as follows:

- the guests arrive,
- the guests are seated by the food service staff,
- the menu is presented and orders for the first course and beverages are taken,
- the first course and beverages are served and cleared,
- the guest make their own selection from the carvery for the main course,
- the table is cleared by food service staff,
- the sweet course is served by the food service staff, and
- the coffee is served by the food service staff.

When running a carvery service, the traditional restaurant brigade system is not usually required. Often the staff are allocated a particular course of the menu to serve, so that all the first courses are served by one team of staff, all the sweets are served by another team, and so on. Another method is to use the restaurant station system with each station team serving and clearing their own section of the restaurant.

Accompaniments for dishes

Some dishes have particular accompaniments offered with them to enhance the overall flavour and enjoyment of the dish. The food service staff need to know which dishes need accompaniments so that these can be prepared before the service time and offered at the correct time during the meal.

Table 4.8 Accompaniments for dishes

Dish	Accompaniments	Notes
Chilled		
Hors d'oeuvre	Oil and vinegar	Fish knife and fork
Tomato juice	Worcester sauce	Teaspoon on side of plate
Grapefruit cocktail	Caster sugar	Teaspoon or grapefruit spoon
Chilled melon	Ground ginger, caster sugar	Dessert spoon and fork.
Fish (cold)		
Smoked trout	Horseradish sauce, brown bread and butter	Fish knife and fork
Smoked salmon	Cayenne pepper, peppermill, brown bread	Fish knife and fork
Potted shrimps	Cayenne pepper, peppermill, hot toast	Fish knife and fork
Miscellaneous		
Oysters	Cayenne peppermill, chilli vinegar, Tabasco sauce, brown bread and butter	Serve on crushed ice, fingerboard and spare napkin
Snails	Brown bread and butter	Snails tongs and snail fork

Table continued

Table 4.8 continued

Dish	Accompaniments	Notes
Vegetables		
Asparagus	Hot – Hollandaise sauce/buerre fondue, cold – vinaigrette	Asparagus tongs and large fingerbowl, spare napkin
Corn on the cob	Buerre fondue	Corn on cob holder
Globe artichoke	Hot – Hollandaise sauce/burre fondue	Large fork, spare finger bowl
Soups		
Petit marmite	Grated Parmesan cheese, flutes	Dessert spoon
French onion soup	Grated Parmesan cheese, flutes	Soup spoon
Minestrone	Grated Parmesan cheese, flutes	Soup spoon
Bouillabaisse	French bread slices	Fish knife and fork and dessert spoon
Purée soups	Croutons	Soup spoon
Crème soups		Soup spoons
Farinaceous food		
Ravioli	Grated Parmesan cheese	Dessert fork and spoon
Spaghetti	Grated Parmesan cheese	Large fork and dessert spoon
Gnocchi	Grated Parmesan cheese	Dessert fork and spoon
Fish (hot)		
Whitebait	Cayenne pepper, brown bread and butter	Fish knife and fork
Grilled herring	Mustard sauce	Fish knife and fork
Poached salmon	Hollandaise sauce	Fish knife and fork
Deep fried fish in batter	Tomato sauce	Fish knife and fork
Fried fish in egg and breadcrumbs	Tartar sauce	Fish knife and fork
Kedgeree	Mild curry sauce	Fish knife and fork
Poultry		
Roast chicken	Bread sauce, roast gravy	Large knife and fork
Roast duck	Apple sauce, roast gravy	Large knife and fork
Roast turkey	Cranberry sauce, bread sauce, roast gravy	Large knife and fork
Meats		
Roast beef	Horseradish sauce, roast gravy	Large knife and fork
Roast lamb	Mint sauce, roast gravy	Large knife and fork
Roast mutton	Onion sauce, roast gravy	Large knife and fork
Roast pork	Apple sauce, roast gravy	Large knife and fork
Boiled mutton	Caper sauce	Large knife and fork
Boiled gammon	Mustard	Large knife and fork
Boiled salt beef	Natural cooking liquor, mustard	Large knife and fork
Boiled beef	Natural cooking liquor, mustard	Large knife and fork
Curried meats	Poppadoms, mango chutney	Large knife and fork
Mixed grill	English and French mustard	Large knife and fork
Grilled fillet steak	English and French mustard	Large knife and fork
Savories		
Welsh rarebit	Cayenne pepper	Side knife and dessert fork
Devils on horseback	Peppermill	Side knife and dessert fork

WELCOME

TRADITIONAL ROASTS

Toby Carving Rooms offer superb fresh roast meats and seasonal vegetables from the Carving Table.

EVEN MORE CHOICE

And now there's a range of exciting new starters, sizzling main courses and mouthwatering speciality desserts.

SAME GREAT VALUE

They're still great value for money - you can help yourself to freshly cooked vegetables, followed by a dessert, included in the price.

AND FOR CHILDREN

Young children can now enjoy a three course Shoe People meal for just
£1.95

If you wish to reward the people who have served you please do so. Service charge is not included. All prices include VAT.

All items subject to availability.

STARTERS

SOUP ~ 95p
SAVOURY MUSHROOMS ~ £1.85
PRAWN COCKTAIL ~ £1.95
MELON ~ £1.50
MUSHROOM CORIANDER ~ £1.85
DUCK AND ORANGE PÂTÉ ~ £1.70
CAPELLETTI CARBONARA ~ £1.85
MARINA HERRINGS ~ £1.65
SOLE AND CRAB COMBO ~ £2.45

MAIN COURSES

A TOTAL PRICE FOR THE MAIN COURSE & DESSERT

From the Carving Table

Single £7.85 *serving*

Children (12 years and under) £4.15 Ask our chef for a child's portion.

ROAST BEEF
ROAST TURKEY
ROAST JOINT OF THE DAY

Every evening at least one other joint is added for your enjoyment. Help yourself to a selection of freshly cooked vegetables.

Something Different

STEAK SIZZLER ~ £9.25
SALMON FILLET ~ £9.25

Vegetarian Choice

CHEESY LEEK & POTATO BAKE ~ £6.50
MUSHROOM STROGANOFF ~ £7.95

Side Dishes

GARLIC BREAD ~ 90p
SIDE SALAD ~ 90p
BUTTON MUSHROOMS ~ 90p

DESSERTS

The following desserts are included in the price of your main course. Choose from:

OLD FASHIONED SHERRY TRIFLE
TRADITIONAL APPLE PIE
REAL DAIRY ICE CREAM
TROPICAL CHEESECAKE
FRESH FRUIT SALAD
SPOTTED DICK
CHEESEBOARD

Speciality Desserts

Alternatively why not spoil yourself with one of our tempting speciality desserts, for just 95p extra on the price of your meal.

TOFFEE SUNDAE
RASPBERRY & BLACKCURRANT FOOL
ALABAMA CHOCOLATE FUDGE CAKE
TREACLE SPONGE
KNICKERBOCKER GLORY
CHOCOLATE MINT PARADISE

Coffee, Tea and Liqueurs

All accompanied by chocolate mints.

FRESH COFFEE & CREAM ~ 60p
TEA ~ 55p
FRESH COFFEE WITH DOUBLE CREAM ~ 75p

Speciality Coffees

THE PERFECT WAY TO END A MEAL ~ £1.50

Wine List

Why not complement your meal with a bottle of wine? Our extensive list of wines offers quality and value for money. A range of beers, spirits and low & non-alcoholic drinks is also available.

Bookings

We are pleased to accept bookings. Telephone, or call in and the manager will reserve your table.

This Toby Carving Room is one of over 100 throughout the United Kingdom.
For more details, please ring 021 452 1911 or write to the Operations Director, Toby Carving Rooms at Hagley House, 83 Hagley Road, Edgbaston, Birmingham B16 8QG.
All prices correct at time of going to press (October 1990).

S/10/90

4.7 Carvery menu.

COFFEE SHOPS

Coffee shops are frequently provided in hotels to enable the guests and members of the general public to obtain snacks, light meals and beverages. This facility relieves pressure on both the main restaurants and the room service facilities. It also provides a social meeting point within the hotel for guests and their friends. The menu is individually priced so that guests may order a complete meal or just a coffee and pastry as they please.

Coffee shops are normally run using a less formal style of food service than is found in the main restaurants, with an attractive display of foods in a heated/chilled area. They are open all day, every day and evening. Some encourage members of the general public to use the facility and thereby see the inside of the hotel. This provides a useful form of indirect advertising.

If the coffee shop is sited at an airport or railway station, there is potential for exploiting trade from passengers who have time to wait between stages of their journey. Some of these coffee shops, and those in shopping precincts, are self-service because of the large numbers of customers per hour.

In all cases, coffee shops are attractively and cheerfully decorated with the use of indoor plants, pictures, screens and benches to break up the seating area. There should always be adequate space between the seats for hand luggage, briefcases, shopping and umbrellas.

ACTIVITIES

- Compile a full English breakfast menu.
- Calculate the food cost of two fried eggs, two rashers of back bacon, one tomato, one round of fried bread, 2 oz mushrooms.
- List the items that you would expect to find on a continental breakfast menu in a hotel.
- Discuss the reasons why carvery service is becoming increasingly popular in the UK. If possible, visit a carvery and note the differences between a traditional kitchen and restaurant and the carvery kitchen and restaurant.

CHAPTER 33

Cafeteria service, vending and function catering

Self-service is a quick form of service used in cafeterias, where customers select their food from the display area and pay for it before taking their tray to a vacant table in the adjacent dining area. Food service personnel perform very different tasks and roles from those performed in the more traditional forms of service. Their duties are primarily concerned with the serving of food requested by the customer and the selling of food directly to customers who often need encouragement to purchase items from the menu.

It should be remembered that there must be an efficient system for replenishing the supplies of foods on the menu from the kitchen, where foods must be kept at room temperatures of below 5°C or above 65°C.

This type of service is found in numerous establishments including schools, colleges, halls of residence, staff cafeterias in industry, hospitals and roadside catering establishments. For success, it is essential that the speed of service is quick and efficient. Delays may be caused for various reasons, the most common causes being:

- the cashier is unable to take money and give change fast enough,
- the customer is unable to decide which foods to purchase,
- trays are not available for customers to use,
- foods are not available because service staff have not restocked in time,
- the service staff are too slow in serving the customers.

It is essential that prompt and efficient clearing is maintained in the dining area at all times. Tables that are cluttered with used crockery indicate a lack of supervision, training and hygienic practice. This will discourage customers from entering and cause a slack attitude among the staff.

LAYOUT AND SEATING

There are numerous types of cafeteria layouts. A combined hot and cold plan would normally be arranged in a basic order as shown in figure 4.8.

The seating capacity must be estimated in order that all the customers are able to sit down at any one time. This is estimated by dividing the service time by the average eating time, and then dividing the resulting figure by the number of people served. The following details will serve as an illustration:

Service time	12 minutes
Average eating time	20 minutes
Number served	720 customers

From these details, it can be calculated that the tables are used 6 times in a 2 hour period (120 ÷ 20). The numbers seated at any one time is 120 (720 ÷ 6). Therefore, minimum seating needed is 120 seats. With a suggested overflow of 30 seats, the total seating required, 150 seats.

MENU → TRAYS → COLD FOOD → SWEETS → HOT FOODS

MENU: *Large display*
TRAYS: *Cleaned and stacked*
COLD FOOD: *Salads and sandwhiches*
SWEETS: *Cold/hot selections*
HOT FOODS: *Hot main meals and vegetables*

↓

BEVERAGES → CASHIER → CUTLERY → CUSTOMER

BEVERAGES: *Soups, teas. coffee and cold drinks*
CUSTOMER: *Goes to dining area*

4.8 Basic layout of a cafeteria style outlet.

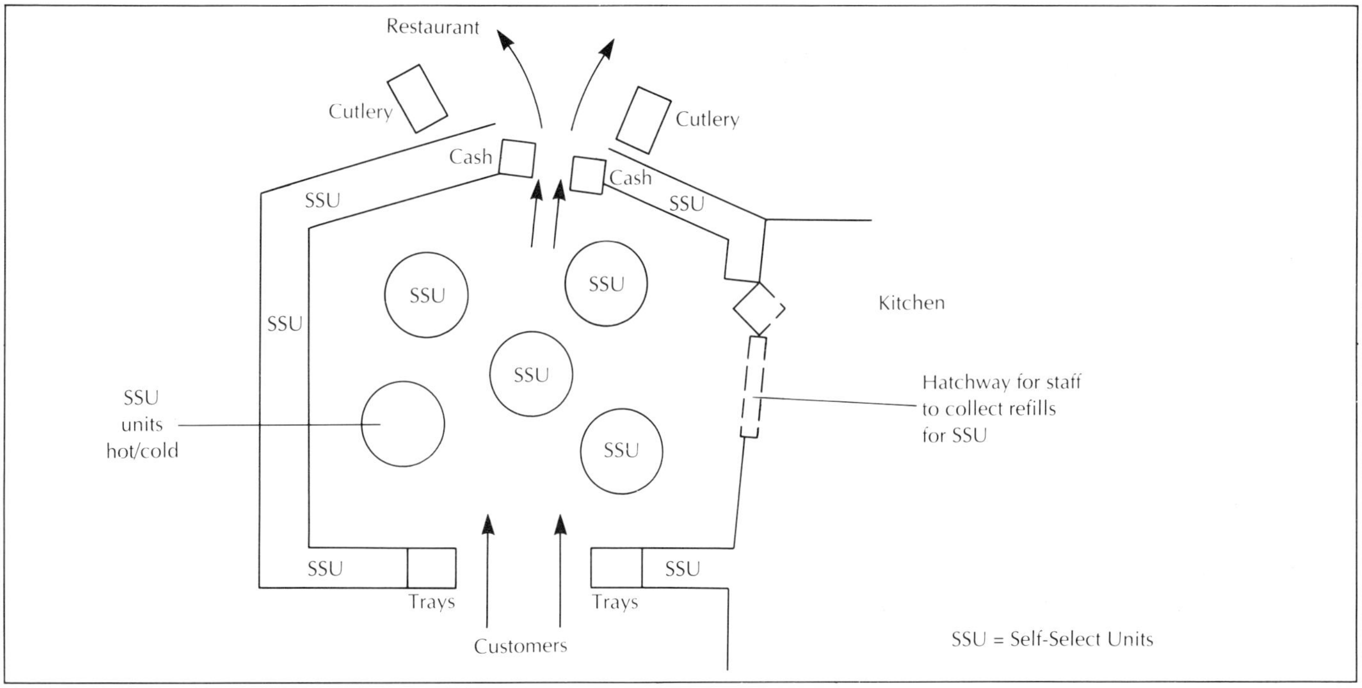

4.9 Plan of a self-service cafeteria.

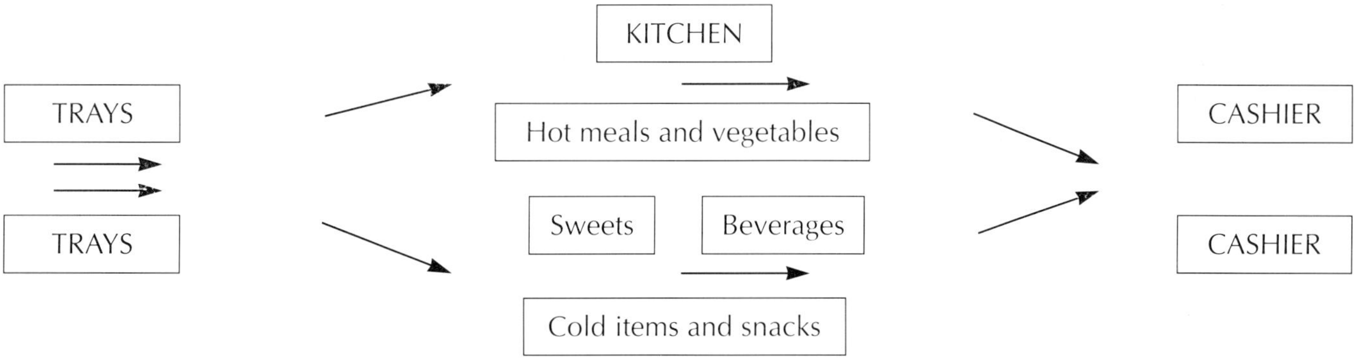

4.10 Layout of a freeflow, self-service cafeteria.

FREE FLOW SELF-SERVICE

This is where the customer collects a tray and then walks around self-service counters that are placed in an informal arrangement that permits the customer to select the food of his or her choice before arriving at the cashier to pay for the meal (figures 4.9 and 4.10).

This type of service relies upon the following points:

- clearly displayed menus at the entrance area,
- clearly named and priced foods on display,
- displays always well stocked and clean and tidy,
- a logical layout of the food,
- constant replenishment of popular dishes,
- cashiers who can memorize the prices and give correct change quickly.

Another self-service selection arrangement that is frequently used, relies upon the customer having a clear idea of their intended purchases before they enter the service area. These are arranged with the hot foods likely to constitute a main meal being served from one area, while the cold items and snacks being served from other areas.

PORTION CONTROL

This is an important consideration and the various methods discussed on page 136 are particularly relevant in this type of operation. Here we see the use of measured capacity ladles and scoops, individual pies and puddings and bowls and dishes used to denote the portion size for soups and salads. Profits may soon be reduced if careful control of the portion sizes are not constantly monitored.

THE CAFETERIA MENU

To ensure an efficient, successful cafeteria service there must be adequately trained staff and suitable menus and prices for the potential customers. The catering manager must consider the following aspects before deciding on a cafeteria menu.

General

- The preparation time available, the staff capabilities, the staff skills and the equipment available.
- The ability of the chefs to keep food hot or cold and at top quality, and their ability to cook/chill and reheat food upon demand in a microwave, convection oven or high-pressure steamer.
- The need to create customer appeal, customer recognition of dishes and the acceptability of dishes; the variety of colour, texture, taste and shape of foods.
- The prices being acceptable to customers, and the selection offering a good variety of dishes to make acceptable combinations for meals.

Once these aspects have been considered, there are two types of menu that can be used.

Standard

- A standard menu gives standard cost levels, preparation times, procedures and equipment requirements.
- It allows for minimum items in stock and limited staff skills.
- The customer becomes familiar with the menu which thereby ensures acceptance.

Skeleton menu

- A skeleton menu has the advantages of the standard menu plus a variety of one or two items in each menu module.
- It increases customer satisfaction and staff interests in their situation by providing variety.

Displaying the menu is the main method of attracting customers into the cafeteria area. It should be displayed in an attractive, clearly priced manner with pictures or photographs to help sell the products. Parts of the menu that apply to a section of the services area are often displayed on the back wall to aid customer selection.

CAFETERIA PREPARATION

Preparation starts at the end of the previous day's work. The cashier empties the cash register and banks the cash, records the takings, and deposits the 'float' in the safe. The general assistants, return food items to the kitchen; clear the beverages from the hot and cold units; clean all the service areas, display cabinets and cold shelves; and clean the service area including the floor areas. Before service starts in the morning the following tasks must be completed:

- the hot plates must be switched on, check containers within, and the crockery numbers,
- complete the menu boards and check the prices,
- the beverage units must be rinsed through, the power switched on, and refilled,
- the customer seating area should be cleaned and checked,
- the cutlery, trays, napkins and water glasses are put in place,
- the condiments and sundries are prepared and placed in position,
- the cold food is placed in the chilled display areas and the reserves in cold cabinets,
- the beverages and cups are prepared,
- the hot foods are placed on the hot plate and bain-marie,
- the service utensils are collected,
- the cashier prepares the till and checks the menu, prices and float,
- the staff position themselves ready for service.

CAFETERIA CLEARING METHODS

There are two main clearing methods used in self-service dining areas. Cafeterias that are open to the general public are normally cleared by a catering assistant and trolley. The self-clearing method is used in welfare units such as industrial canteens and staff cafeterias.

Catering assistant and trolley

- The trolley is pushed between tables and the catering assistant clears the used crockery items and trays.
- With a disposable cloth and hand-held spray filled with a germicidal detergent solution, the assistant cleans the table for the next customer, without passing bacteria from one area to another.
- An adequate number of trolleys is essential so that

one full trolley may be unloaded as a clear one is taken out by the assistant.

- It is necessary to wash the trolleys after each use to prevent bacterial growth.

Customer self-clearing

- Customers are asked to clear their own used crockery and trays to a hatchway or conveyor belt when they have finished their meal.
- The clearing area must be placed near the exit door and positioned to prevent any cross-flow of traffic between new customers and those who are leaving.
- Staff are required to keep this clearing area free from piles of dirty crockery.
- Scraps of uneaten food must be removed to prevent bacterial growth.
- The tables need to be cleaned for new customers.

VENDING

This is where the customer selects food from a vending machine - the required sum of money is paid and he or she instantly receives the goods. The choice of goods will be governed by one of the following factors:

- recognition of a brand name - on confectionery and carbonated, nonalcoholic drinks,
- recognition of the product though not the brand name - on items such as biscuits and snack products,
- the attractiveness of the product itself - making it an immediate sale because it looks irresistible.

The last factor is the one that will concern the caterer, as it is attractiveness that governs the sale of light meals, sandwiches and pastries from vending machines. Most vending machine manufacturers will recommend the type of packaging that is suitable for their machines. All packaging should be noncrushable, clear, clean-looking and designed to give maximum visual appeal to the food inside. The pricing of vending machine products should be competitive so that the customer is not disinclined to purchase on price grounds alone. The presentation is vital and needs to be planned with the maximum shelf life of the product in mind. For example, it is no use planning a garnish that will spoil after 3 hours if the product has a 12-hour shelf life.

Specialist product vending machines are now in use, for example, machines for chips, hot dogs and beefburgers. These are not in general use in traditional food service outlets, but can be found in amusement centres and outdoor areas.

Types of vending machines

Beverages

Hot beverage vending machines work on the principle of customer selection, the cup being released already containing the appropriate dry ingredients and boiling water being dispended into the cup. The types of beverages that are commonly for sale in these machines are, coffee (black and white, decaffeinated and sweetened), tea with milk (unsweetened and sweetened), lemon tea, soup, blackcurrant juice and chocolate. The caterer may alter the range to suit the customer preferences - these change in accordance with the seasons of the year (more soup and hot chocolate are sold in the winter and more lemon tea and blackcurrant are sold in the summer months).

Cold beverage vending machines dispense chilled beverages that may be still or sparkling. Here there is no water heating mechanism, but a chilling device instead. The flavours of beverages likely to be dispensed from these type of machines are orangeade, lemonade, blackcurrant, strawberryade and lime juice cordial. Prepacked beverage vending machines dispense cartons and cans of brand name beverages from a chilled holding cabinet. These are very popular with younger customers, but they are bulky because of the large volume of space required to store the cans and cartons.

In some hotels, these machines are available to residents only so that they may purchase alcoholic beverages at times when the bars are closed. These machines need proof of resident status, usually provided by a disc attached to the room key tab.

Chilled cabinet vending machines

Here the individually packaged items are stacked in a chilled cabinet that will maintain the food at a safe temperature of 3°C, and can be acceptable and safe for a maximum of 24 hours, though 12–16 hours is the normal time practical. The items should be clearly labelled and date stamped to enhance sales and customer acceptability. Many of these machines work on the rotating drum principle.

These types of vending operations may include a microwave oven so that it is possible to purchase a chilled meal and then, using a token provided, place the meal in the microwave where it will be regulated to an appropriate and safe eating temperature. This type of machine is often found alongside other types of machines which dispense chilled snacks and sandwiches, chilled meals, hot beverages and cold beverages. A range of machines are sometimes found in a work place that runs 24 hours a day, or in hospitals and some roadside service areas. (see figure 4.11)

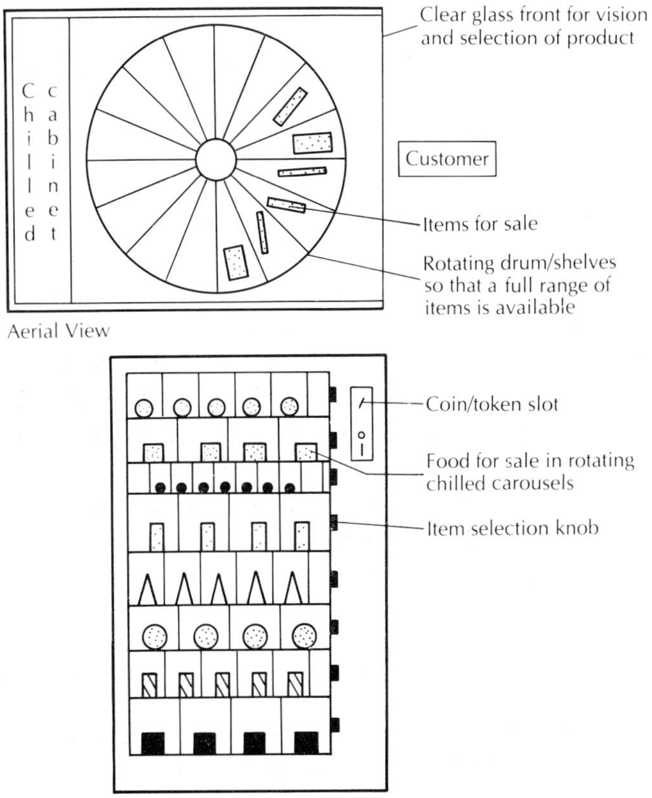

4.11 *Chilled cabinet vending machine with rotating drums.*

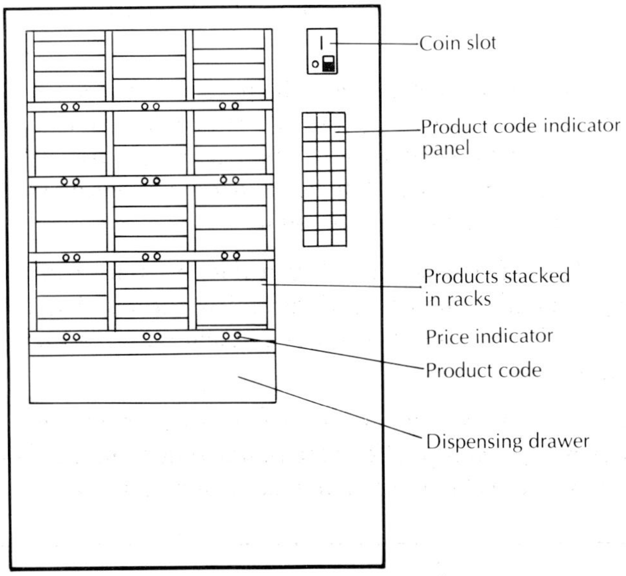

4.12 *Ambient temperature vending machine.*

Ambient temperature cabinets

These dispense any nonspecialist item that the caterer wishes to sell, such as potato crisps, biscuits, sweets, chocolate and nuts. The price of each item is indicated at the front of the row of stock with its choice letter (see figure 4.12).

The advantages and disadvantages of vending machines

There are some establishments that use vending machines without any cause for dissatisfaction, and many establishment use vending machines in conjunction with more labour intensive forms of food service. Nevertheless, there are some points that should be considered before a decision is made to increase or decrease the volume of vending sales in any establishment.

Advantages

- The items for sale are always available.
- A standard range of items is held in stock (or the leaser will restock machines in some instances).
- The leaser will undertake break down repairs and servicing on a prompt call out basis.
- The machines provide a comprehensive service to customers.
- They are easily installed and minimum staffing is required.
- They are cost saving because they are labour and time saving.

Disadvantages

- They tend to be impersonal and noisy.
- In some public areas, they tend to attract vandals and litter.
- There is some customer resistance experienced to the concept of vending.
- There is some customer resistance to the quality of the product for sale, particularly hot beverages.
- They can be expensive to licence and maintain under contract.
- There may be only minimal profit.
- They require the supply of electricity and, in some, mains water as well.

Staff training

The refilling and cleaning of machines will be the responsibility of the food handling or food service personnel. Staff need to know how to:

- open the machine,
- empty coins and free stuck coins,

- clean the inside of the machine to avoid cross-contamination,
- switch off electricity and water supplies in an emergency,
- restock the machine,
- change the pricing structure of the products,
- rectify simple faults such as wrongly stacked products, coins or cups.

Any break downs will cause a loss of sales, and so it is essential to clean and restock machines properly and regularly to maximize sales.

FUNCTION CATERING

Function catering is a busy sector of the catering industry which has enormous potential for growth and development. There are always meetings, exhibitions, receptions and private parties being held in every town and city in the UK. It is the caterer's job to ensure that his or her company is the one that is awarded the catering contract. Table 4.9 lists the five planning stages for successful function catering and gives examples of the types of operations and tasks to be considered at each stage.

BUFFETS

Buffets on the caterer's premises may be for very small private functions for less than a dozen covers, or they could be for several hundred persons. The overriding concern with the provision of buffets is the temperature at which prepared foods are held prior to their service to the guests. All cold foods for a buffet must be kept refrigerated at under 5°C until the moment that they are to be served. Buffet items take up a large amount of refrigerator space – an aspect that must be allowed for while a buffet is being planned. Equally, any hot items for the buffet must be served at a temperature above 65°C to ensure safety from harmful bacteria.

Buffet service

A buffet can be either a finger buffet, where all the foods may be eaten without the need of cutlery, and each item is 'bite size', so it can be easily handled and eaten by the guests; or a fork buffet, where the foods are slightly more substantial in size and are designed to be eaten using a fork. These items are often accompanied by salads and, sometimes, small, buttered potatoes. In either case the guests come up to the buffet area and help themselves, or are served by staff as the style of the function dictates.

Table 4.9 The planning process

1. *Decide*	The numbers involved, the space required, safety requirements, the menu(s) and beverages, type of service and entertainment (cost per head/function).
2. *Arrange*	Room bookings/marquees and awnings. A written contract for client and book entertainment. Check safety regulations (fire, electricity, emergency exits). Apply for a liquor licence/extension. Arrange for transportation of equipment, fittings and food.
3. *Liaison*	Plan menus with the executive chef. Plan tableware with the restaurant manager. Plan beverages and wines with the beverage manager. Plan for table linen and cloths with the housekeeper. Contact the personnel department for hire of extra staff. Contact a florist for floral decorations. Consult the security officer for safety aspects. Contact the billing office to open a function account.
4. *Advance organization*	Printing of menus and name cards. Staff are briefed and trained if necessary. Check entertainment, amplifiers and permitted sound levels. Check microphones, lighting and other electrical requirements. Plan the furniture layout in relation to safety and emergency exits. Plan staff duties and rotas.
5. *Organizing the event itself*	The setting up of the function area. Liaise with the housekeeper, florist, executive chef, restaurant manager and beverage manager. Keep the billing office informed of all charges. Be on hand to liaise with the client to ensure that the function is a success.

The main disadvantages of this style of service is that guests frequently end up in a queue for some time before they can be served. It is therefore essential that this potential problem is addressed during the planning stages so that it is eliminated or reduced as much as possible. Very often, the layout of the buffet service area itself can be planned in such a way that more than one area can be used for guests to help themselves (see figure 4.13). Alternatively, for larger buffets a rectangular service area will provide even more space.

Back up for buffets is essential. The service areas must be well stocked throughout the service time. Often, refrigerated boxes below the service tables are used to house replenishment trays of food. These can be utilized immediately, and are useful where the staff have little access to the kitchen once the service has started. Food looks more attractive if more than one level of service is used, and if the tables are clothed to the floor level.

OFF-SITE CATERING

Off-site catering is a specialist operation and could be simply the provision of a few sandwiches and a pot of coffee in a company office, or a full silver service dinner for hundreds of people in a marquee. Not only must the caterer follow the five planning stages (see Table 4.9), but he or she must also arrange for the transportation of every item of tableware and food. Outdoor functions, in particular, require meticulous planning, and a hard floor and shelter in the form of marquees and awnings should be arranged in case of bad weather. The food must be transportable, presentable upon service to the customers and, above all, safe to consume. The food temperatures must be considered during preparation, transportation, presentation and service. This sounds easy but it requires properly refrigerated transportation facilities, and adequate chilled storage or reheating facilities.

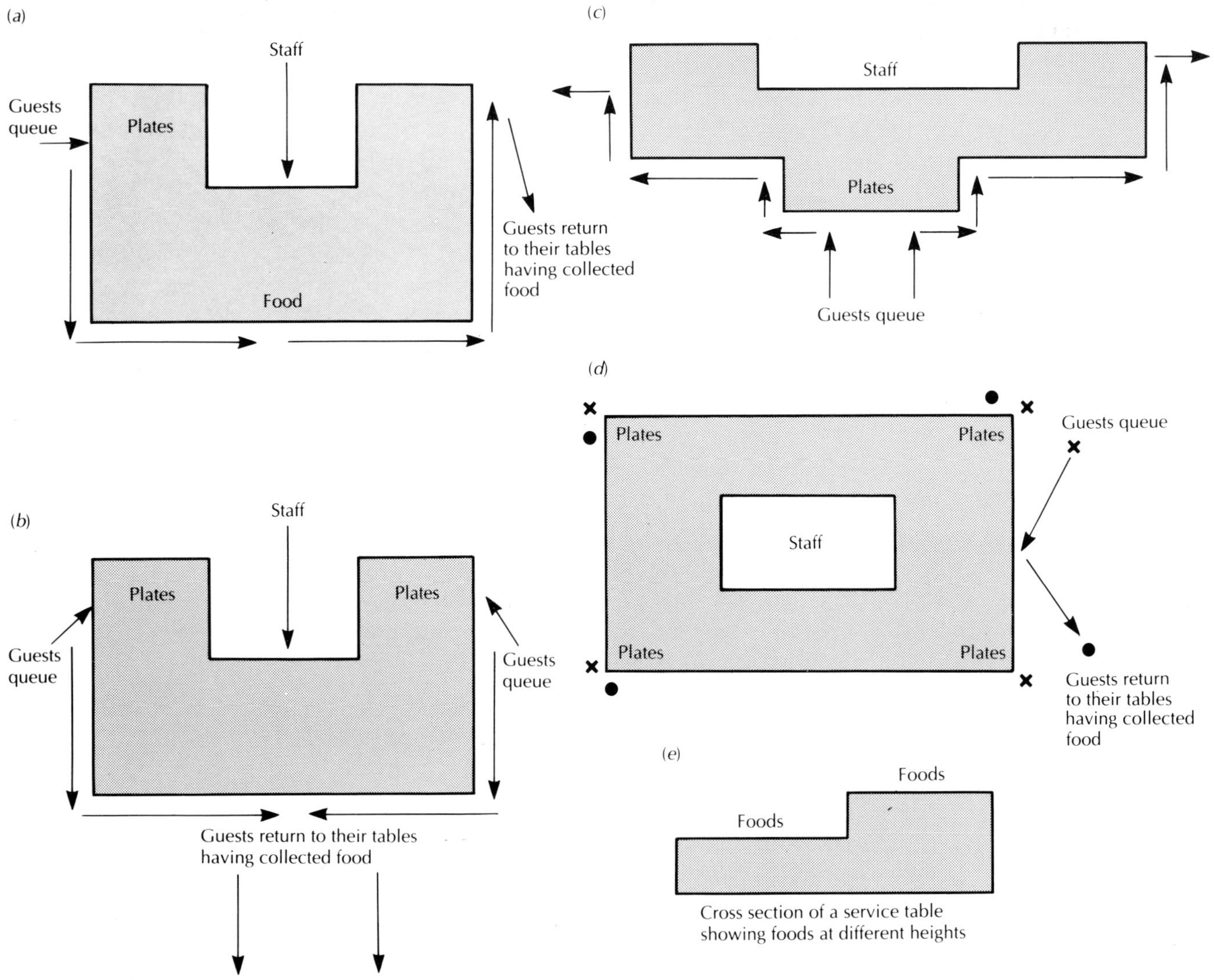

4.13 *Layouts for buffet service.*

ACTIVITIES

- Compile a list of rules for cafeteria service staff. What factors do these staff have to consider when they are serving their customers?
- Discuss why good food hygiene is so imporant in cafeteria service units.
- Consider the use of the following. Why are they important?
  1. chilled display shelves,
  2. chilled back up cabinets,
  3. back bar cooking to order,
  4. individually wrapped snacks,
  5. thermostatically controlled hot cabinets.
- Study the table d'hôte menu below and plan
  1 the mise-en-place necessary to silver serve this menu,
  2. list the cutlery that should be laid for each dish,
  3. using terminology that food service personnel would use to describe the menu to guests in the restaurant, describe each item on the menu.

Crème Crecy
Oeufs Florentine
Gnocchi Romaine

***

Filet de Pile Bonne Femme
Poulet Sauté Paysanne
Carbonnades de boeuf

***

Pommes Croquettes – Pommes au Lard
Choux Fleur Polonaise – Courgettesà la Crème

***

Crepes Normande
Coupe Jacques
Soufflé Milanaise

***

CHAPTER 34 Beverages

Beverages are served in formal restaurants and lounges, and in less formal coffee shops and cafeteria style catering units. Beverages can be divided into three quite distinct categories: hot beverages, nonalcoholic beverages and alcoholic beverages. Nonalcoholic beverages include fruit juices, squashes, nonalcoholic wines and mineral waters. These are frequently served with alcoholic beverages either as mixers, for example a tonic water is often served with a measure of gin. These types of beverages are also served in lounges and restaurants as well as in coffee shops, roadside catering units, cafeterias and vending machines. The essential here is all these nonalcoholic beverages must be stored in cool areas and chilled before service. To increase this chilled effect ice is often added at the point of service. Complementary flavoured slices of fruit are also common. Alcoholic beverages each have their own specialist form of additions at service. These are indicated after the details about each group of alcoholic beverage. See page 214.

HOT BEVERAGES

In a hotel, hot beverages will be prepared in the still room for the lounge and restaurant service, and in the floor or room service pantry for room service. Some hotels now provide tea and coffee making facilities for guest to use in their rooms - it frees the room service department of this task, and it is preferred by the hotel guests who can then make their own hot beverage when they like. In the less formal establishments such as coffee shops, roadside units and cafeterias, hot beverages are served in large quantities. Vending machines play an important part in hot beverage sales.

Tea

Tea is made from an infusion of leaves of an evergreen shrub. Climate, altitude and soil affect the growth and quality of the plants and the taste and fragrance of the tea. The best quality shrubs grow at an altitude of 2,000 m and are harvested in the spring. A hot, humid climate is required and the main tea growing countries are India, Sri Lanka, China, Japan, Indonesia, South Africa, Latin America and the Soviet Union.

There are three types of tea: green tea is roasted as soon as it is harvested to produce a strong, bitter tea; it is the speciality of China. Black tea is fermented and then dried; it is the most common type of tea used in the UK and comes from Sri Lanka (Ceylon), India or China. Oolong tea is fermented for a very short period of time and is a tea that has the characteristics of both the green and the black tea. This is a speciality of Taiwan and very popular in USA. In the UK, Ceylon, Indian and China teas are popular.

Ceylon teas

This is the most popular type of tea as it has a clear, natural taste and may be easily infused to the strength required. The varieties of Ceylon teas that are best known include Superior Orange Pekoe, which has large fragrant leaves and may be served without milk; Flowery Orange Pekoe, which has rolled leaves and makes a tea ideally served with lemon; UVA Highland, which is produced from large leaves and makes a tea that is ideal for service without milk; and medium grown Broken Orange Pekoe, which give a full flavoured tea that is ideal for service with milk.

Indian teas

The most well-known and used Indian tea in the UK is Darjeeling, which makes a well-flavoured tea suitable for service with milk. Assam is another well-known type of tea that is made from the small broken leaves. This full-bodied tea is usually served with milk.

China teas

China teas are either classed as ordinary or steamed. Ordinary types include Yunnar, a popular, well-flavoured tea, Caravan, a low-caffeine tea, Keemur and Great Mandarin, a jasmine-flavoured tea. The steamed types include Imperial Souchong, Lapsang

Souchong and Tarny Souchong. Earl Grey is a popular, black, Westernized China tea with oil of bergamont added making it fall into the scented tea category.

Tea making

Tea is a mild stimulant because it contains a small amount of caffeine. If tea is allowed to infuse for too long (for more than 5 minutes), the tanning in the leaves will produce a strong taint to the brew. It is for this reason that most establishments, including roadside catering units, serve tea in individual pots, brewed to order. Individual pots of tea are appreciated by the customers and a choice of the type of tea also increases customer satisfaction and hence tea sales. Tea is easily brewed provided the following points are followed:

- The water must be freshly boiled - a boiling water service point is an essential piece of equipment if any number of pots of tea are to be made.
- The pot must be scrupulously clean and prewarmed.
- One teaspoon or tea bag per person is the measure normally recommended (this depends upon the strength and type of tea being used).
- Infusion time is recommended at between 3 and 5 minutes.
- Just before pouring, the infusion should be stirred to enhance the total flavour.
- In the UK, it is common practice to serve tea on a tray with the following: a jug of cold fresh milk, a bowl of cubed white sugar, and a jug of clear, hot water (in case the tea infusion is too strong for the customer). Tea is always served in a tea cup with a saucer and teaspoon. A strainer will be required if loose tea leaves are used.
- If tea is to be served in a formal setting, such as a restaurant or lounge area in a hotel, the tray should be taken to the customer's table. The teapot, hot water jug, milk jug, and sugar bowl, tea cups and saucers should be positioned appropriately, so that the customer can pour out his or her own cup of tea.

Different forms of tea service include:

- The service of slices of lemon in place of cold milk (sometimes preferred with China tea).
- The service of strong Russian tea, in tall glasses, hot but without milk. Here the cube sugar is not put in the glass but placed onto the tongue and the strong tea sipped through.
- Tea with ice - a cool, long drink in a glass served with mint leaves and lemon for very hot weather.
- Decaffeinated tea, made as traditional tea.
- Instant tea - not a very popular type - is used in some poor-quality vending operations.
- One cup tea filters, which are circular plastic tubs with a filter base and a one cup measure of tea leaves. Boiling water is poured into the tub allowing the water to filter through the leaves and into the cup below. These are similar to the very popular one cup filter coffees and are used in fast-food and popular eating units. They are more expensive to purchase than tea bags, but no tea pots are required, thereby saving stock, space and washing up.

Coffee

Coffee trees grow small red berries that contain seeds - coffee beans. There are two species of coffee tree: Coffea arabica and Coffea robusta. Arabian coffees have long, oval, slightly flat beans that produce mild coffee. These are cultivated in Brazil, Arabia, Ethiopia, India, Mexico and Costa Rica. Robusta coffee beans have an irregular, convex shape and are smaller than the arabica bean. They also contain two and a half times more caffeine than the arabica bean. Robusta beans are cultivated in the Côte d'Ivory (formerly the Ivory Coast), Angola and Zaire.

At harvesting, the berries are collected, all the skin and pulp removed, and the coffee beans hulled before being stored in dry conditions. These newly harvested and untreated beans are known as 'green beans', and are sold to coffee merchants in their 'green state'. To produce the coffee flavour, the beans are roasted (they become swollen in size, a darker colour and produce the characteristic coffee aroma. After roasting, the beans are ground so that the coffee can be infused. Once roasted and ground, the aroma soon disperses into the atmosphere. It is for this reason that all ground coffee is either vacuum packed or sealed in airtight containers until required.

Caffeine found in coffee is an alkaloid, and has the ability to aid digestion and to keep the brain alert and functioning, even in the middle of the night. It can be the cause of insomnia. Because caffeine has such pronounced effects upon the body, there is a growing demand for decaffeinated coffees. These are processed to remove the caffeine while leaving the coffee flavour unaffected.

Large quantities of instant coffee (in a powder or granules) are sold in the UK. To produce this, 'liquid' coffee is spray or freeze dried, packaged and sealed in tins and jars, or individual sachets. Like ground coffee, instant coffee will quickly lose its aroma once it is exposed to the atmosphere. The individual sachets are used in hotel bedrooms where hot beverage making facilities are provided. Each sachet contains sufficient instant coffee for one cup. Another popular way of using coffee is the individual cup filter tubs that contain sufficient ground coffee for one cup. These may hold either ordinary or decaffeinated ground coffee. Liquid coffee essence is concentrated and sweetened coffee extract. This is used extensively in cookery to produce coffee-flavoured items such as ga^teau mocha and coffee creams. When making coffee, the following should be observed:

- Freshly drawn water should be used.
- The ground coffee must be kept sealed in as airtight a condition as possible.
- Accurate measuring of the coffee grounds is essential.
- Once prepared, the coffee must be served as soon as possible - the flavour deteriorates if the coffee is allowed to stand.
- If hot milk is to be offered in preference to chilled cream, it should never be boiled and should be just above blood heat (if it is too hot, it will curdle when it is added to the coffee). It should always be strained into the milk jug before being served to the customer.

A coffee still machine

This machine is heated by electricity or gas and is an expensive piece of equipment, normally used in hotels where coffee is made in large quantities in the still room. It has a central boiler that produces boiling water either via a tap for direct use (this is used for tea making), or to make coffee in either of the two containers on each side of the boiler. The boiling water runs onto the ground coffee and filters into the heated container below. It is drawn from the base of the container via a small tap. This machine also has a steam injection point for the warming of milk to be served with the coffee (or for other hot milk beverages).

Percolation method

For this method, a base jug is filled with cold water and placed on a hot plate to boil. A second container with a filter is positioned on top of the jug with the measured quantity of coffee. When the water boils, it rises up a funnel into the upper container and mixes with the coffee. The temperature of this infusion will automatically reduce slightly, causing it to filter back into the base jug leaving the coffee grouts behind. This is a good visual and aromatic method of making coffee, but tends ot be slow and messy as the top container has to be removed prior to service.

Filter method

Here an electric filter coffee making machine is used. The machine contains a water tank that will heat up the amount of water required to make one jug of coffee. An indicator light will signal when the correct temperature has been reached, and if a measured jug full of cold water is poured into a slot at the top of the machine, the hot water will filter through a measured quantity of coffee into the jug below. So there is always one measure of water in the machine. This is an efficient and quick method of making jugs of ground coffee and is used in restaurants, coffee shops, and in conference and function suits.

La cafetiere

In this method, a heat resistant glass jug of two or four cup capacity with a well-fitting lid with a central plunger attached is used. The plunger has a mesh filter at its base which is the same diameter as the jug interior. A measured quantity of coffee is placed in the base of the empty jug and boiling water is poured onto the ground coffee. The lid, with the plunger in the 'up' position, is carefully placed on the jug. The coffee is then left for 2–3 minutes to infuse in the water before the plunger is slowly and steadily pushed down, which traps the grouts below the mesh filter. The coffee is then ready to be poured directly from the jug - normally by the guest at his or her table in a formal restaurant, a hotel lounge or in a hotel room.

Individual disposable filters

These make one cup of coffee. They are plastic tubs with ground coffee sealed in the base. The tub is placed over a cup and boiling water poured in the top. A disposable lid is placed on the top to aid heat retention (the lid also acts as a saucer for the used tub). Once the water has filtered through, the cup of coffee is then ready to drink. These are popular in roadside and work catering areas, public houses and many popular and fast-food units where good, ground coffee is appreciated. With this method, the customer receives freshly filtered good quality coffee, there is no wastage of brewed coffee, and the coffee can be made instantly to order and at little cost.

Expresso

In this method, steam is passed over finely ground coffee beans to produce individual cups of coffee to order. The machines are fairly bulky and expensive to install and are seen in specialist coffee and patisserie shops. Milk may be heated by a steam injection nozzle to produce the well-known cup of cappuccino.

The following points should be considered when deciding upon which coffee making method is to be employed:

- the type of customer service offered,
- the number of cups or pots to be served per hour,
- the cost to be charged to the customer,
- the customer expectations,
- the equipment, space and time available to make the coffee.

Speciality coffees

These are some speciality coffees that are sometimes offered for sale to the customer. They tend to be included on menus where there is a higher than average spend per customer, as they are labour intensive to prepare and therefore costly. They should also be sipped slowly by the consumer because they are liqueur based.

These coffees are served in a warmed goblet (18cl size) placed on an underplate with a doiley. A small quantity of sugar added to increase the density of the coffee so that the cream will float over the surface. To prepare speciality coffee the following method is used:

- prepare the warmed goblet and place it on a doiley on an underplate,
- add the required quantity of sugar,
- pour strong black coffee into the glass and stir to dissolve the sugar,
- add one measure of spirit, for example, whisky,
- pour double cream carefully over the back of a spoon so that it touches the surface of the coffee slowly and floats on the surface.

These coffees are becoming increasingly popular at the end of a meal, sometimes served in place of a sweet course at the guest's request. They are named according to the type of spirit used, for example, Irish coffee contains Irish whisky, Highland coffee contains Scotch whisky, Russian coffee contains vodka, and Jamaican coffee contains Jamaican rum.

Another type of coffee is iced coffee. This is white coffee that has been chilled and served in tall glasses. It is a popular summer beverage in cafés and coffee shops. It is prepared from coffee essence, cold milk and, sometimes, vanilla ice cream, and blended in an electric beverage blender immediately prior to service.

Milk-based drinks

With the increased awareness in healthy eating, milk-based beverages are increasing in popularity. Hot chocolate and malted milk are the two popular milk-based beverages frequently sold. They are offered in fast-food, popular food and work-based catering units, roadside catering units, hospitals, as well as in the lounge and room service of hotels.

To prepare these beverages, the milk should be heated to a temperature just below boiling point (once boiled the flavour changes and spoils the beverage taste), and then the powdered drinking chocolate is whisked in. Finally, the beverage is sweetened to taste an poured into a warmed beaker or cup ready for service. It should be served to the customer in the beaker or cup, never in a pot or jug. Some hot chocolate drinks are served with whipped cream and flaked chocolate on top for added effect and flavour. It is now possible to purchase instant hot chocolate and malted milk preparations, which only require the addition of boiling water. They are purchased in tins or individual sachets and may be used in hotels for guests to make their own milk drinks in their rooms in preference to tea or coffee. These instant mixes are dried milk based, so no additional milk is required.

Milk shakes are another form of milk-based beverage. They are quickly prepared by using a specially designed electric whipping machine. This will whisk up the cold milk and the chosen flavoured fruit syrup to produce the frothy milk shake. Some establishments use a soft ice cream in the milk shake to increase its creamy texture, and it is possible to make milk shakes by using a sweetened and flavoured powder mix instead of a syrup. Milk shakes can be made from whole, semi-skimmed or skimmed milk, but the whole milk will produce the traditional creamy texture. Some powdered milk shake mixes have dried milk added, so that it is only necessary to add cold water to produce the milk shake.

Milk shakes are served in tall glasses and drunk with a wide straw and cream floated over the surface. In fast-food establishments, they are served in disposable, lidded beakers with a straw so that they can be drunk off the premises.

NON-ALCOHOLIC BEVERAGES

There is an increasing demand for nonalcoholic beverages to be available in all types of food service establishments. These nonalcoholic beverages may be divided into three main types: bottled waters, fruit juices and squashes, and fruit syrups.

Bottle waters

These fall into three main types:

Aerated waters

These are flavoured or plain waters that have been carbonated with carbonic gas, which gives them their effervescent qualities. Flavourings are added by using essences so that these waters are always clear as well as sparkling. The waters are sold in cans or bottles and are intended either to be drunk on their own or as a mixer with the appropriate spirit. Examples include soda water, tonic water, bitter lemon, dry ginger, lemonade, ginger beer, coca cola and orangeade. All these beverages should be served chilled; in a service situation they are kept in a cold cabinet or on a cold shelf.

Mineral waters

These are natural waters that have a mineral content. They may be sparkling (carbonated) or still and are sold in glass or plastic bottles. Once opened, they should be drunk within 24 hours and stored in a refrigerator. They should always be served chilled. Examples of mineral waters include Perrier water from France, Appollinovis water from Germany, and San Pellegrino water from Italy.

Spring waters

These are pure spring waters and are sold in glass or plastic bottles as mineral water; once unsealed they should be stored in a refrigerator and consumed within 24 hours. These may be purchased as still or sparkling and are always served chilled. Examples include Ashbourne, Buxton and Malvern waters from England, Highland Spring from Scotland, and Evian and Badoit from France.

Fruit juices and squashes

Fruit juices are becoming increasingly popular in the UK, and are not only served as a chilled preliminary course at a meal, but in place of more traditional beverages in bars. They are also popular in hotel bedroom mini bars, as well as in cafeteria and fast-food outlets. Fruit juice may be just that - the juice obtained from freshly squeezed citrus fruit but, increasingly, it includes long-life (UHT) concentrated fruit juice and pasteurised, concentrated juice, which has a shelf life of five days.

Fruit squashes are fruit-flavoured, concentrated drinks that are not made from 100 per cent juice. Water is used to dilute squashes and examples include lime cordial, lemon squash and orange squash.

Fruit syrups

Flavoured syrups are used to create milk shakes and a thicker variety is used to top ice cream sundaes and other sweets. The topping syrups have a concentrated flavour and sometimes very bright colour, and are ready to use direct from their containers, they are a very useful commodity to keep for quick sweets in popular and fast-food establishments. Some have the addition of a fast setting agent so that they solidify when in contact with the cold ice cream.

ALCOHOLIC BEVERAGES

Beverages that have an alcoholic content are produced by introducing yeasts to ferment such substances as malt extract and grape juice. The yeast converts the glucose sugar present into alcohol and carbon dioxide, though there are other reactions taking place at the same time which determine the character of the finished beverage. The amount of alcohol produced depends upon the amount of glucose in the liquid.

Beers

Malt is made by moistening barley which is then evenly spread out and maintained at a moderate temperature to encourage the grains to germinate. When this occurs the grain's diastase (an enzyme) converts the starch in the barley to dextrins and maltose (a glucose). The malt is then dried (the lower the drying temperatures, the darker the beer). Next the malt is ground and made into a mash with water. Then the liquid is drained off (this is known as the wort). The wort is boiled with hops to give it flavour and then cooled and fermented in vats with yeast.

After three days of fermentation, the beer should be of the desired alcoholic strength. At this point, the surplus yeast is skimmed off leaving a head of dark foam behind. Finally, the beer is transferred to casks, kegs or barrels where finings are added to aid the clarification process. Beer contains dextrins, sugar, alcohol, riboflavin and nicotine acid. These nutrients are present in varying proportions in draught, bottled and canned beers.

Storage of beers

Beers should be served in well-ventilated, clean cellars at a constant temperature of 13°C. If traditional casks are used, they should be placed on trestles and allowed to rest for 24 hours after transportation to settle the beer. The barrels are tapped and the spile (wooden peg or spigot) is inserted to control the flow of air when the contents are withdrawn. In modern metal kegs, the beer is pressurized with carbon dioxide to assist the flow of the beer. All pipes and pumps should be absolutely clean to avoid contaminating the beer and so altering the flavour. Bottled beers should be allowed 24 hours to settle before being opened for service. Beer is served in imperial pint or half-pint measures. The beer should be carefully poured against the side of the glass to ensure that only a small head of froth is created.

Cider

Cider is produced from apples in the South West of England, Devon, Somerset and Herefordshire. In the production process the apples are cleaned, crushed and the juice fermented; then the cider is bottled, canned or barrelled. As fermentation continues in the bottle or cans, this cider is aerated.

Wines

The quality of a wine depends upon the weather, the quality of the grape harvest, the type of soil used to grow the vines, the type of vines used and the skill of the grower, and so these variables will produce many different types, colours, strengths and flavours of wine in any one year. Wines are produced annually from grapes that are specially cultivated. In the winter the vines are pruned and cuttings taken; in the spring the soil is cultivated, weeded and the vines sprayed; in summer further spraying of the vines occurs and the growing grapes will be protected from harsh weather if necessary, finally, in the autumn the ripe grapes are picked.

A dry year with constant sunshine will produce a good quality, vintage wine. A cold spring and a wet summer will result in the grapes having a high acid level and a low sugar level. The flesh of the grape contains water and sugar, and needs to be combined with yeast to cause the fermentation process to commence. Wine may contain all or some of the

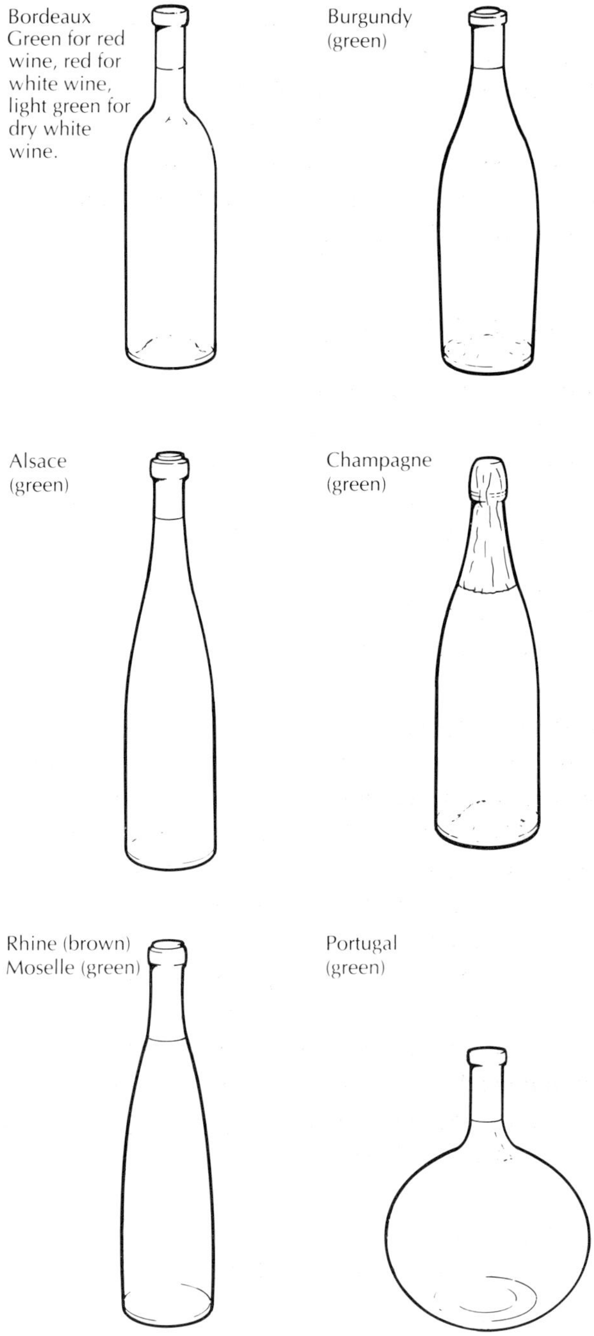

4.14 Bottle shapes.

following substances: glycerol, sugars, pectins, oxygen, colour pigments, traces of nitrogenous compounds, tannins and acids. It is this mix of contents that creates the character of the resulting wine.

The conditions that suit the vine best are found in two areas: between 30 and 50°C north and south of the equator. There are many varieties of wine and the skilled grower will select the type of vine that suits the soil and regional climate.

Black grapes have colouring matter in their skin; red wine may only be produced from these grapes. However, white wine may be produced from black or white grapes. A summary of the wine making process is shown in Table 4.10.

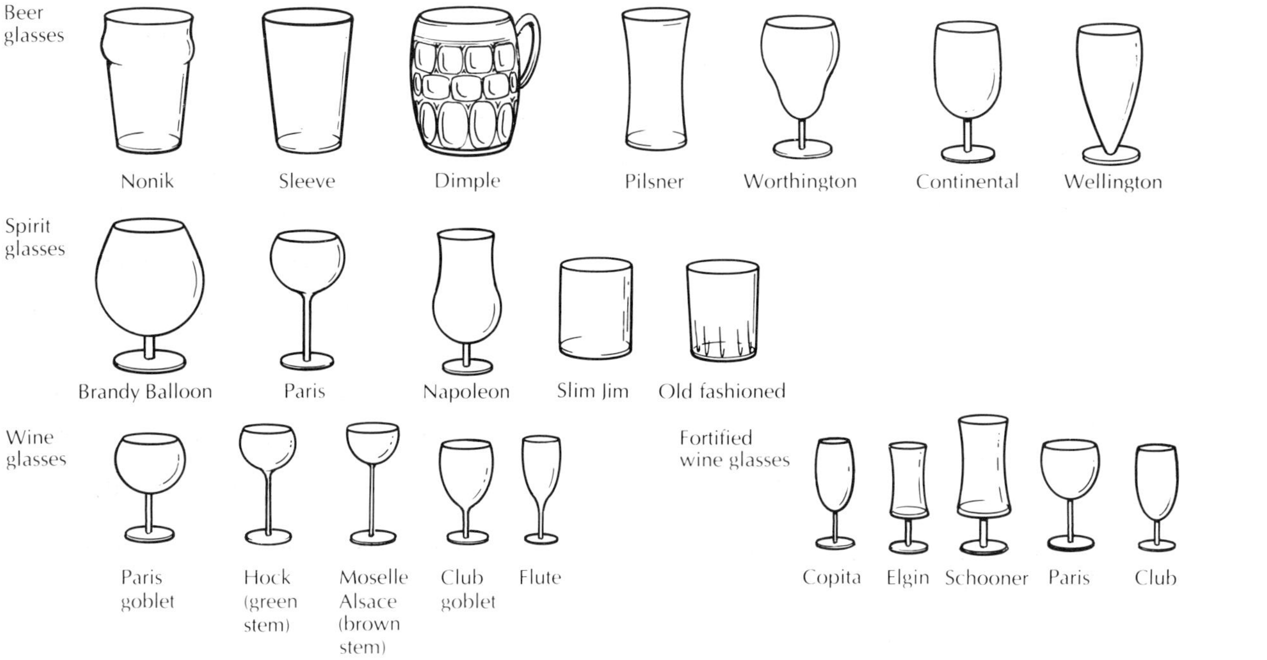

4.15 Glass shapes.

Table 4.10 Summary of the wine making process

Red wine Red grapes → Crusher: stalks may be removed → Fermenting vat: grapes crushed into vat and fermented 5–14 days

↓

Wine runs out from fermenting vat

White wine Red or white grapes → Crusher: grape stalks are removed → Horizontal press: grapes are pressed to allow any juice to flow away

↓

Fermenting vat juice passes into fermenting vat

→ *Sweet wine* is removed from the vat before all the sugar is converted into alcohol

→ *Dry wine* left in vat until fermented

↓ *Sparkling wine:* fermentation continues in bottles

Rosé wine Red grapes → Crusher: stalks may be removed → Fermenting vat 1: When the juice has taken sufficient colour from the grape skins, it is run into the second vat

Fermenting vat 2: wine continues to ferment.

↓

A pale pink coloured wine is produced.

Wine storage

The following important points must be observed when storing wine:

- The cellar must be cool and kept at a constant temperature of 12–14°C.
- Wine should be stored away from direct sunlight, as strong light can have a harmful effect upon the quality and colour of the wines.
- The cellar must be clean and protected from any strong smelling items such as fuel or paraffin.
- The wines should be stored horizontally on racks that are not affected by vibrations from traffic or lifts.
- The racks should be labelled or coded to coincide with the wine list number to avoid any unnecessary handling or movement of the bottles.

Table 4.11 Wine classification

Country	Classification	Meaning
France	Vin Ordinaire/Vin de Pays	Nonquality
	Vin Delimité de Qualité Supérieure (VDQS)	Middle quality
	Appellation d'Origine Controlée (AOC)	Top quality.
Germany	Tafelwein	Nonquality
	Qualitatswein bestimmte Anbaugebiete (QBA)	Middle quality
	Qualitatswein mit Pradehat (AMP)	Top quality
Italy	Denominazione di Origine Controllata (DOC)	Middle quality
	Denominazione di Origine Controllata e Garantita (DOGG)	Top quality

Wines and food

There are a number of general guidelines for the serving of wines with food (see Table 4.12). However, the wine waiter must never object to the combination chosen (some people like to drink a red wine with fish or a sweet white wine with meat).

Table 4.12 Recommended wine and food combinations

Food	Wine
Aperitif	Fortified wine (sherry, vermouth)
Hors d'oeuvre	White wine (dry or medium dry)
Fish, shellfish	White or rosé wine (dry, medium dry, sparkling or very dry)
Meat, poultry	Red wine (rich, full-flavoured)
Sweet course	White or rosé wine (medium, sweet or sparkling)
Cheese: strong	Red wine ((full-bodied, port)
Cheese: mild	White or rosé (dry or medium dry)
Fruit	White or rosé (naturally light and sweet)

Fortified wines

A fortified wine is one to which brandy or alcohol has been added – this stops the fermentation process and strengthens the wine. Fortified wines are blended to create an acceptable product that may be either dry or full-bodied and sweet.

Sherry

The best quality sherry comes from the Andalusian region of Spain, although other sherries are produced in Australia, South Africa and Cyprus. It is made from white wine produced from the Palomino grape. Once the young wine has been lightly fortified, the blending of the sherry is completed using the solera system, which enables a constant taste and colour of any blend to be produced. The different types of sherry available:

- Fino is a very dry, pale sherry and is an ideal apéritif.
- Manzanilla is a pale, delicate sherry which is dry in taste.
- Montillado is a matured sherry of a dark colour and is full and sweet in flavour.

- Oloroso is a rich, dark, sweet sherry.
- Milk is a smooth, sweet, dark-coloured sherry.
- Cream is an older, smoother and richer sherry than milk sherry.
- Brown is a dark and sweet sherry but it not as popular as other sherries.

Sherry is traditionally served as an aperitif before a meal and helps to stimulate the digestive juices. Pale, dry sherries may be served slightly chilled, but sweeter, cream sherries should be served at room temperature. Sherry is served in small, straight glasses of a 10 cl capacity.

Port

Port is produced in the Duoro region of Portugal. The grapes are crushed and fermented and the juice is then blended with brandy. This strengthened wine is blended and matured. There are five main types of port available:

- Ruby is four or five years old and is a blend of new and old wines.
- Tawny is matured longer than ruby port and is lighter in colour.
- Old Tawny is a smooth and matured port of a light colour.
- White port is made from white grapes; it is dry and served as an apéritif.
- Vintage is a mellow and well-matured port; it is very smooth and must be decanted.

Port is traditionally served at the end of a meal, poured from a decanter into small sized glasses of a 10 cl capacity. Port should be served at room temperature.

Spirits

Brandy

Brandy is a spirit distilled from wine and matured in wooded casks where it takes on its characteristic golden colour. Bottles of brandy should be stored upright in a mild temperature between 15°C and 18°C and always away from direct heat and sunlight. This is why the best brandies are bottled in coloured glass bottles. Cognac is a brandy of excellent quality and is produced in a region near the south west of France. Brandy is served in specially designed balloon glasses, which can be cradled in a warm hand to encourage the bouquet.

Whisky

Whisky is a spirit that is distilled from fermented barley, rye or maize (natural, pure spring water is a necessary ingredient).

- Malt whisky is made from malted barley. The starch in the grain is turned to sugar, then mixed with liquid, fermented and distilled.
- Grain whisky is mainly made from maize, with a small proportion of malted barley. It is distilled and matured for at least three years before being blended.
- Blended Scotch whisky is a mix of 60 per cent grain and 40 per cent malt, and is blended using whisky of different ages and origins to produce the required taste.
- De luxe Scotch whiskies are older and more mellow in flavour. They are smooth whiskies that have been matured for at least seven years (some for as many as twelve years).

Whisky is served in $^1/_6$ gill measures and may be drunk neat, with ice (when it is known as 'on the rocks'), or sometimes with a mixer, such as dry ginger ale. Malt and de luxe Scotch whiskies are drunk neat in straight or cut glass spirit tumblers.

Rum

Rum is a spirit distilled from sugar cane and is made in the Caribbean, Jamaica, Guyana, Trinidad and Barbados. As a dark, rich, highly-flavoured spirit, it may be used in cakes, Christmas puddings and mincemeats, fruit babas and other sweets. It used to be drunk daily by the Royal Navy as their 'rum ration'.

Bacardi is a clear, light spirit without the highly flavoured taste of dark rum. This light rum is served with fruit drinks and ice. Rums are served in $^1/_6$ gill measures.

Gin

This is a spirit made from grain and molasses, and flavoured with juniper berries and coriander seeds.

- London gin is made by redistilling a neutral spirit with flavourings; it may be clear or straw coloured.
- Plymouth gin is served in 16 gill measures, often with tonic water or fruit juice to flavour and to create a long drink.

Vodka

This is distilled from grain or molasses, and is a pure spirit with no flavourings added. Vodka is tasteless and so may be mixed with any fruit based juice or mineral water, and is served in 16 gill measures in a tall glass to create a long drink.

Liqueurs

Liqueurs are spirits flavoured with fruits, herbs, roots and plants and, sometimes, sweetened or blended.

- Advocat is a thick, yellow and creamy, and is made from eggs and spirit.
- Apricot brandy is a mix of apricots and brandy.

BEVERAGE SERVICE AREAS

There is a certain amount of preparation required before the service time in any licensed bar area. Whatever the type of establishment, style of decor or time of opening, the same tasks must be completed to ensure that once open, the bar service will run smoothly and efficiently. Below is a check list that is used in this preservice time.

Customer area

- Floors should be dust free and in a clean condition.
- Furniture should be free from glass rim stains, smears, dust and sticky residues.
- Furniture should be positioned suitably to provide access to doors and the service area and be so arranged that customers can sit in groups.
- Adjustments should be made to heating levels and ventilation to provide a pleasant atmosphere.
- Ash trays, drip mats, promotional material, plants and flowers need to be placed in the appropriate places.
- Cloakrooms should be in a clean, neat and hygienic condition with adequate ventilation. Coat hanging areas should be tidy and accessible.
- External access and outdoor seating areas should be clean and tidy and look inviting.

Service area

When closing the service area:

- Wipe and tidy behind the bar counter and wash the measurers.
- Wipe the counter service area and remove the bar towels, trays, etc.
- Leave the display areas tidy and refill the wine cooler cabinet.
- Stack all the clean glasses in tidy rows.
- Leave the bottle storage shelves tidy and place all empties in the skip. Refill the cold shelf.
- Place cut fruit in cling film in the refrigerated area. Wash the cutting board and knife.
- Switch off the glass washing machine and leave it open. Place all cloths and swabs in lined bins.
- Wash and replace all drip trays and empty the wire baskets.
- Empty the till and take the final reading. Place the takings and float in the safe. Leave the till unplugged and open.
- Close the bar shutters and turn off unused electric plugs. Close and lock the windows and check around the entire area, looking for cigarette ends, dripping taps, open windows and unlocked doors. Unplug the TV, gaming machine, etc. When completely satisfied that all is in order, switch off the lights and lock the door.

When opening the service area:

- Polish the display area, shelves and mirrors.
- Clean the bottle shelves and restock the shelves (bringing forward all existing stock). Check the temperature of chilled storage area and empty and clean the skip.
- Fill the ice buckets and water jugs. Prepare the fruit slices and get the sticks, stirrers and shakers ready.
- Switch on the glass washing machine and clean the outside and top of the machine. Make sure a clean dust pan and brush are available and there is an adequate supply of glass cloths and swabs. Soap, nail brush and hand drying equipment in position.
- Check the pumps and dispensers and check the supplies of beverages, and that the wine buckets, waiters' friends and trays are placed ready.
- Switch on the till, check the float, set the date and time on the till and change the till roll if necessary. Check the time on the clocks in the customer area and that pens and pads are available. Switch on the gaming machines.

SELLING ALCOHOLIC BEVERAGES

Licences

In order to sell or supply alcoholic beverages it is necessary to obtain a licence under the Licensing Act 1988. There are two types of licence – an on-licence allows the vendor to sell alcohol to customers to drink on or off the premises. This is the type of licence required to run a public house, restaurant or hotel. An off-licence only allows the vendor to sell alcohol to customers to be drunk off the premises.

This is the type of licence required by wine and spirit merchants who run off-licence shops.

An on-licence premises has permitted trading hours, which means that the licensee can choose his or her trading hours to fall between 11 am and 11 pm in England and Wales, and between 11 am and 2.30 pm and 5.30 and 11 pm in Scotland. Sundays, Good Friday and Christmas Day have short permitted hours. Licences vary according to the type of establishment and its requirements. In addition, there are age restrictions that are designed to control the consumption of alcohol to young adults on licensed premises:

- persons under 18 years may neither purchase nor consume alcohol,
- persons over 16 years may purchase and drink beer or cider with a meal in a part of the premises exclusively for the service and consumption of food,
- young persons over 14 years are permitted into licensed premises, but they may not purchase or drink alcoholic beverages.

Identification schemes

There are identification schemes in several inner-city areas where those persons over 18 years of age may apply for an identification card to be used when frequenting licensed premises. Only those customers who can produce the card will be served with alcoholic drinks. This scheme will help to reduce the underage drinking dilemma for the licensee who, at times, finds this impossible to control. The penalty for serving drinks to people who are underage is a fine of £400 maximum.

Prices

Licensees must display a clearly printed price list of beverages so that customers standing at the bar should be able to read the list of prices. This is enforced by the Price Marking (Food and Drink on Premises) Order 1979. The size of measure for those drinks should be indicated on the price list. In lounge service and room service, these lists take the form of a card 'menu' where the drinks, measures and prices are listed.

Measures

The standard measures are listed below. It is permitted to sell beverages in multiples of these measures:

- spirits - $^1/_6$ gill = 2.33 cl,
- fortified wines - $^1/_3$ gill = 4.66 cl,
- liqueurs - $^1/_6$ = 2.33 cl,
- wine - 125 or 175 cl,
- beers and cider - $^1/_3$ pint or 1 pint,

(Note that 1 gill = 5 fl oz = $^1/_4$ pint =14 cl).

Sealed optic

A sealed 'optic' is designed to dispense the precise measure of $^1/_6$ gill of spirit. Newly designed optic measures permit the beverage to be dispensed when a small lever is pressed by the bowl of the glass, which is more hygienic than the older 'three prong' design optic that could cause cross-contamination.

Thimble measures

These will measure a standard quantity of spirit, but they do not allow for spillages and are not, therefore, totally accurate. They are also a slow, messy, method and are normally only used for less common liqueurs.

Pourers

These incorporate a sealed optic and a spout so that the single standard measure can be dispersed by upturning the bottle over the appropriate glass. These are not as fast as the wall-mounted spirit optics and are therefore normally used for less popular types of beverages.

Draught dispenser units

These are used to dispense beers, lagers and ciders, as well as some carbonated mineral waters such as lemonade. The designs of these units will differ. Some may have a large traditional pump handle, others will have a small tap of press-button mechanism. As a general rule, to achieve an acceptable glass of beer, the glass should be held at a 45 angle with the dispenser nozzle clear of the inside of the glass. As the glass fills up, it is lowered and levelled thus producing a full measure without too large a 'head' (froth).

Bottled beverages

Bottled beers, lager and ciders are measured and sold by the bottle capacity. When ordered, the bottle should be opened and poured slowly into the glass holding both the glass and bottle at a 45 angle, gradually levelling the glass as it fills. Some bottled beers need a good degree of skill to dispense correctly. These include Guinness and special brews that tend to contain a sediment.

SERVICE OF DRINKS

The apéritif

Apéritifs are served before the meal to whet the appetite. These drinks are usually served in the lounge area or lounge bar. However, in the restaurant, the customer often likes one at the table while ordering and waiting for the meal. The waiter should therefore ask the host if an appetiser is required as soon as the party has been seated. These should be served as soon as possible, before the order for the meal is taken. The guests may then enjoy their drinks while waiting for the meal.

Apéritifs are served to guests by the glass, and should be presented on a cloth-covered wine salver in case of spillage. The glass should be placed on the table just below and to the right of the table glass, which should not be removed. The most common appetizers served before a meal are shown in Table 4.13.

Table 4.13 Common apéritifs

Types	Service
Sherry	In a plain, undecorated glass, often tulip shaped, narrow at the top and broad at the base, 10 cl capacity
Madeira	In a plain glass of 5–10 cl capacity
Vermouth	In a plain glass of average capacity 8–12 cl

At a reception

When the guests are standing talking in the lounge area, the waiter or waitress should invite them to help themselves to an apéritif from his or her tray. All empty glasses should be collected during this reception time, and ash trays exchanged as soon as they have been used. Small dishes of nuts or potato-based small snacks are sometimes offered with premeal drinks.

Bottled beers are usually served in 1/2 pint glasses or at the bar in 1 pint capacity glasses. Some customers prefer beer to wine to accompany their meal at midday. The waiter should collect the beers from the dispenser bar and ensure that they are served to the customer immediately so that the head is still on the top of each glass of beer.

Spirits may be requested in the restaurant or lounge areas. They are served by using a standard measure of $^1/_6$ gill or a double measure, $^1/_3$ gill. Many guests request a particular brand of spirit, if this is not available the waiter should suggest comparable brands. The spirits will normally be in glasses of 10–12 cl capacity, together with any bottles of minerals or mixes requested. The glasses should be placed on the table and a splash of the mixer carefully poured into the glass. The bottle of mixer should then be placed on the table next to the glass of spirit. Ice should always be available for these drinks.

Selling wines in the restaurant

Bottles of wine are sold to the customer by the wine waiter. A wine list is taken to the host of the group and any advice offered only if requested.

- Wine lists should be presented to the host/guest and advice given if requested.
- The wines on display must be attractive and the labels clearly visible.
- White wines are often sold by the glass from a special chilled dispense cabinet. Full-bodied red wines by the glass should be served at room temperature. Some lighter and younger red wines are served chilled.
- The wines available should meet the needs of the customer.
- A short description of the character of the wines on the wine list will help to increase sales.
- Wine sales will increase the guest's enjoyment of his or her meal and should, for this reason, be encouraged.

Service of wine

- The chosen wines are presented to the host guest by the wine waiter before being corked with the label uppermost.
- Red wine and sweet white wines are then corked and left to breathe until the service time.
- A medium or dry white wine is corked and placed in an ice bucket to chill prior to service.
- When the wine is to be served, the waiter should pour a little taste in the host guest's glass. When this taste has been approved, the waiter should then proceed to serve the guests seated round the table, taking care never to fill each glass more than half full; the host should be served last.
- Further wines requested for other courses of the meal should be served in the same manner. Each new wine should be served in a clean glass.

CHAPTER 35 Food and beverage control

Adequate control of takings is essential as it will enable the management to:

- forecast future trade trends,
- ensure customers are billed correctly,
- ensure costings are correct and therefore charges are correct,
- reduce the amount of wastage by not over producing,
- reduce or eliminate pilferage by keeping a tight check at all times.

BILLING AND CONTROL SYSTEMS

Fast-food billing

In fast-food establishments the menu is limited to a standard selection. The ordering procedure is illustrated in figure 4.16. With this system, the customer has to wait while the foods that have been ordered are prepared and served. The customer will not wish to wait long, so preparation is normally standardized to ensure that it is kept to an acceptable minimum time. Having eaten, the customer can leave since the bill has already been paid. This system prevents any customer from leaving the premises without paying for the meal.

CUSTOMER
Enters
Chooses a table and sits down
Selects food and beverages from a standard menu

↓

WAITRESS
Records the order on a printed pad
and orders the food

↓

CASHIER
Prepares a bill from the top copy of the pad

↓

WAITRESS
Serves the food to the customer
Presents the bill for payment

↓

CUSTOMER
Pays the bill
Eats the food
Departs when he or she is ready

4.16 Ordering procedure in a fast-food outlet.

Take away billing

Here the customer orders the food and pays for it, and then waits just a few minutes while his or her order is prepared and packaged. Once the customer has received his or her order he or she may either take it to a tabled area and eat it there, or leave the premises to eat the food elsewhere. Examples of these types of establishments include fish and chip shops, hamburger bars, pizza houses and fried chicken bars.

Triplicate system

The triplicate system is the standard manual system of control used in many restaurants (see also page 197). When the waiter takes the guest's order for food or beverages, he records it on a triplicate pad, the duplicate copy which goes to the cashier then enables the guest's account to be opened. Thereafter, every change that the guest incurs during the meal will be recorded by the waiter, who passes the duplicate copies to the cashier.

Each guest's account consists of two parts, the account itself and, attached in front of it, the guest's statement (bill). As charges are posted, the two parts are written up simultaneously so no discrepancies between them should be possible. After posting each charge, the guest's account is balanced automatically. Thus, provided that all the duplicate orders have been posted, each account should be balanced ready for presentation to the guest at the end of the meal.

When the account is paid, the amount is recorded by the billing machine and the top copy of the account (the statement) is given to the guest. The second copy is then filed for management control checking.

The bill will be presented to the guests, at the end of the meal, folded on a small plate or tray by a member of the food service team. There are many types of cash registers and billing machines available for use in every type and size of outlet. Each establishment will select the cash control system that best suits their requirements.

Billing machines are in use in many restaurants with the following advantages:

- records are neater and legible with fewer errors,
- accounts are always up to date,
- accounts are more acceptable to the guest when totalled and presented,
- most machines have mechanized, automatic date printing and abbreviated descriptions of items
- they can prepare related records simultaneously,
- they save on labour costs.

Computerized billing

This follows the standard billing procedure, only instead of writing out the orders and bills by hand, the food service staff either use a soft key pad or a simple keyboard to record the customer's order. This may include the customer's name and table number, so that a personalized account can be prepared at the end of the meal. Each time further items are ordered, these are keyed into the system, so keeping a full record of orders at all times. The advantages of this system are that it provides an instantaneous print out of every order, which saves time in the restaurant; it provides an instant account for each customer; and it produces menu sales analysis figures on a daily, weekly or monthly basis.

RESTAURANT, LOUNGE AND BAR CONTROL

A hotel's cash terminal tills are sited in the restaurant, lounge, and bar retail areas, making it possible to obtain an instant check on a customer's credit worthiness. Charges may then be posted automatically to the appropriate account.

The sequence of action and control that is carried out at a cash terminal is outlined in Table 4.14. This action occurs at any cash terminal in the premises where a guest wishes to place a charge for goods or services on his account. It is a very quick and easy system to operate, but is expensive to install, and staff have to be adequately trained to use the tills and to read the data that is produced. There are many types of systems available, but they all have the same basic function of accounting for cash, and services as they are sold.

Table 4.14 The sequence of action and control at a cash terminal

Action	Control
The guest's cheque is presented for payment	
The guest requests that the bill is placed upon his or her account	Proof is required that the guest has an account
The guest is asked to sign his or her name and room number on the back of the cheque	Proof in case of a query at a later date
The cashier inserts the cheque into the terminal till and presses the appropriate key	The terminal memory recalls the cheque total from the original order
The cashier finalizes the transaction by activating the charge settle key	The display reminds the cashier to enter the guest's room number and his or her initials; this confirms the guest's credit worthiness
The cashier checks the display reading of the guest's name, room number and amount entered onto that account	The amount is posted to the guest's account
The cashier gives a receipt to the guest	The guest checks that receipt against the account before settling the final amount due

METHODS OF PAYMENT

The different methods of payment most frequently used by guests once a bill is presented are shown in figure 4.17. Some are more popular than others, but each is explained below.

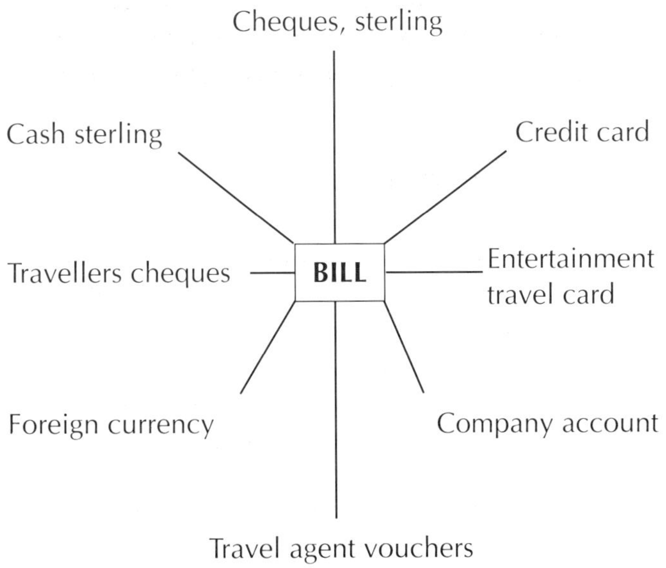

4.17 Different methods of payment.

Cash, sterling

Payment by sterling is the best method for the hotel since the cash is instantly available for use.

Cheque, sterling

Bankers' cheque cards only guarantee cheques up to £100 in value, so this may restrict the use of cheques by guest's whom the hotel does not know. However, if a guest wishes to pay by this method, certain steps must be followed:

- the cheque should be signed by the guest in front of the receptionist, cashier or waitress,
- the cheque card's date and signature should be valid,
- the cashier or receptionist should write the cheque card number on the reverse of the cheque,
- the figures should agree with the written cash amount.

Since cheques take a minimum of three days to clear through the bank, the hotel is deprived for this time of its money.

Credit cards

Credit cards issued to bank customers are used to purchase goods or to pay for services:

- the cashier takes in imprint of the credit card and the guest signs the triplicate sales voucher (the bank then undertakes to credit the hotel's bank account with the amount of money owed, minus a small commission charge).
- the cashier may obtain telephone authorization from the credit card company to verify the guest's credit worthiness.

There are machines which provide a quick and efficient means of handling credit cards.

- the customer is presented with his or her bill in the normal manner,
- the customer presents the credit card for payment,
- the cashier accepts the card and 'swipes' it through a payment terminal where it is automatically checked for credit worthiness (see (a) below),
- the cashier then enters the bill total on the terminal and the customer's receipt is automatically printed,
- the receipt is taken to the customer for him or her to sign, which authorizes payment,
- the credit card and the receipt are handed to the

(a) The transaction amount is entered and a customer receipt is automatically printed.

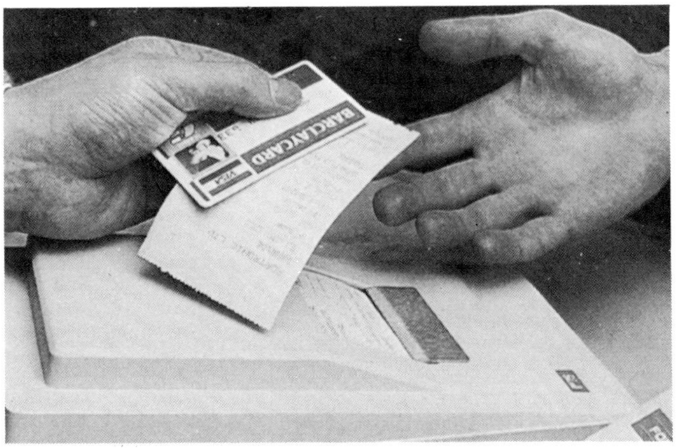

(b) The customer's card and transaction receipt are returned and the customer leaves with minimum delay.

A credit card payment system.

Entertainment and travel cards

The procedures for accepting entertainment and travel cards and for checking the credit worthiness of the guest are similar to those used for bank credit cards:

- the sales vouchers are sent to the card company at regular intervals,
- the card company then sends a cheque to the hotel for the total amount given on the vouchers, minus a small commission charge,
- the hotel may have its money tied up in these vouchers for weeks, which is a distinct disadvantage to the cash flow figures.

Company accounts

Regular guests from companies who have arranged to have a ledger account with the hotel, may simply sign their bill upon departure. The hotel then sends the bill to the company, who in turn will settle the account. This arrangement encourages regular business from executive guests.

Travel agent vouchers

Travel agents sometimes issue vouchers to a guest upon receipt of the cost of their standard stay or visit. Therefore when the guest arrives at the hotel the voucher is presented as full payment for the agreed services, such as bed and breakfast terms or half board. With this type of payment, care needs to be taken to ensure that all extra services are paid for by the guest, these may be meals in the restaurant or coffee shop at the hotel. Furthermore, the credit worthiness of the travel agent needs to be checked before the hotel accepts the booking form from the travel agent.

Foreign currency

Some large hotels will exchange a guest's foreign currency for sterling currency. However, the rate at which the currency is exchanged will normally be slightly lower than that available at a bank to cover the hotel handling charges.

Exchange rates are constantly changing, therefore an exchange rate table is usually displayed in the hotel for the guests' infomation. Guests will normally have changed their currency before using the hotel restaurant.

Traveller's cheques

The customer purchases traveller's cheques in his or her own country before commencing his journey - these may be in any currency the customer requires, that of his own country, or sterling. The traveller's cheques are signed when they are purchased and the serial numbers are noted by the issuing bank. When using traveller's cheques to pay an account, the guest will sign and date each cheque. The cashier should watch the guest sign the cheques and ask for proof of the guest's identity by checking his or her passport. The traveller's cheques are exchanged for their sterling value by the bank and then credited to the restaurant or hotel's account.

CHAPTER 36

Computerization in food services

Computerization in food services
Computerization in hotels allows guests to charge food and beverage items directly to their hotel accounts. The food service staff simply require proof of the guests' residency (usually this is their key card), and their name. The charge details are then types into the computer terminal in the restaurant or hotel bar and the charges will be immediately added to the guests' accounts.

This instantaneous billing facility saves many mistakes and omissions from the guest's account which frequently happens when using the triplicate billing system. There are also benefits in labour costs as there is no longer a large team of billing clerks sitting for hours writing up all the bills by hand. This system is also less labour intensive than the bulky and noisy point-of-sale billing machines that will print out the customer's bill in the restaurant. Computerized systems will itemize charges as well as give the final total of a guest's amount.

HAND-HELD, POINT-OF-SALE SYSTEMS

These are becoming increasingly used in popular restaurants, coffee shops and roadside restaurants. They are hand-held, soft key order pads with a small visual screen for order confirmation messages, or dish no longer available information. The advantages of these systems are that they give instant confirmation of the order from the kitchen, thus saving time and mistakes; the food service staff can spend more time talking to and serving the customer instead of writing out the orders and delivering the copies of the triplicate system; they are linked up with the billing clerk's terminal so that the bills are automatically prepared; and they can be used for beverage sales in formal restaurants and in lounge bars, cocktail bars and lounge service areas (see opposite).

MINI-BARS

These are small, refrigerated units containing beverages that are fixed into a cupboard in the guest's hotel bedroom. The mini bar is stocked up daily by the room service staff, and contains a selection of miniature bottles of spirits, small cans of lager, beer, fruit juices, mineral water and other soft drinks. When the guest removes a beverage, a sensor positioned beneath the computerized control system will automatically charge the cost of the item to the guest's account. (If the mini bar is not fully computerized, then it is checked daily by room servicing staff and any sales are noted and then added to the guest's account.)

A hand-held, point-of-sale system.

COMPUTERIZED BEVERAGE DISPENSING SYSTEMS

Computerized beverage dispensing systems are either based on electronically controlled sensor pads, hand-held spirit dispenser guns, cocktail dispenser units, or mixed beverage dispenser units. The skill is to select the system that will improve efficiency without losing customers and the unique atmosphere of the establishment.

All the control systems have an automatic billing and stock control system. They also have the ability to record operator sales individually so that the manager can analyse the performance of each member of staff in terms of sales, errors and cash taken.

Electronically controlled sensor pad

This is available in a mobile trolley format so that it can be used in function suites, receptions, and off-site catering functions. It requires only mains electricity and can quickly calculate the correct charges billed at the end of a function. Another advantage is that these units can be coin or card operated so that they can be used for self-service areas at trade conventions and exhibitions, thus reducing labour cost. They are very hygienic as the glass does not have any contact with the dispenser, all sales are recorded, and a bottle replacement light automatically signals an empty bottle. The system will take 75 cl to 1.5 litre bottles of spirits, fortified wines and vermouths and up to six glasses can be dispensed automatically (see photo (a)).

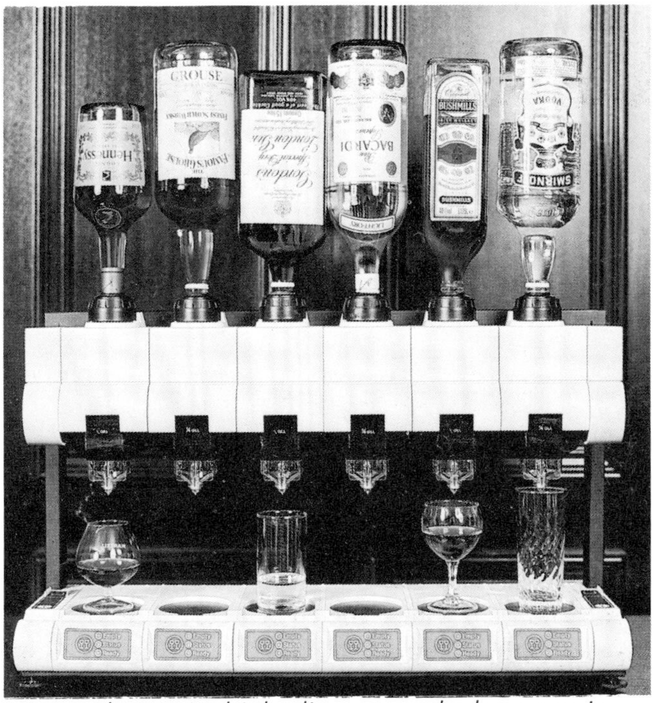

(a) An electronic drinks dispenser – the beverage is delivered when the glass is placed on the sensor pad.

Portioner magnetic ring

This has a code reading system similar to the computerized bar code reading system now used in supermarkets and chemists. In the beverage dispense system, the code is fixed to the portioner unit that is placed in the top of the opened bottle. To dispense a tot, the bottle is inverted into a magnetic dispensing ring where an 'eye' recognizes the code and opens the portioner, allowing the drink to pour into the glass placed below the magnetic ring. No contact is necessary between the power unit and the glass, and so this is a hygienic method of dispensing drinks. The magnetic ring is fixed to a control unit that is key activated by the operator. The size of the measure may be determined by the staff by pressing the appropriate key on the control unit. This allows for single and double tots, vermouth measures and glasses of house wines, to be dispensed from the same magnetic ring and control unit. This system is ideal for dispenser bars where a large variety of beverages must be available, but where the volume of sales may at some times of the day be relatively small (see photo (b)).

(b) a portioner dispensing ring.

Hand-held dispenser gun

This system is designed to increase the speed of beverage dispensing in busy bar areas. It is already in use in may bars, particularly for the dispensing of mixers and cordials, such as soda water, lime cordial, lemonade and cola. However, the system can be used just as successfully for alchoholic beverages. The hand-held dispenser unit can dispense up to twelve different beverages just by the touch of a soft key pad on the dispenser head. This releases the appropriate beverage and records the sale automatically. Each measure takes approximately 1 second to dispense the beverage outlet is never in contact with the glass. However, the dispenser unit does require cleaning at regular intervals and should be hung on its rack when not being used to avoid contamination of the beverage outlet (see photo (a)).

(a) a hand-held dispenser gun.

Cocktail dispenser unit

This is designed to dispense any of up to ninety types of cocktails or long drinks from thirty basic components, which are all freely programmable. The machine will mix cocktails and any number of different drinks in different quantities as long as the recipe has been correctly programmed into the dispenser's computer. The unit is made of polished stainless steel and operates with the individual staff's key and a soft pad for the selection of cocktails or long drinks. The dispenser spout is washed after each use by soda water to eliminate taste spoilage. This machine is invaluable in the busy night club bar where fast service and efficient control of stock is essential (see photo (b). However, this machine is impersonal and so is not suitable for smaller, quieter units where watching the cocktails being mixed and poured is part of the pleasure experienced when purchasing a cocktail.

(b) a cocktail dispenser unit.

Mixed beverage dispenser unit

This unit is becoming increasingly popular, and is used on cross channel ferries, in popular restaurants, steak houses, and bars in busy hotels.

The unit can dispense up to twelve types of beverage, each in appropriate volume measures such as $^1/_4$ pint, $^1/_2$ pint and 1 pint measures. The types of beverage that this is designed for include beer, lagers, ciders, fruit juices, carbonated mineral water and soda waters.

It can take up to sixty-four different operating keys, so that staff can have their own identification key. Another control facility is that there can be a two tier price structure programmed into the computer to allow for sales in the lounge bar at normal prices, and for sales in the restaurant at a higher price. It is possible to have this unit connected to cash registers or guest accounts so that automatic billing becomes possible.

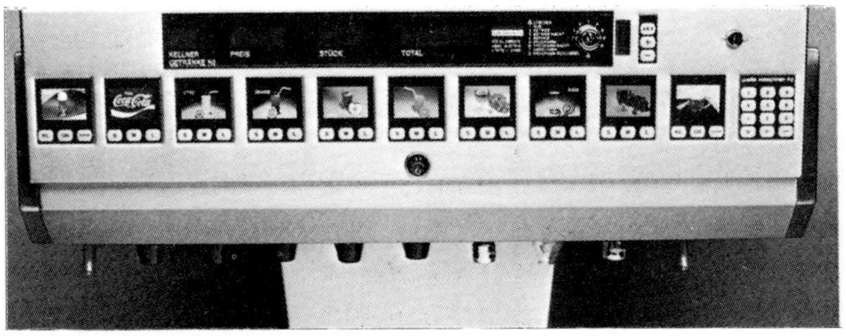

(c) a mixed beverage dispenser unit.

Each beverage has a separate dispense nozzle so that no taste contamination can occur, and the rim of the glass does not touch the nozzle, making it hygienic and free from potential cross contamination.

The beverages are pumped through conventional pipes and coolers from the beer cellar to the dispenser unit as required (see photo (c)).

Glossary

Accompaniments Items offered to accompany a dish.
Additives These are substances added to food to preserve the food, to slow down the decaying process, to stabilize the consistency of the food, or to colour the food.
À la carte Type of menu that offers individually priced dishes. Dishes are prepared to order, and are, therefore, a more expensive and slower form of service.
Al denté Cooking vegetables, cereals and rice until they are soft but retain a slight resistance or 'bite'.
Anaerobe An organism that can grow in areas without oxygen.
Antioxidants These are added to foods to slow down the normal process of food decay that occurs when food is exposed to oxygen in the air.
Aperitif drinks served prior to a luncheon or dinner to stimulate the digestive juice.
Aromatic merchandising The encouragement of sales by using the aroma of the foods.
Baking powder A chemical raising agent made from two parts sodium bicarbonate and one part cream of tartare.
Basic sauces These are sauces from which other sauces may be derived.
Basting The action of moistening meat during cooking with hot fat and the juices produced.
Beurre manié Equal quantities of butter or flour kneaded together into a paste. Small pellets of this may be whisked into a simmering liquid as the butter melts, the flour grains are distributed into the liquid and the starch thickens the liquid.
Blanch To immerse meat in cold water and bring to the boil for one or two minutes in order to seal food that is to be kept white, to start the cooking process, or to destroy food spoilage organisms in vegetables prior to deep freezing.
Bouquet garni The name given to a bundle of herbs that are used to infuse flavour into foods while they are being cooked.
Brand name The manufacturer's trade name, mark or logo. Customers often select a product by its brand name.
Brunoise A fine dice of vegetables. Used in broths and for garnishing.
Cereals Cultivated grasses whose seeds are edible, for example oats, barley wheat, maize and rye.
Chemical food poisoning Poisoning caused by contamination from chipped enamelware, worn copperware, lead pipes, insecticides, pesticides, rodenticides or cleaning chemicals.
Chilled shelf A fixed, refrigerated shelf in a service area at 5°C - 8°C.
Clouté onion An onion peeled and studded with cloves and used to infuse flavour into liquids.
Coagulate To cause a 'fluid' to change into a semisolid or solid mass.
Collagen White transparent connective tissue found in the muscles of meats. It turns to gelatine when cooked.
Cooked out This is the term used to describe the process of applying heat to a roux; the heat will cook the flour causing the grains to burst and allow the starch to be released into the liquid and thus thicken the sauce.
Court bouillon A flavoured liquid used when poaching fish.
Cross-contamination The term used to describe the transference of pathogenic contaminates from one source to another.
Customer flow The volume or numbers of customers that pass through an area.
Darne A slice taken across the back bone of a round fish, sometimes called a 'cullet'.
Delice Small fillets that are folded before being poached or steamed.
Demi-tasse cup A small coffee cup used for the service of coffee in a formal restaurant at the end of a luncheon or dinner.
Detergent A cleansing agent made from varying combinations of soap, carbonate silicates and phosphate.
Dietary fibre The cellulose in foods that absorbs

water and expels waste from the body.
Direct sales Sales made directly to another person, for example, the food service personnel sell direct to their customers.
Disinfect To rid of micro-organisms that are potentially harmful.
Dry methods of cooking Methods where the food is cooked in dry heat, for example roasting, baking and shallow frying.
Duxelle A mix of finely diced mushroom and shallots used as a stuffing.
Elastin Inedible connective tissue that forms the walls of tubular fibres.
Emulsifiers These mix together ingredients, such as water and oil, that would normally separate.
Emulsify To form an emulsion - a mixture in which two components are held in suspension.
Exchange rate The value of the UK sterling currency in relation to other currencies in the world.
Extractives These are the meat juices found in the muscle fibres of animal flesh that give the flesh its distinctive flavour.
Family service A type of service used in less formal restaurants where the food is served to the customer already plated, and the vegetables are placed in dishes so that the customer may help him or herself.
Farinaceous This refers to products that have a high starch content such as pastas.
Fillet The boned side of a fish or a strip of boneless meat.
Fire certificate A certificate given by the Local Fire Authority to indicate that a premises is safe, has sufficient fire fighting appliances and staff training.
Fish kettle A deep, oblong lidded pan with a perforated rack that is used to cook a whole fish without damaging the skin or flesh.
Flash point The temperature at which heated oil will ignite and burst into flames.
Food poisoning Illness caused by eating contaminated foods giving some of the following symtoms - diarrhoea, vomiting, abdominal pain, headaches and fever.
Food spoilage organisms cause food to decay but do not themselves cause food poisoning.
Garnish To add food to a dish to increase its visual appeal or flavour.
Gastronome tray A deep tray with a lid used for cooking foods.
Glaze This means either to brush foods with clarified butter to give them a sheen, or to brush raw pastry with beaten egg so that when baked it has a glossy, golden finish.
Gluten The name of the protein found in flour.
Griddle A preheated, lightly-oiled cooking plate used for the cooking of beefburgers, eggs, bacon, etc, in fast-food and back-bar, call order areas.
Gueridon service Personalized service used in formal restaurants. The dish is finished or carved at a trolley or table adjacent to the customer in the restaurant. Some dishes are finished with a flambé technique on a gas or spirit lamp.
High-risk foods Foods that are particularly susceptible to pathogen growth.
Homogenize To break up the fat globules in milk so that they are evenly distributed.
Hors d'oeuvre A small, light course served at the beginning of a formal meal.
Hydrogenated This is the process used to harden oils during the manufacture of margarines. The harder the margarine, the more saturated fat is present.
Jardinère Small batons of vegetables used for vegetable mixes.
Julienne Thin strips of vegetables that are used for garnishing.
Kneading The second stage of the bread making process where the dough is handled to strengthen the gluten and to gain a good rise and an even textured dough.
Liaison A mixture of egg yolks and cream used for thickening and enriching sauces, soups and white stews.
Linen room A central store room for all linens within an establishment.
Maçedoine A dice cut of vegetables.
Marinade An acidic liquid (containing either wine or vinegar) used to soak joints or portions of meat to tenderize and flavour prior to cooking.
Marketing mix The combination of the following factors when deciding upon the best marketing mix for the product: the product, the price, the promotion and the place of sale.
Meat extenders This is the term used to TVP when it is used with meat to extend or suppliment the dish.
Meat substitutes The use of TVP in place of fresh meat.
Microbes Single cell organisms that are microscopic - unable to be seen by the naked eye. They are sometimes referred to as micro-organisms or bacteria.
Mirepoix Roughly chopped prepared carrots, celery, onions, parsley and thyme, used in many basic kitchen processes.

Mise-en place Literally it means to put into place. A french term used to indicate the preservice preparation time or tasks in the restaurant.
Modified starch A commercially prepared thickening agent that will thicken a liquid upon the application of boiling liquid or cold liquid.
Moist methods of cooking Methods of cooking that use moisture around the food as it cooks, for example boiling, poaching and steaming.
Nutrients The name given to the five basic constituents of food: carbohydrates, proteins, minerals, vitamins and fats.
Offal The term is used to describe all edible parts of an animal carcass that are not muscle fibre (meat), for example liver, tongue, tripe, sweetmeats, tail and heart.
Organic farming Farming without the use of any chemical fertilizers or pesticides, and without the use of livestock growth regulators and food additives.
Panada A sauce that is made twice as thick as usual and used as a binding sauce.
Pasteurized To subject a liquid or food to heat in order to destroy harmful micro-organisms.
Pathogens Microbes that will cause illness to humans.
Paupiette A skinned fillet of fish that is spread with forcemeat, rolled, then oven poached or baked. Or a thin slice of topside beef, spread with stuffing, rolled, sealed and braised.
Paysanne A cut of vegetables that is used for soups. The vegetables are cut into thin shapes of circles, squares or triangles.
Perishable A term used to describe fresh food commodities that are liable to rot easily.
Personal disposable income The amount of money available to an individual to spend as he or she wishes over a given period.
pH scale The scale used for the acidity or alkalinity of a substance: pH7 is neutral, less than 7 is acid and greater than 7 is alkali.
Preservation of foods The treatment given to fresh foods to prolong their safe life, for example, by canning, deep freezing and dehydrating.
Preservatives Substances added to foods to prevent the growth of bacteria yeasts and moulds which cause fresh foods to decay and spoil.
Proving A process used in bread making and yeast dough preparation where the mixture is allowed to double its size by 'resting' in a warm, moist atmosphere.
Pulses The term used to describe the dried seeds of plants made from pods, for example, beans and lentils.
Raising agent An agent or combination of ingredients that causes a baked product to rise when heat is applied and thereby created a light texture.
Rechauffé A term used to describe the creation of a dish from items of food that have already been cooked.
Reconstitute The mixing of a dried food commodity with a liquid prior to cooking.
Refreshing The plunging of cooked vegetables, pasta or rice into cold water to cool the food rapidly, thus arresting the cooking process and washing away excess starch.
Roux A thickening agent used to create the basic sauces made from equal quantities of fat and flour.
Sauté To toss foods quickly in a little fat.
Savouries Small, tasty items served after the sweet course of a formal dinner menu. They are designed to take the sweetness off the palate.
Smoke point The temperature at which heated oil will produce a smoke haze. This is lower than the flash point.
Spoken merchandising Speaking to customers using a positive sales style to create a selling environment.
Stabalizer A substance added to a food to maintain it in an unchanging state.
Staff appraisals The process used by managers to look in depth at staff performance and standards achieved, as part of the general management process.
Staffing ratio The number of staff employed in relation to the number of meals sold.
Sterilized A process used for milk which will preserve the milk for up to 3 months if kept in sealed bottles. Also used to destroy harmful micro-organisms on equipment, flatware, cutlery and glasses, after the washing process.
Stir-fry To cook vegetables and cooked meats and rice in a deep sloping sided pan over a fierce heat.
Stock An unthickened, unseasoned liquid that contains nutrients and flavour extracted from the bones and vegetables that have been simmered.
Suprême A fillet of large fish cut on a slant before being poached or grilled, or a cut of chicken with the breast flesh and the wing together in one piece. This may be poached or grilled.
Sweated vegetables Vegetables are diced or rough chopped and then placed in a hot pan containing a small amount of melted butter. The vegetables are tossed and left on a medium heat for a few minutes in the lidded pan so that they soften without colouring.

Table d'hôte Type of menu that offers customer choice in each course and a set number of courses for a predetermined, inclusive charge.
Textured vegetable protein (TVP) is made from soya beans and is used in many forms as a high protein meat substitute or supplement.
Thermostat that maintains a system at a constant temperature.
Tomato concassé The term used to describe coarsely-chopped, skinned and deseeded tomato flesh.
Tourné vegetables Vegetables that have been cut and shaped into small, barrel-shaped pieces and used as a garnish.
Toxin The poison produced by pathogens in food.
Traditional restaurant brigade The team of food service staff in a traditional, formal restaurant.
Triplicate billing Manual order system using triple carbonated order pads, one copy of each goes to the kitchen, cashier and food service staff.
Tronçon A slice taken from across the back bone of a flat fish.
Turnover The amount of business transactions per day.
Ultra heat treatment (UHT) A process used for milk and cream to destroy harmful micro-organisms. Milk and cream treated in this way will last in sealed cartons for up to six months.
Visual merchandising Encouraging sales by using visual methods - the use of coloured photographs to illustrate dishes on a menu, for example.
Written merchandising Encouraging sales using the written word - the menu and wine list are examples.
Yeast A form of fungi that is used as a raising agent in the baking of breads and cakes and for fermenting beer, whisky, etc.

INDEX